Biotechnology for Environmental Protection in the Pulp and Paper Industry

Springer

Berlin
Heidelberg
New York
Barcelona
Budapest
Hong Kong
London
Milan
Paris
Singapore
Tokyo

Dr. PRATIMA BAJPAI
Dr. PRAMOD K. BAJPAI
Thapar Corporate Research & Development Centre
Chemical Engineering Division
PO Box 68
147001 Patiala
India

Professor Dr. RYUICHIRO KONDO
Kyushu University
Department of Forest Products
Faculty of Agriculture
Hakozaki, Higashi-ku
812-8581 Fukuoka
Japan

ISBN-13: 978-3-642-64271-5

Library of Congress Cataloging-in-Publication Data

Bajpai, P. (Pratima)
Biotechnology for environmental protection in the pulp and paper industry / Pratima Bajpai, Pramod K. Bajpai, Ryuichiro Kondo.
p. cm.
Includes bibliographical references.
ISBN-13: 978-3-642-64271-5 e-ISBN-13: 978-3-642-60136-1
DOI: 10.107/978-3-642-60136-1
1. Wood-pulp - Biotechnology. 2. Wood-pulp industry - Environmental aspects. 3. Paper industry - Environmental aspects. 4. Paper chemistry. I. Bajpai, Pramod K. II. Kondo, Ryuichiro, 1949- .
III. Title.
TS1176.6.B56B35 1999
676'. 042 - dc21 99-30761

Softcover reprint of the hardcover 1st edition 1999

Cover design: Design & Production, Heidelberg
Typesetting: Best-set Typesetter Ltd., Hong Kong
SPIN 10693384 30/3136 - 5 4 3 2 1 0 - Printed on acid-free paper

P. Bajpai · P.K. Bajpai · R. Kondo

Biotechnology for Environmental Protection in the Pulp and Paper Industry

With 27 Figures and 104 Tables

Springer

Acknowledgements

We are grateful to Professor Kokki Sakai for his constant support and encouragement in the execution of this project. Our heartfelt gratitude goes to Mr. Koki Fujita for his assistance in a number of tasks in the preparation of this manuscript. We would also like to express our thanks to Dr. I. Reid for suggesting a few good topics for this book. We offer our sincere thanks to the friends and organizations who provided us with information on the latest advances in their respective areas.

Sincere appreciation is extended to Ms. Chie Takatsuka, Ms. Chie Ishihara, and Ms. Hirano Makiko for their help in word processing in the early stages of the preparation of the manuscript and to Mrs. Katsuko Mori for secretarial assistance. We would like to thank all the students/research scholars of the Wood Chemistry, Forest Products Department of Kyushu University for their help and cooperation.

PKB and PB wish to express their particular gratitude to Professor M.P. Kapoor for granting them sabbatical leave to work at Kyushu University, and for supporting them on every step of the way. We were lucky to have numerous peoples' help with various parts of this work. We are grateful to Mr. S.S. Gill for his invaluable help in word processing in the final stages of preparation. Also to Mr. A. Sappal and Mr. S.N. Sharma for their help in typing some chapters. The valuable help provided by Mr. S.S. Saini and Mr. B.C. Saini in preparing the drawings and figures and by Mr. S.P. Mishra, Mr. O.P. Mishra and Mr. Sanjay Kumar in proofreading and general organization are gratefully acknowledged. Our thanks also go to many others who gave us permission to use drawings and other illustrative material.

Finally, we thank Springer-Verlag for their efforts in publishing the book.

P. Bajpai, P.K. Bajpai, and R. Kondo

Contents

1 Introduction

The pulp and paper industry comprises a large and growing portion of the world's economy. Pulp and paper production has increased globally and will continue to increase in the near future. Approximately 155 million tons of wood pulp are produced worldwide and about 260 million are projected for the year 2010.[1] However, the industry is very capital-intensive, with small profit margins, which tends to limit experimentation, development, and incorporation of new technologies into the mills. To be able to cope with increasing demand, an increase in productivity and improved environmental performance is needed, as the industry is also under constant pressure to reduce and modify environmental emissions to air and water.

In the US, the pulp and paper industry ranks third, after primary metals and the chemical industry, in terms of freshwater withdrawal. It is believed that by the year 2000, the paper industry will have become the largest manufacturing user of water.[2] However, the industry ranks fifth among the major industries in its contribution to the water-pollution problem, having thus had significant impact on the nation's streams; but the situation is improving rather than growing worse. Since 1976, many mills have built secondary biological waste treatment plants, by 1991, 99% of the pulp and paper industry in US had secondary treatment or its equivalent. The fact that the majority of the mills have wastewater treatment has had a visible beneficial effect on water quality.

Like any other large-scale industry, the pulp and paper industry exerts its own impact on the environment. Although the situation is improving year by year, pulp mills still release unpleasantly smelling sulfur compounds into the air and discharge wastewaters which enhance eutrophication and have toxic effects on the surrounding biota. Organochlorine compounds formed during the bleaching of chemical pulp have attracted most attention in recent years.

Environmental Regulations

While not all AOX (adsorbable organic halides) compounds are harmful to health or the environment, a number of them have been determined to be so. Rather than single out the harmful constituents of AOX for regulatory purposes, regulatory agencies in Canada and Scandinavian countries have used AOX as the measure by which to set limits on different discharges. Such regulations have forced the pulp and paper industry in Europe, Canada, and Japan, and recently in the USA[3] also, to reduce AOX discharges through a combination of process modifications and effluent treatment technologies.

Europe

While environmental regulations provided the impetus for the drive towards AOX reduction, today it is market force which is sustaining it. The Scandinavian industry today is moving beyond AOX reduction as a solution, and working towards total elimination of effluent discharge as the long-term solution to environmental problems. The Scandinavian pulp and paper industry is thoroughly convinced of the need to reduce and even eliminate AOX discharges. It does not require pressure from environmental regulations to spur it.

In **Sweden**, the Licensing Board for Environmental Protection is an autonomous body responsible for setting the terms and conditions for granting of environmental permits. The board takes into account ecological, technological, and economical factors in setting the terms for permits. Regulatory limits for pulp and paper mills are set on an individual, case-by-case basis. As such, there is no set of effluent limits for the industry as a whole. However, the regulations are fairly similar for mills that produce a similar product; consequently, from an exhaustive survey of discharge limits in place at the various mills, one can determine allowable discharge levels for mills producing different products. In general terms, Table 1.1 lists the limits for discharges from pulp mills in Sweden.[4] Regulations required that by the end of 1995, all mills decrease their emissions of organic chlorine to 1 kg/ton for softwood kraft pulp, and 0.5 kg/ton for sulfite or hardwood kraft pulp, which has been further reduced to 0.5 and 0.3 kg/ton, respectively.

Table 1.1. Regulatory emission limits for Swedish pulp mills

Parameter	Discharge limit (kg/ton pulp)		
	1991	1995	1999
BOD_5	10–20		
COD	10–20		
TSS	40–70		
TOCl Softwood	2.0	1.0	0.5
Hardwood	1.0	0.5	0.3
AOX Softwood	2.5		
Hardwood	1.3		
Nitrogen	0.3		
Phosphorus	0.1		

Based on data from Ref. 4.

In **Finland**, regulatory trends in general have closely followed those in Sweden. The AOX discharge has steadily decreased from 3.7 kg/ton in 1989 to less than 1 kg/ton by 1995.

Effluent discharge limits in **Norway** are imposed and permits granted by the State Pollution Control Authority on a mill-by-mill basis, based on the conditions of the receiving body of water, with limits being stricter for discharge to lakes and fjords than to the sea.

Effluent regulations in **Germany** are set up by the Ministry of Environment, based on recommendations by various experts. The enforcement of the regulations is through the levying of what is known as discharge fees. Mills that exceed the discharge limits set by the regulations are liable to pay the discharge fees. The effluent limits for pulp and paper mills in Germany are given in Table 1.2.[5]

French regulations do not have AOX standards or discharge limits; however, mills are expected to measure their AOX discharge levels periodically. The regulations do set discharge limits for TSS, BOD_5, and COD for pulp and nonintegrated paper mills.

Portugal has specific emission limits for discharges to water from pulp and integrated paper mills as shown in Table 1.3.[5] Regulations for general chemical industries are applicable to nonintegrated paper mills.

In **Austria**, the AOX discharge limit was fixed at 0.75–1.5 kg/ton in 1994, which has been further reduced to 0.5–1.0 kg/ton after 1995.[6]

Table 1.2. Regulatory emission limits for pulp and paper mills in Germany

Mill type	Discharge limit (kg/ton)		
	BOD_5	COD	AOX
Paper Mills			
Wood-free unsized	1	3	0.04
Wood-free sized	2	6	0.04
Wood-free highly refined	3	9	0.04
Wood-free coated	–	2	0.02
Wood-containing	–	3(5)[a]	0.01
Recycle fiber-based	–	5	0.012
Chemical pulp mills	5[b]	70[b]	1.0

[a] COD of 5 kg/ton when >50% of pulp is TMP, or when a substantial part of the pulp is bleached with peroxide.
[b] 24-h composite samples.
Based on data from Ref. 5.

Table 1.3. Discharge limits for pulp and integrated paper mills in Portugal

Mill type	Discharge limit (kg/ton)			
	TSS	BOD_5	COD	AOX
Unbleached kraft pulp				
1991	2.5	5	–	–
1996	1.5	3	35	–
Bleached kraft pulp				
1991	10	9	–	–
1996	3	6	50	1.5
Bleached sulfite pulp				
1991	12.5	45	–	–
1996	6	25	120	1.5
Integrated kraft liner				
1991	4	6	–	–
1996	2	4	30	–

Based on data from Ref. 5.

Table 1.4. AOX discharge limits in Canada

Province	Year	AOX limit (kg/ton)
Ontario	1991	2.5
British Columbia	1992–1993	2.5 (Softwood)
		3.8 (Hardwood)
Quebec	1992	2.5 (Softwood)
		1.5 (Hardwood)
	1995	1.5 (Softwood)
		0.5 (Hardwood)

Based on data from Ref. 7.

North America

Canada, though behind Scandinavia in achieving AOX reduction goals, is aggressively setting targets and demanding that its pulp and paper industry meet them. Tables 1.4 and 1.5 summarize the salient points of regulations by the various provinces.[7] The regulations developed by British Columbia (Table 1.5) are the most stringent in the country. Very similar regulations are being developed by Ontario.

The **United States** did not have specific discharge limits on AOX, but did have a limit on dioxin discharge. The Clean Water Act of 1987 listed dioxin as a toxic compound. In an exhaustive study of 104 US bleached pulp mills in 1988–1989, the EPA measured concentrations of 2,3,7,8-TCDD and 2,3,7,8-TCDF in the effluents. A further in-depth study of 25 mills resulted in the con-

Table 1.5. AOX regulations in British Columbia

1. Mills need to meet the AOX discharge limit of 1.5 kg/ton by 31 December 1995
2. Mills will be required to eliminate AOX generation in the bleaching process before 31 December 2002
3. Mills eliminating AOX generation by 31 December 2000, will be exempted from the 1.5 kg/ton requirement

Based on data from Ref. 7.

Table 1.6. Cluster Rule – bleach plant effluent discharge limits

Effluents	
1. AOX	0.623 kg/ton (monthly average) 0.512 kg/ton (annual average) 0.951 kg/ton (daily maximum)
2. Dioxin and 12 chlorinated phenolics	Non – detect
3. TCDF	31.9 pg/l (daily maximum)
4. Chloroform	4.14 g/ton
5. Color, acetone, MEK, methylene chloride	No limit

Based on data from Ref. 8.

clusion that the chlorine bleaching stage was responsible for dioxin generation.[5] According to the EPA's Clean Water Act, pulp mills are required to regulate their discharges to ensure that the dioxin levels in receiving streams do not exceed 0.013 ppq (parts per quadrillion), which is well below the detection limit of the detection devices – ranging from 1 to 10 ppq. Various states have proposed their own limits on dioxin discharge, usually higher than the EPA standard; for example, the states of Maryland and Georgia have a dioxin limit of 1.2 ppq.

Recently, the so-called Cluster Rule, first proposed in December 1993, was signed by EPA on 14 November 1997 after some modifications.[8] The following are the highlights of the Cluster Rule. The compliance period for effluent is up to 3 years from the date of publication of the rule in the Federal Register. The best available technology (BAT) for the bleached paper grade kraft and soda subcategory is complete substitution of chlorine dioxide for chlorine. The BAT for calcium-, magnesium-, and sodium-based paper grade sulfite mills is TCF bleaching. The BAT for an ammonium-based and specialty grade paper sulfite mill is ECF bleaching. The limitations on COD were deferred by EPA and there were no changes in the limitations of BOD and TSS for all the categories. The specific limits of AOX and other organochlorine compounds are summarized in Table 1.6.

Regarding the air standard, any mill using a chlorine- or chlorine dioxide-based bleaching system (towers, seal tanks, washers, etc.) must scrub the emissions. The compliance period is 3 years from publication in the Federal

Register, except for emissions from brownstock washer systems, oxygen delignification systems, and certain other high-volume, low-concentration (HVLC) system emission points, which have a compliance period of 8 years. Mills participating in the voluntary incentive program have up to 6 years to comply with the bleaching requirements if they meet certain conditions.

Asia

Japan has the second largest pulp and paper industry in the world. In 1996–1997, Japan's 30 million tons of paper and board production ranked second internationally after that of the US, while pulp production was third at 21 million tons.[9] Japan has 117 major pulp and paper sites spread throughout the country. Most of the pulp and paper produced in Japan is consumed internally. A relatively small amount of paper is imported, while greater quantities of imported pulp (4 million ton/year) are required to satisfy internal paper-making demand. Seventeen Japanese companies operated about 40 bleached kraft mills with a combined production of 7 million tons/year in 1996–1997. Substantial quantities of pulpwood in log and chip form (28 million m^3) were imported in 1996–1997 to supplement local wood resources. In Japan, waste paper comprises a significant proportion of fiber input (16 million tons/year).

The increased focus on organochlorine discharges was precipitated by a single event in Japan. In October 1990, fish caught in the Kinsei River near Kawanoe City were reported to have a high concentrations of dioxins.[10] Public blame was laid on the pulp and paper industry, since the Daio Paper Corporation operates a large mill in the area. The Environmental Agency and Ministry of International Trade and Industry (MITI) took immediate action to investigate the amount and composition of organochlorines in Japanese pulp mill effluents. The Japanese industry, through the Japan Paper Association (JPA), also took positive steps to deal with the dioxin issue; it set up a special committee to investigate the technical means to eliminate dioxin discharges and announced in December 1990 that all Japanese pulp mills must voluntarily limit organochlorine discharges (in terms of AOX) to 1.5 kg/ton by the end of 1993. These events have placed the environmental performance of the entire pulp and paper industry in Japan under closer scrutiny. Today, most of the bleached kraft mills are discharging less than 1 kg/ton AOX after modifying the pulping and bleaching processes. The discharge limits for pulp and paper mills in Japan are given in Table 1.7.[11]

The **Indian** pulp and paper industry produced 2.51 million tons of paper and board plus 300 000 tons of newsprint in the year 1994–1995 although its installed capacity was 3.95 million tons and 400 000 tons, respectively. Actual consumption of paper and board was 2.5 million tons and that of newsprint 690 000 tons. The balance of the quantities was imported.[12] There are more than 350 small paper mills in addition to about 20 so-called large mills. About two-thirds of India's paper mills have a daily capacity of 50 tons or less. In addi-

Table 1.7. Discharge limits for pulp and paper mills in Japan

Parameter	Value	
	Maximum	Daily average
BOD (mg/l)	160	120
COD (mg/l)	160	120
TSS (mg/l)	200	150
AOX (kg/ton pulp)	1.5	

Based on data from Ref. 11.

Table 1.8. Effluent standards for the Indian pulp and paper industry. (The Gazette of India, No. 174, 19 May 1993)

Mill type	Discharge limits				
	BOD_5 (mg/l)	COD (mg/l)	TSS (mg/l)	TOCI (kg/ton prod)	Flow[a] (m^3/ton)
Large pulp and paper	30	350	50–100	2	200(100)
Newsprint/rayon grade	30	350	50–100	2	150
Small pulp and paper					
- Agrobased	30	–	100	–	200(50)
- Waste-paper based	30	–	100	–	75(50)

[a] The values in parentheses are applicable to mills established after January 1992.

tion to bamboo and hardwood, several nonwood fiber sources, such as bagasse, wheat straw, and wild grasses are used as raw material. Due to small-scale operations and utilization of a broad spectrum of raw materials, extensive automation has not been feasible, which has resulted in a large specific water consumption and pollution problem. The Ministry of Environment and Forests (Government of India), through the Central Pollution Control Board and State Pollution Control Boards, is regulating the effluent discharges. The present standards are summarized in Table 1.8.

In **China** also, there are many small pulp and paper mills spread throughout the country and utilizing various types of nonwoody raw materials like agricultural residues. The limitations on the discharge of mill effluents depend on the type of receiving body, as shown in Table 1.9.

In the countries of **South-East Asian** region, there is no standard for AOX discharge, However, for other effluent parameters, there is a wide range of limits, as shown in Table 1.10.[13]

Table 1.9. Chinese mill effluent limitations

Parameter (kg/ton)	First-class receiver	Second-class receiver	Third-class receiver
BOD_5	7.2	28.8	120
COD	16.8	48.0	96
TSS	24.0	84.0	192
AOX	1.5	2.5	–

Based on data from Ref. 13.

Table 1.10. Effluent discharge standards in SE Asia

a) Thailand

Parameter (mg/l)	Pulp mill	Paper mill
BOD_5	20	20
COD	120–500	120
TSS	50–150	50–150
TDS	3000–5000	3000–5000
TKN	100–200	100–200

b) Indonesia

Flow (m^3/ton)	50–95	35–50
BOD_5 (mg/l)	50–100	60–90
(kg/ton)	3–9.5	2–6.5
COD (mg/l)	120–350	100–175
(kg/ton)	7.2–29.75	5.6–10
TSS (mg/l)	75–100	80–100
(kg/ton)	3–9.5	2–8.5

c) Malaysia

Parameter (mg/l)	Standard A	Standard B
BOD_5	20	50
COD	50	100
TSS	50	100

d) Philippines

Parameter	Protected water category II	Inland waters class C
BOD_5 (mg/l)	30	50
COD	case by case	case by case
TSS (mg/l)	50	70
TDS (mg/l)	1000	–
Temperature	3 °C rise	3 °C rise
Color (PCU)	100	150

Based on data from Ref. 13.

Australasia

By international standards, pulp and paper production in Australia and New Zealand is relatively small. In Australia, virtually all of the 0.8 million tons of pulp produced in 1990 were used in the production of 2 million tons of paper. A further 1 million ton of paper and board were imported to satisfy the country's demand.[14] New Zealand, by contrast, is a net exporter of pulp and paper products. About 45% of the 0.8 million tons of paper and 1.4 million tons of pulp produced in 1990 was exported.[14]

The regulatory environment in Australia and New Zealand has changed markedly over the past few years. The present standards for the effluent discharge are (in kg/ton product): BOD 1.6, TSS 5.3, COD 22, AOX 0.37, and dioxins at non-detectable levels.

South America

Amongst South American countries, Brazil produces substantial quantities of pulp and paper from tropical hardwoods. The effluent discharge limits of Brazil (in kg/ton product) are given below: BOD_5 1.3–6.5, TSS 1.4–60, COD 4.6–45, and AOX 0.2–1.0.

Importance of Biotechnology

Although biotechnology has been in use in the paper industry for quite some time, recently the awareness has grown to employ biotechnological processes in pulp and paper manufacturing. Wastewater treatment systems for removal of oxygen-demanding substances and suspended solids, and preparing starch-using enzymes for paper sizing have long been part of the industry. Improvement in fiber supply by selection of superior trees is still being done by forest product companies. Even the control of slime and deposits on paper machines can be considered as an aspect of biotechnology. However, within the past several years, biotechnologists have sought specific applications for microorganisms/enzymes in the pulp and paper industry.

The need for sustainable technologies also brought biotechnology into the realm of pulp and paper-making about a decade ago. The application of xylanases in pulp bleaching was an essential first step in demonstrating that enzymes are efficient technological means and can be introduced into existing plants without large investments. Recently, other enzymes like lipases or glucanases were tested for large-scale applications.[1] Enzymatic as well as fungal processes are being developed to increase pulp brightness, to reduce troublesome pitch, to improve the quality of wastewater and to purify bleach plant effluents. International conferences on Biotechnology in the Pulp and Paper Industry have been organized every 3 years since 1981 to review the state of research and development in this field.

Table 1.11. Biotechnology in the pulp and paper industry – current status

Process	Status
Biological depithing	Commercial scale
Removal of pitch in pulp by enzymes	Commercial scale
Xylanases for pulp bleaching	Commercial scale
Removal of pitch in wood chips by microorganisms	Commercial scale
Improvement of pulp drainage by enzymes	Commercial scale
Enzymatic deinking	Commercial scale
Pulp bleaching with laccase mediator system	Pilot scale
Biomechanical pulping	Pilot scale
Purification of bleach plant effluents	Pilot scale
Modification of fiber properties by enzymes for improving beatability	Laboratory scale
Production of dissolving pulp	Laboratory scale
Use of enzymes for debarking	Laboratory scale
Use of enzymes for retting of flax fibers	Laboratory scale

Table 1.11 presents the current developmental stages of various biotechnological approaches for use in pulp and paper industry. While many applications are still in the R & D stage, several new applications have found their way into the mill in an unprecedented short period of time. In addition, some of these new developments in biotechnology, if successful, could have a profound impact on the future technology of the pulp and paper manufacturing process.

As future generations of environmentalists look at the world, new ideas emerge that encompass the concept of previous investigators. This book, entitled *Biotechnology for Environmental Protection in the Pulp and Paper Industry*, gives updated information on various biotechnological processes useful in the pulp and paper industry which could help in reducing the environmental pollution problems, in addition to other benefits. Various chapters deal with the latest developments in areas such as raw material preparation, pulping, bleaching, water management, waste treatment, and utilizations. A last chapter on biofiltration has also been added, which discusses the potentials of the removal of odorous emissions from the kraft pulping process. The major benefits/advantages, limitations, and future prospects of these processes have also been discussed.

References

1. Messner E, Srebotnik E, Fiechter A: Editorial. J. Biotechnol. 1997; 53: 91–92.
2. Yulke SG, O' Rouke JT, Asano T: Water reuse in the pulp and paper industry in California. Proc. Water Reuse Symposium II, AIChE, New York 1981.
3. Vice K, Trepte R, Stuart P, Johnson T: The cluster rule – the road to promulgation. Tappi J. 1997; 80(12): 34–36.
4. Hartler, N: Regulatory issues in Sweden. Tappi Non-Chlorine Bleaching Conference, Hilton Head, SC, March 2–5, 1992.

5. Shrinath SA, Bowen IJ: An overview of AOX regulations and reduction strategies. In: Environmental issues and technology in the pulp and paper industry. Joyce, TW, ed. Tappi Press, Atlanta, GA 1995; 31–46.
6. Rennel J: TCF-An example of the growing importance of environmental perceptions in the choice of fibres. Nord. Pulp Pap. Res. J. 1995; 10(1): 24–32.
7. Pryke DC: Regulatory issues in Canada. Tappi Non-Chlorine Bleaching Conference, Hilton Head, SC, March 2–5, 1992.
8. Anonymous: Industry news: cluster rule signed. Tappi J. 1997; 80(12): 14.
9. Anonymous.: Export and import of paper and pulp; pulp wood consumed and production of pulp and paper. Jpn. Tappi J. 1998; 52(3): 443–444.
10. Allison RW, Sakai K: The status of non-chlorine bleaching in Australia and Japan. Presented at Tappi Non-Chlorine Bleaching Conference, Hilton Head, SC, March 2–5, 1992.
11. Japan Paper Association (JPA): Effluent discharge standard. Annual Report, April 1997.
12. Meadows DG: The pulp and paper industry in India. Tappi J. 1997; 80(8): 91–98.
13. Schmidt A: Towards more environmentally friendly mills. Presented in the Federation of Asean Pulp and Paper Industries-12th Board of Directors Meeting, Chiang Mai, Thailand, Dec. 1–3, 1997.
14. Anonymous: Annual review: world trends and trade. Pulp Pap. Int. 1991; 33(7): 63, 71, 79.

2 Wood Pretreatment to Remove Toxic Extractives

Environmental research on the aquatic impacts of pulp mill effluents has been concentrated in Scandinavia and North America. In the 1970s, the regulatory framework covered conventional parameters such as biochemical oxygen demand, suspended solids, and nutrients. From the 1980s until today, concern has been directed towards chemical toxicity. It has been claimed that the organic materials in the effluents from pulp mills cause a number of biological effects. These suspicions have prompted intensive research to study relationships between exposure and effects, as well as profound process changes within the pulp industry. Chemical toxicity analysis has been performed at various levels of biological organization: community, population, individual, organ, tissue, cell, subcellular, and molecular. Many studies performed on pulp mill effluents have focused upon within-organism effects on physiology and biochemistry.[1] Despite the profound process changes in the industry, there is still evidence of physiological changes in organisms exposed to mill effluent. However, the ecological significance of these responses is unknown; in large part they are provoked by lipophilic compounds present in the wood entering the mill (extractives) rather than by materials formed or modified during pulping or bleaching.

The amount and composition of lipophilic extractives vary within the tree itself, between wood species, geographical location, and seasons of the year. The lipophilic extractives in hardwood consist mainly of fatty acids esterified with glycerol, fatty alcohols, sterols, and terpene alcohols. In addition, there are small amounts of free fatty acids, free alcohols, and hydrocarbons. High proportions of saturated and dienoic fatty acids are found among the acids. Triterpenoids are present in birch in a great variety. Birch contains more neutral extractives than pine.[2] Resin acids, the tricyclic, diterpenic, and carboxylic acids are natural wood extractives and are predominantly found in the bark of pine trees, where they play a role in plant defense. Structural variability between six of the most common resin acids is related to the degree of unsaturation and the location of the double bonds (Fig. 2.1). Two classes are recognized: the abietane types, which have conjugated double bonds, and the pimarane types, which do not.

These compounds are environmentally significant because of their relative persistence and toxicity to fish and are responsible for a large part of the acute toxicity of pulp mill effluents.[3] The main resin acids contributing to toxicity in pulp mill effluents are dehydroabietic, abietic, isopimaric, and pimaric acids.

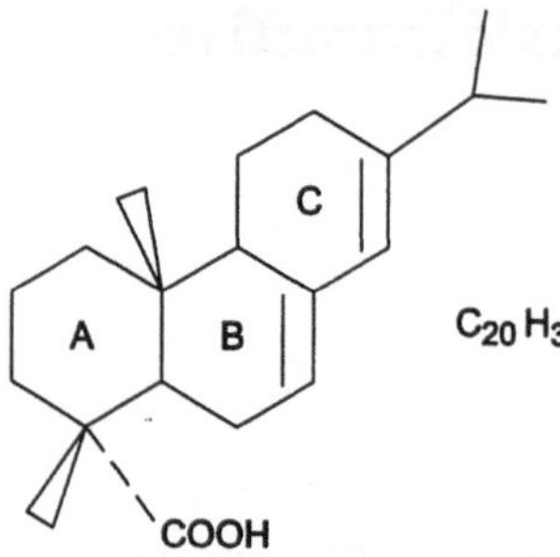

$C_{20}H_{30}O_2$

Fig. 2.1. General structure of resin acids commonly found in pulp mill effluent. Variability between the most common congeners is related to the number and (or) location of the double bonds. The structure illustrated here is of ABA. DHA differs from this in that the C-ring is aromatic and the B-ring has no double bond

Acute toxicity levels for these resin acids are between 0.4 and 1.8 mg/l.[4] The works of Rabergh et al.,[5] Pesonen and Andersson,[6] Tana,[7] Oikari et al.,[8] and Oikari and Nakari[9] give strong evidence that extractives such as resin acids are cytotoxic and enzyme inhibitors and that defensive responses would be expected in fish exposed to effluents containing high levels of extractives. Tana[7] reported inhibition of the conjugation enzyme uranosylglucuronosyl-transferase (UDP-GT), which eliminates substances from the body, in rainbow trout exposed to 5 µg/l dehydroabietic acid (DHAA) for 60 days, and also reported decreased liver glycogen values in fish exposed to DHAA for 80 days. Pesonen and Andersson[6] tested DHAA on primary cell cultures of rainbow trout hepatocytes. They found that this resin acid decreased the enzyme 7-ethoxyresorufin-o-deethylase (EROD) activity at concentrations between 0.1 and 40 µg/l. In a study by Rabergh et al.,[5] the cytotoxic action on isolated rainbow trout hepatocytes by resin acids was studied. Exposure to dehydroabietic and isopimaric acid inhibited bile acid uptake, confirming that resin acids cause impaired liver function in fish and can be major contributors to pulp mill effluent toxicity. The resin acids concentrations used by Rabergh et al.[5] were much higher, 30–97 mg/l, than those used by Oikari and Nakari[9] in a 3-11-day experiment with rainbow trout and simulated unbleached kraft mill effluent (70–150 µg/l).

Time-dependent stimulation and inhibition of the UDP-GT activity in rainbow trout simultaneously exposed to DHAA and trichlorophenol was observed by Tana.[7] Such a response may be caused by compounds with different modes of action at the subcellular level so that some compounds such as trichlorophenol stimulate conjugation processes, whereas DHAA damages membranes and partly inhibits enzyme activities.[9]

Resin and fatty acids levels are normally reduced to sublethal concentrations with the implementation of secondary treatment.[10] Besides being the major contributors to toxicity to aquatic life, these lipophilic extractives (pitch), cause a number of serious problems in the production process.[11,12] During the pulping process, these resinous materials are released from wood and later stick to tile and metal parts including the rolls and wires of the paper-making machines. The pitch also stains the felts and canvas and eventually reaches the drier section. This pitch accumulation can cause paper spotting and web breaks on the machine, which are severe problems in production. Wood extractive components such as triglycerids, resin acids, and steryl esters are major components of paper-machine pitch deposits.[13–16] In addition, pitch outbreaks are more common when resinous wood species are used and during seasons when wood resin content is particularly high. Pitch of pines, including loblolly, slash, and red pines, is known to cause serious problems. Hardwood pitch, particularly from tropical hardwood species and eucalyptus, can also be detrimental. In addition, these extractives are thought to increase the yellowing, i.e., brightness reversion of the pulp and paper products, to cause odor problems, and to increase the necessity for chlorination of the pulp.[12]

Due to the toxicity of these lipophillic wood extractives, it is necessary to have a pretreatment that reduces the liberation of these compounds, especially in mills lacking secondary effluent treatment.

Biotechnological Methods for Reducing Toxic Extractives

The following approaches can be used to reduce the liberation of toxic extractives.

1. Use of Fungi. A variety of wood-inhabiting fungi including sapstain fungi, basidiomycetes, and molds are capable of degrading wood extractives.[17]

Sapstain fungi rapidly colonize the sapwood of logs and wood chips. These fungi grow mainly in ray and parenchyma cells and are capable of deeply penetrating logs and wood chips. They can also grow within resin canals, tracheids, and fiber cells, penetrate simple and bordered pits, and sometimes form boreholes through the wood cell wall. Extractives and simple sugars found in the parenchymal cells are the major nutrient source for sapstain fungi. These fungi are not capable of degrading cellulose and lignin; hemicellulose is degraded to a very small extent. Common species of softwood sapstain include *Ophiostoma ips, O. piliferum, O. piceae, Aureobasidium pullulans, Leptographium lundbergii, Alternaria alternata, Cephaloascus fragrans, Cladosporium* spp., *Lasiodiplodia theobromae* and *Phialophora* spp. Common species of sapstain fungi on hardwood include: *Ophiostoma pluriannulatum, Ceratocystis moniliformis, L. theobromae,* and *Ceratocystis rigidum.*[18] Many of these species are capable of degrading wood extractives. Extractive degradation by *Ophiostoma* spp., particularly *O. piliferum* and *O. piceae*, has been most widely studied. A wide variety of sapstain fungi have been found to degrade wood extractives (Table 2.1).[17,19] *Ceratocystis adiposa, O. piceae,* and *O piliferum* were found to be the best species for extractive reduction. About 41, 32, and 26% reduction in extractives of Southern yellow pine was obtained in 2 weeks at room temperature with *C. adiposa, O. piceae,* and O. piliferum, respectively. Screening of nine different strains of *O. piliferum* on sterile Southern yellow pine showed that five strains reduced dichloromethane extractives by 25–35%; three strains

Table 2.1. Extractive degradation by sapstain fungi on nonsterile Southern yellow pine

Fungal species	Control extractives (%)	Treated extractives (%)	Reduction (%)
C. adiposa	2.13	1.26	41
O. piliferum	3.34	2.27	32
C. adjuncti	1.98	1.44	27
C. minor	2.13	1.57	26
O. piceae	2.13	1.57	26
O. populina	2.13	1.62	24
O. abiocarpa	2.13	1.61	24
C. tremuloaurea	2.13	1.65	23
O. fraxinopennsylvanica	2.13	1.71	20
O. plurianulatum	2.13	1.73	19
E. aereum	1.98	1.61	19
C. ponderosa	2.19	1.86	18
C. penicullata	1.98	1.58	15
O. olivaceum	2.27	1.93	15
E. clavigerum	2.13	1.84	14
C. hunti	2.13	1.89	11
C. ambrosia	2.13	1.92	10
O. distortum	2.27	2.07	9
E. robustum	2.27	2.06	9
C. virescens	2.27	2.11	7
O. galeiformis	2.13	1.98	7
C. coerulescens	2.13	2.13	0
O. dryocetidis	2.24	2.27	0
O. stenorcerns	2.13	2.21	0
X. conudamae	2.44	3.16	0
X. hypoxylon	2.44	2.88	0

Reproduced from Ref. 17.

did not degrade the extractives. Screening of 45 different strains of *O. piceae* on sterile aspen chips showed that one strain reduced the extractives by 60%, 13 strains by 16–35%, 21 strains by 1–15%, and 10 strains did not degrade the extractives. Chen et al.[20] also reported degradation of extractives in aspen and lodgepole pine with five different sapstain fungi. Triglycerides were found to be the most abundant component of both aspen and softwood extractives. The wax and steryl ester content of aspen was about three times that of lodgepole pine. Fatty acids and resin acids were the second most common component of lodgepole pine extractives but were present in very small amounts in aspen. The extractives were reduced by 28–33% in aspen and 10–17% in lodgepole pine (Table 2.2). Analysis of extractive components showed that all five fungi decreased the triglyceride content and that four of the five fungi increased the free fatty acid content.

The fungus *O. piliferum* is marketed as Cartapip by Clariant, UK, to the pulp and paper industry, and removes extractives by metabolizing them. This is a colorless strain of *O. piliferum*, an ascomycete that often dominates in natu-

Table 2.2. Extractive content of sterile lodgepole pine and aspen treated with sapstain fungi

Treatment	Aspen extractives (%)	Lodgepole pine extractives (%)
Control	3.09 ± 0.07	2.31 ± 0.03
Aged control	2.88 ± 0.04	2.26 ± 0.02
Cartapip 97	2.15 ± 0.01	1.94 ± 0.03
Strain A	2.13 ± 0.02	2.08 ± 0.01
Strain B	2.22 ± 0.01	NA
Strain C	2.07 ± 0.04	1.92 ± 0.01
Strain D	2.08 ± 0.05	1.92 ± 0.01

NA: Data not available.
Reproduced from Ref. 17.

Table 2.3. Pitch components in pine chips after 2-week incubation with *O. piliferum*

Component[a]	Acid content (mg/g of o.d. wood)				
	T_o[b,d]	Nonsterile[c]	Strain A[d]	Strain B	Strain C
Fatty acids					
Free	0.9 ± 0.09	5.6 ± 0.4	4.4 ± 0.7	5.4 ± 0.1	4.2 ± 0.04
Esterified[e]	*10.3 ± 1.1*	*2.5 ± 0.1*	*1.2 ± 0.2*	*1.3 ± 0.2*	*1.0 ± 0.2*
Total fatty acids	11.2 ± 1.2	8.1 ± 0.4	5.6 ± 0.5	6.7 ± 0.2	5.2 ± 0.1
Resin acids	6.9 ± 0.6	3.8 ± 0.1	4.8 ± 0.3	4.0 ± 0.2	3.8 ± 0.4
Unidentified[f]	14.9 ± 0.3	10.2 ± 2.1	6.0 ± 0.5	10.5 ± 1.6	8.9 ± 0.3

Reproduced with permission from Tappi Press, Ref. 15.
[a] Acid component determined as the methyl ester.
[b] Chips frozen at start of experiment.
[c] Nonsterile chips aged 2 weeks to allow growth of natural background organisms.
[d] Average of two samples.
[e] Esterified component is the difference between free acids and total acids.
[f] Unidentified component is the difference between the total extractives and the total fatty acids and resin acids.

rally seasoned piles. Cartapip is marketed as a dry powder, which is diluted with water to a 1–3% solids solution and sprayed onto wood chips. It rapidly proliferates and removes pitch/resin from wood chips within 4 to 10 days. Chip piles show good coverage of fungal growth, with pitch, as determined by dichloromethane extractables, being reduced 50% or more.[21-24] In general, fungal treatment results in an overall decrease in total fatty acids and total resin acids (Tables 2.3, 2.4). Triglycerides are almost completely hydrolyzed. The following resin acids are decreased by more than 50% after 2 weeks of treatment: levopimaric, abietic, palustric, pimaric, and neoabietic. Linoleic and palmitic fatty acids are decreased by more than 50% after 2 weeks of treatment, and

Table 2.4. Major fatty and resin acids in ether extracts of pine chips after 2-week incubation with *O. piliferum*

Component[a]	Acid content (mg/g of o.d. wood)				
	T_0[b]	Nonsterile[c]	Strain A	Strain B	Strain C
Resin acids					
Levopimaric	2.4 ± 0.2	1.1 ± 0.03	1.2 ± 0.1	1.2 ± 0.06	0.9 ± 0.09
Abietic	1.2 ± 0.2	0.5 ± 0.05	0.7 ± 0.007	0.5 ± 0.06	0.6 ± 0.1
Palustric	1.2 ± 0.1	0.6 ± 0.05	0.9 ± 0.07	0.7 ± 0.02	0.6 ± 0.08
Dehydroabietic	0.6 ± 0.06	0.8 ± 0.07	1.1 ± 0.07	0.8 ± 0.005	0.8 ± 0.05
Pimaric	0.5 ± 0.09	0.3 ± 0.01	0.3 ± 0.02	0.3 ± 0.01	0.3 ± 0.03
Neoabietic	0.6 ± 0.002	0.3 ± 0.02	0.3 ± 0.04	0.3 ± 0.05	0.3 ± 0.04
Isopimaric	0.1 ± 0.03	0.1 ± 0.01	0.2 ± 0.04	0.1 ± 0.01	0.1 ± 0.008
Fatty acids					
Oleic	4.3 ± 0.5	3.8 ± 0.2	2.5 ± 0.2	3.2 ± 0.08	2.4 ± 0.09
Linoleic	5.6 ± 0.6	3.0 ± 0.3	2.2 ± 0.2	2.4 ± 0.08	2.0 ± 0.03
Palmitic	0.9 ± 0.05	0.8 ± 0.1	0.5 ± 0.05	0.7 ± 0.09	0.5 ± 0.01
Stearic	0.2 ± 0.02	0.2 ± 0.02	0.2 ± 0.01	0.2 ± 0.01	0.2 ± 0.008

Reproduced with permission from Tappi Press, Ref. 15.
[a] Represents total acid fraction determined as the methyl esters.
[b] Chips frozen at start of experiment.
[c] Nonsterile chips aged 2 weeks to allow growth of natural background organisms.

Table 2.5. Use of a depitching organism in a TMP mill

DCM extractive content of secondary refiner pulp	−37.5%
Alum usage	−31.7%
Bleach usage	−36.9%
Brightness	+0.9%
Tensile index	+5.4%
Tear index	+3.4%
Burst index	+3.3%

Reproduced from Ref. 17.

oleic acid is decreased by 41%. The results of Cartapip treatment, such as pitch removal and maintenance of chip brightness and improved paper machine runnability, have been documented by use in mills. In a thermomechanical pulp mill, using Southern yellow pine, a trial was performed comparing a 2-week period using the Cartapip product on their wood chips to a 2-week period without Cartapip; the results are shown in Table 2.5.[17] Reductions in the dichloromethane extractive content of secondary refiner pulp caused expected reductions in use of alum, a pitch control chemical. Use of Cartapip resulted in 36.9% reduction in bleach usage, along with an increase in paper brightness of 0.9%. In addition, strength properties were increased due to the lower extractive content of the paper. A 2-month Cartapip 97 trial at a US thermomechanical pulp mill using Southern yellow pine showed significant reduction in the dichloromethane extractives of wood chips and an increase in burst index.[20] A 1-week trial was performed at a mill in northwestern USA using a blend of 60% lodgepole pine and spruce, and 40% fir and hemlock. Cartapip 97 treatment reduced the average dichloromethane extractive content of the chips reclaimed from the storage pile by 25%.[25] Both fungal treatment and natural microbial activity increased the free fatty acid content of the extractives. The increase in free fatty acid content results from initial hydrolysis of esterified fatty acids to free fatty acids. The free fatty acid content of the Cartapip-treated chips is lower than that of the naturally aged chips, indicating further metabolism and removal of these components by the fungus. To date, the wood species that have been effectively treated for pitch removal include many different pine species, spruce, birch, eucalyptus, aspen, and mixed tropical hardwoods. The dosage of the fungus product and storage time in the chip pile may vary depending upon temperature and wood species. Cartapip is a unique solution to pitch problems before the wood enters the mills.

Basidiomycetes also degrade extractives effectively.[17] These fungi extensively colonize wood. Brown rot fungi preferentially degrade wood polysaccharides, including cellulose, and cause rapid depolymerization. Brown rotted wood usually shows virtually no decrease in total lignin content. These fungi do not degrade lignin but modify it by oxidation and demethylation of methoxy groups. White rot fungi are the predominant degraders of lignin in nature. Some species of white rot fungi preferentially degrade lignin to wood polysaccharides and other species degrade all wood components simultaneously. The basidiomycetes have been shown to degrade pitch extractives. It has been reported that ethanol/benzene (1:2) extractive content of oak, treated with *Phaenerochaete chrysosporium*, decreased by 46% after 12 weeks.[26] Treatment of sterile Southern yellow pine wood chips with *P. chrysosporium* for 2 weeks resulted in a 21% reduction in dichloromethane extractives.[17] Two biopulping fungi, *Cereporiopsis subvermispora* and *Phanerochaete chrysosporium*, have been examined to lower the resin content of wood chips.[27,28] The comparison was made with the commercial depitching fungus *O. piliferum*. *C. subvermispora* and *O. piliferum* lowered the resin content of loblolly pine (2.55–2.64%) by 18–27% in 2 weeks and 33–35% in 4 weeks. In spruce wood, all three fungi lowered the resin content from 1.2 to 0.8–0.9% in 2 weeks (Tables 2.6, 2.7). Various Basidiomycetes have shown significant reduction in extractive content (Table 2.8).[17,19,29] About 41% reduction in dichloromethane extractives in Southern yellow pine was obtained with *P. chrysosporium* in 2 weeks at room temperature. *Hyphodontia setulosa*, *Perennipora subacida*, *P. gigantea*, and *Phlebia tremellosa* also performed well, reducing the extractives by about 40%. Treatment with two brown rot fungi, *Coniophora puteana* and *Gloeophyllum saepiarium*, and one white rot fungus, *Phellinus igniarius*, resulted in large reductions in extractive content.[17]

Molds are capable of degrading wood extractives to a lesser extent because they grow less prolifically than do other wood-inhabiting fungi. Molds grow

Table 2.6. Resin content (% of dry wood) of loblolly pine chips treated with *C. subvemispora* or *O. piliferum* after 1 to 4 weeks' incubation

	Start	1 week	2 weeks	3 weeks	4 weeks
Control	2.55	2.62	2.64	2.55	2.63
C. subvermispora		2.04	1.93	1.75	1.75
O. piliferum		2.39	2.16	1.81	1.70

Reproduced with permission from Walter de Gruyter, Ref. 27.

Table 2.7. Resin content of spruce chips treated with various fungi after 2 weeks' incubation and kappa numbers after sulfite cooking

Fungus	Resin content (%)	Kappa number
Control	1.2	25.9
C. subvermispora	0.8	20.1
O. piliferum	0.9	25.5
P. chrysosporium	0.9	23.8

Reproduced with permission from Walter de Gruyter, Ref. 27.

Table 2.8. Extractive content of sterile Southern yellow pine treated with various Basidiomycetes

Fungal species	Control extractives (%)	Treatment extractives (%)	Reduction (%)
Phanerochaete chrysosporium	2.19	1.30	41
P. subacida	3.34	2.01	40
P. gigantea	3.34	2.03	39
P. tremellosa	1.98	1.21	39
H. setulosa	1.98	1.20	39
Coriolus versicolor	1.98	1.28	36
Inonotus rheades	3.34	2.18	34
Trichaptum abietinium	4.70	3.13	33
C. subvermispora	3.34	2.18	29
Trichaptum biforme	4.70	3.13	24
Schizophylum commune	2.50	2.03	17
Sistotrerma brinkmanii	2.44	2.17	11
Pleurotus ostreatus	2.44	2.30	6
Alurodiscus sp.	2.44	3.70	0
Ganoderma collosum	2.29	1.75	23
Phellinus igniarius	2.44	2.48	0

Reproduced from Ref. 17.

best on wood that is very wet or that has been exposed to very high humidity for a long time. On softwoods, molds grow mainly on wood surfaces. On hardwoods, they can enter the wood at exposed parenchyma, vessels, or ruptured cells, and can move throughout the wood by penetrating pit membranes.[17,19]

Table 2.9. DCM extractive content of nonsterile Southern yellow pine treated with various molds

Fungal species	Extractives (%) control[a]	Extractives (%) treatment	Reduction (%)
Phlebia roqueforti	3.34	2.16	35
Leptographium terrebrantis	2.27	1.92	15
Verticicladiella truncata	2.27	1.96	14
Diplodia pinea	2.27	2.01	11
Codinaea sp.	2.27	2.07	9
Aureobasidium pullulans	3.34	3.26	2

Reproduced from Ref. 17.
[a] The control was chips that had been frozen at −20 °C since the start of the experiment.

Penicillium roqueforti, Penicillium funiculosum, Rhizopus arrhizus, and *Trichoderma lignorum* were found to degrade wood waxes in liquid culture.[30] *R. arrhizus, Gliocladium viride, Penicillium rubrium, T. lignorum,* and *Aspergillus fumigatus* were found to reduce the ethanol/benzene (1:2) extractive content of wood chips.[17] The wood chips were stored at 35 °C and sampled after 30 days. The type of wood chips tested was not given. Screening of various molds for their ability to degrade wood extractives revealed that the best fungus was *P. roqueforti* (Table 2.9).[19] It reduced the dichloromethane extractive content by 35% in 2 weeks in Southern yellow pine chips at room temperature.

2. Use of Enzymes. Lipase enzymes can also be used for removal of pitch. These enzymes catalyze the hydrolysis of triglycerides, which have been identified as one of the key components for pitch troubles. Lipases from *Aspergillus oryzae, Candida cylindraceae,* and *Pseudomonas* sp. have been used.[13,14,31–34] Some thermostable lipases have been also evaluated.[14] A few commercial preparations of lipases for pitch removal are available in the market.[14]

Mechanical Pulps

Lipase from *Candida cylindraceae* was used for pitch removal from groundwood pulp from Japanese red pine.[13] In lab-scale studies, when the resinous materials in the groundwood pulp slurry were treated with lipase, triglyceride (TG) was hydrolyzed and the pitch deposits were remarkably reduced (Table 2.10). Since the effect of lipase on reducing pitch deposits was confirmed, the technology was applied to the actual paper-making process.[13] The effects of enzyme concentration, reaction temperature, reaction time, and agitation mode on the hydrolysis of triglyceride were investigated in order to select optimum conditions of the lipase treatment for the mill trial.[13] It was found that it was necessary to have a strong mixing system to keep contact between

Table 2.10. Effect of lipase treatment on pitch deposition

Resinous solid[a]	Pitch-water suspension		Pitch-pulp suspension[b]	
Polar compounds/Nonpolar compounds	Control (mg)	Lipase treatment (mg)	Control (mg)	Lipase treatment (mg)
9/1	76	49	30	trace
7/3	137	53	46	trace
5/5	204	79	50	trace

Reproduced with permission from Tappi Press, Ref. 13.
[a] Contained 0.6 g of both compounds totally.
[b] Pulp consistency: 1%.

Table 2.11. Effect of lipase concentration on hydrolytic rate of TG

Lipase concentration ppm/4% pulp slurry	Hydrolytic rate of TG (%)
5	74.2
3	50.0
1	12.9

Reproduced with permission from Tappi Press, Ref. 13.

enzyme and triglyceride for the effective reaction. Under sufficient mixing conditions, lipase 5000 U/kg groundwood pulp could hydrolyze more than 80% of the triglyceride in the surface pitch within 2 h. No effect of the lipase treatment on the brightness and strength properties of the pulp was observed. The effect of lipase treatment on the pitch deposition was investigated at the No.9 paper machine in the Ishinomaki mill of the Jujo Paper Company. The lipase was added to the groundwood pulping line and the hydrolytic rates were determined. In the three trials, the lipase was added into 4% groundwood pulp slurry at the rates of 5, 3, and 1 ppm, respectively. The hydrolytic rates of triglyceride were 74.2, 50, and 12.9%, respectively (Table 2.11). During the trials, pitch deposits were rarely observed at the stock chest and the drier felt rolls. Also, the frequency of pitch holes in the paper web was notably reduced and the amount of pitch control agents, such as talc, was reduced. A long-run mill trial of 1 month was also conducted on the same paper machine. The lipase was added to the groundwood stock chest at the rate of 3 ppm. The content of surface pitch and triglyceride in the pulp stock, amount of wet pitch (pitch deposit in the wire and press section), number of holes in the paper web caused by the pitch, and visible pitch deposition on the wall of the chest were measured during the production of yellow telephone-directory paper and

Table 2.12. Amount of pitch and TG in the lipase treatment operation

Date	Content of unseasoned wood (%)	GP chest		Mixing chest		Stock inlet		Unseasoned wood to GP	
		Pitch[a]	TG[b]	Pitch	TG	Pitch	TG	Pitch	TG
5/5	0	2.18	6.9	1.50	0.8	1.70	Trace	–	–
5/6	0	1.48	8.5	1.79	0.8	1.85	Trace	–	–
5/10	20	1.94	13.5	2.07	Trace	2.18	Trace	–	–
5/21	20	2.37	11.4	2.45	1.2	1.41	Trace	2.67	21.3
5/26	30	2.09	15.7	1.96	1.3	2.01	Trace	–	–
5/30	30	2.02	19.2	2.38	2.0	1.61	Trace	2.76	17.6
6/10	40	1.86	16.2	2.49	5.9	1.76	4.6	–	–
6/18	40	2.02	17.3	2.53	5.1	1.64	5.8	2.84	20.1
6/20	50	1.94	16.1	2.53	2.7	2.01	2.9	–	–
6/21	50	1.87	19.0	2.34	4.4	1.87	2.2	2.57	24.7

Reproduced with permission from Tappi Press, Ref. 14.
[a] All measurements in percent based on oven-dry wood.
[b] All measurements in percent based on pitch.

newsprint. Lipase treatment was found to have a great effect on preventing pitch troubles during the long-run mill trial.

Lipases from *Aspergillus oryzae* for control of pitch were also used.[31] It was found that lipases rapidly hydrolyzed triglycerides. Normally, dosages of 100 to 500 g enzyme preparation per ton of pulp added directly to the whitewater system were used for 90% destruction of triglycerides.

A commercial preparation of lipase (Resinase A 2X) for the control of pitch was also used.[14] The effect of lipase dose on rate of triglyceride hydrolysis was investigated in laboratory experiments. More than 70% of triglyceride was hydrolyzed at 500 ppm of lipase. At the Yatsushiro mill of Jujo paper Co., the plant-scale trial for enzymatic pitch control with Resinase A 2X revealed that the enzymatic treatment reduced triglyceride content at the inlet of the mixing chest to less than 6%, while 50% unseasoned wood was used[14] (Table 2.12). During the entire trial period, the removal frequency was kept at less than 0.3 times per day by enzymatic treatment. Enzyme addition guaranteed a normal and stable operation at the paper machine with a relatively high unseasoned wood mixture. Since the trial (1990), enzymatic pitch control and unseasoned wood have been continuously used at the Yatsushiro mill. The groundwood pulp produced from fresh wood was found to give a brightness 3.5 points higher than that from wood which had been seasoned to reduce the pitch. The mill has reduced the cost of seasoning and bleaching chemicals. At the Ishinomaki mill, the plant-scale trial with Resinase A 2X had a dramatic effect on all pitch-related parameters. The amount of pitch was reduced to about 50% of the normal level. The significant decrease in the number of heavy and light defects resulted in less frequent recycling of substandard paper product. The use of micronized talc was reduced. During the trial, the paper dynamic friction coefficient (DFC) increased and white carbon additions were reduced to 1%. DFC increased with enzyme addition, but its effect leveled off with addition of more than 1000 ppm. The DFC of paper is normally affected by extractives, pigments, and fillers. Triglyceride was found to be one of the key compounds for decreasing DFC, since the enzyme hydrolyzes triglyceride only in pitch, and the other compounds may remain unchanged.

Fujita et al.[14] developed a thermostable fungal lipase which was found to be effective at 80 °C. Mill trial with this enzyme at a concentration of 1000 ppm resulted in 75% hydrolysis of triglycerides. The results were better than those obtained in the laboratory-scale experiments.

Sulfite Pulp

The other common source of pitch deposits, sulfite pulp, also responds to lipase treatment. Fisher and Messner assessed on the laboratory scale a method of triglyceride removal from sulfite pulp that utilizes the hydrolyses of the triglycerides by lipases and the removal of the liberated fatty acids with sodium hydroxide solution.[32] Pitch fractions with and without triglycerides and free fatty acids were chlorinated with chlorine water to simulate chlorination during the bleaching process. Chlorinated pitch without triglycerides and the fatty acids was found to be less adhesive than pitch containing free fatty acids or pitch that was treated with enzyme. It was found that (1) triglyceride content in pulp was a major contributor to pitch adhesiveness and (2) enzyme treatment followed by sodium hydroxide extraction reduced pitch adhesiveness. Fischer et al.[34] also performed pilot-scale experiments concerning the application of lipases to unbleached pulp. Twelve tons of pulp/day were treated with enzyme in a continuous process, the process conditions and the enzyme concentrations were varied, and the resins were analyzed for triglycerides and free fatty acids. The resin content of the pulp was reduced by about 60% during the process, from 0.68 to 0.28% of the dry pulps. The hydrolysis of triglyceride, C 57 fraction, which contains the main part of the most important unsaturated fatty acids (oleic acid and linoleic acid) was about 85%.

Kraft Pulp

Lipases from *Candida cylindraceae* have been shown to be effective in hydrolyzing triglycerides in extractives of fresh birch and birch kraft pulp.[35] The total amount of esterified compounds in fresh birch was decreased by 34%. The decrease was most significant in the fatty acids. About 50% of the esterified fatty acids and even 65% of the saturated fatty acids were hydrolyzed. The esters of fatty alcohols were also hydrolyzed to the same extent (Table 2.13). The amounts of esterified sterols and terpenoids remained the same. For treat-

Table 2.13. Effect of *Candida* lipase treatment on the amounts of free and esterified lipid compounds in birch sawdust

Compound group	Untreated birch sawdust (mg/g dw)			Lipase-treated birch sawdust (mg/g dw)		
	Free	Esterified	Total	Free	Esterified	Total
Fatty acids	*0.85*	*3.05*	*3.90*	*2.94*	*1.64*	*4.58*
saturated C_{12}–C_{28}	1.42	1.84	2.26	1.62	0.65	2.27
unsaturated C_{18}	0.43	1.21	1.64	1.32	0.99	2.31
Fatty alcohols C_{16}–C_{24}	*0.27*	*0.18*	*0.45*	*0.17*	*0.14*	*0.31*
Hydrocarbons C_{23}–C_{31}	*0.15*	*0*	*0.15*	*0.17*	0	*0.17*
Sterols	*0.17*	*0.30*	*0.47*	*0.18*	*0.32*	*0.50*
Terpenes and terpenoids	*0.19*	*0.51*	*0.70*	*0.15*	*0.55*	*0.70*
betulaprenols	0.01	0.13	0.14	0.01	0.17	0.18
triterpenoids	0.17	0.38	0.55	0.13	0.38	0.51
squalene	0.01	0	0.01	0.01	0	0.01
Total	*1.63*	4.04	*5.67*	*3.61*	*2.65*	*6.26*

Reproduced with permission from Tappi Press, Ref. 35.

Table 2.14. Amounts of lipid compounds in the birch sulfate pulp and degree of hydrolysis of esters by *Candida* and *Aspergillus* lipases

Compound group	Untreated sulfate pulp. (mg/g dw)			Hydrolysis of esters (%)	
	Free	Esterified	Total	*Candida*	*Aspergillus*
Fatty acids	*0.58*	*0.50*	*1.08*	50	67
- Saturated	0.26	0.25	0.51		
- Unsaturated	0.32	0.25	0.57		
Fatty alcohols	*0.06*	*0.03*	*0.09*	95	98
Hydrocarbons	0.03	0	*0.03*		
Sterols	*0.24*	*0.12*	*0.36*	45	62
Terpenes and terpenoids	*0.56*	*0.54*	*1.10*	0	18
- Betulaprenols	0.19	0.24	0.43	0	20
- Triterpenoids	0.27	0.30	0.57	0	16
- Squalene	0.10		0.10		
Total	*1.47*	*1.19*	*2.66*	28	43

Reproduced with permission from Tappi Press, Ref. 35.

ment of birch sulfate pulp, the lipases of *C. cylindraceae* hydrolyzed 30% of the esterified lipids as compared to 40% hydrolyzed by Resinase A (*Aspergillus* lipase). Esters of saturated fatty acids, alcohols, and sterols were hydrolyzed by both lipases. The *C. cylindraceae* lipase was incapable of degrading esters of betulaprenols and triterpenoids, whereas the *Aspergillus* lipase hydrolyzed them to the some extent (Table 2.14).

A combination of lipase and cationic polymers has been used in controlling pitch deposits during paper making.[36,37] Lipase was found to be effective in reducing the triglyceride content of the stock slurry via hydrolysis to glycerol and fatty acids and cationic polymer reduced the concentration of fatty acids released during hydrolysis.

Conclusions

Both the fungal and enzymatic methods are effective in reducing the generation of toxic extractives in effluent. Resin acids, free fatty acids, triglycerides, etc., which contribute significantly to pulp mill effluent toxicity, are reduced substantially by these methods. Besides reducing toxic extractives and offering an ecofriendly and nontoxic technology, these methods offer several other advantages, such as improved pulp and paper quality, reduction in bleach chemical consumption, reduction of effluent load, and space and cost savings in a mill wood yard by using unseasoned logs. By reducing the outside storage time of logs, these methods reduce wood discoloration, wood yield loss, and the natural wood degradation which occurs over longer storage time. The white carbon dosage is reduced by the enzymatic method because it hydrolyzes triglycerides and then increases the dynamic friction coefficient of the paper. With chemical pulps (sulfite), the application of lipase improves the properties of resins by lowering their adhesiveness.

The principal problem associated with the fungal pitch control method is the relatively long time required for chip pretreatment after fungal inoculation. There is a need to identify/develop fast-growing fungal cultures using classical or genetic engineering techniques which could do the job in less time. With lipase, improvement of the agitation system is required to improve the effectiveness of the enzyme. This could have a great effect on the saving of the enzyme. Half the amount of the lipase will be enough to obtain the same effect with effective agitation. To reduce the cost of the enzyme, there is a need to improve the productivity of microorganisms.

References

1. Owens JW: The hazard assessment of pulp and paper effluents in the aquatic environment: a review. Environ Toxicol Chem 1991; 10: 1511–1540.
2. Fengel D, Wegener G: Wood: chemistry, ultrastructure, reactions. Berlin, Walter de Gruyter, 1984.
3. Sunito LR, Shiu WY, Mackay D: A review of the nature and properties of chemicals present in pulp mill effleunts. Chemosphere 1988; 17: 1249–1290.
4. Chung LTK, Meier HP, Leach JM: Can pulp mill effluent toxicity be estimated from chemical analyses? Tappi J 1979; 62: 71–74.
5. Rabergh CMI, Isomaa B, Eriksson KE: The resin acids dehydroabietic acid and isopimaric acid inhibit bile acid uptake and perturb potassium transport in isolated hepatocytes from rainbow trout (*Oncorhynchus mykiss*). Aquat Toxicol 1992; 23: 169–180.

6. Pesonen M, Andersson T: Toxic effects of bleached and unbleached paper mill effluents in primary cultures of rainbow trout hepatocytes. Ecotoxicol Environ Safety 1992; 24: 63–71.
7. Tana J: Sublethal effects of chlorinated phenols and resin acids on rainbow trout *(Salmo gairdneri)*. Water Sci 1988; 2: 77–85.
8. Oikari A, Lonn BE, Castren M, Nakari T, Snickars-Nikinmaa B, Bister H, Virtanen E: Toxicological effects of dehydroabietic acid (DHAA) on the trout, *Salmo gairdneri* Richardson, in fresh water. Water Res 1983; 17: 81–89.
9. Oikari A, Nakari T: Kraft pulp mill effluents cause liver dysfunction in trout. Bull Environ Contam Toxicol 1982; 28: 266–270.
10. Servizi JA, Martens DW, Gordon RW, Kutney JP, Singh M, Dimitriadis E, Hewitt GM, Salisbury PJ, Choi LSL: Microbiological detoxification of resin acids. Water Pollut Res J Can 1986; 21: 119–129.
11. Smook GA: Handbook for pulp and paper technologists. Tappi Press, Atlanta, GA 1989; 7.
12. Allen LH: Pitch In wood pulps. Pulp Pap Can 1975; 76(5): 70–77.
13. Irie Y, Hata K: Enzymatic pitch control in papermaking system. Proceedings of 1990 Papermaking Conference, Atlanta, GA, 1990; 1–10.
14. Fujita Y, Awaji H, Taneda H, Matsukura M, Hata K, Shimoto H, Sharyo M, Sakaguchi H, Gibson K: Recent advances in enzymatic pitch control. Tappi J 1992; 74(4): 117–122.
15. Brush TS, Farrell RL, Ho C: Biodegradation of wood extractives from southern yellow pine by *Ophiostoma piliferum*. Tappi J 1994; 77(1): 155–159.
16. Hata K, Matsukura M, Taneda M, Fujita Y: Mill scale application of enzymatic pitch control during paper production. In: Enzymes in pulp and paper processing. Jeffries TW, Viikari L, eds. ACS Symp Ser 1996; 655: 280–296.
17. Farrell RL, Hata K, Wall MB: Solving pitch problems in pulp and paper processes by the use of enzymes or fungi. Adv Biochem Eng Biotechnol 1997; 57: 198–212.
18. Zabel RA, Morrell JJ: Wood microbiology: decay and its prevention. Academic Press, New York 1992; 476.
19. Breuil C, Iverson S, Yong Gao: Fungal treatment of wood chips to remove extractives. In: Environmentally friendly technologies for the pulp and paper industry. Young RA, Akhtar M, eds. John Wiley and Sons Inc., New York, 1998; 541–565.
20. Chen T, Wang Z, Gao Y, Breuil C, Hatton JV: Wood extractives and pitch problems: analysis and partial removal by biological treatment. Appita 1994; 47: 463–468.
21. Hoffman GC, Brush TS, Farrell RL: Cartapip: A biopulping product for control of pitch and resin acid problems in pulp and paper mills. Nav Stores Rev 1982; 102(3): 10–12.
22. Blanchette RA, Farrell RL, Burnes TA, Wendler PA, Zimmerman W, Brush TS, Snyder RA: Biological control in pulp and paper production by *Ophiostoma piliferum*. Tappi J 1992; 75(12): 102–106.
23. Farrell RL, Blanchette RA, Brush TS, Gysin B, Hadar Y, Wendler PA, Zimmerman W: Cartapip: A biopulping product for control of pitch and resin acid problems in pulp and paper industry. Proceedings of 5th International Conference on Biotechnology in the Pulp and Paper Industry Kuwahara M, Shimada M, eds. Tokyo: Unipublishers, 1992; 23–32.
24. Farrell RL, Blanchette RA, Brush TS, Hadar Y, Iverson S, Krisa K, Wendler PA, Zimmerman W: Cartapip, a biopulping product for control of pitch and resin acid problem in pulp mills. J Biotechnol. 1993; 30: 115–122.
25. Haller T, Kile G: Cartapip treatment of wood chips to reduce pitch and improve processing. Proc Tappi Pulp Conf 1992; 1243–1252.
26. Lim CS, Cho NS: Studies on the biological degradation of oak wood by white rot fungus *Phanerochaete chrysosporium*. J Tappi Korea 1992; 22: 32–44.
27. Fischer K, Akhtar M, Blanchette RA, Burnes TA, Messner K, Kirk TK: Reduction of resin content in wood chips during experimental biological pulping process. Holzforschung 1994; 48: 285–290.
28. Akhtar M, Blanchette RA, Myers G, Kirk TK: An Overview of biomechanical pulping research. In: Environmentally friendly technologies for the pulp and paper industry. Young RA, Akhtar M, eds. 1998; 309–340.

29. Schmitt EK, Miranda MB, Williams DP: Applications of fungal inocula in the pulp and paper industry. Presented in 7th International Conference on Biotechnology in Pulp and Paper Industry, Vancouver, BC, June 16–19, 1998.
30. Nilsson T, Asserson A: Treating wood chips with fungi to enhance enzymatic hydrolysis. US Patent 1969; 3 486, 969, 5 pages.
31. Gibson K: Application of lipase enzmes in mechanical pulp production. Tappi Pulp Conf 1991; Book 2: 775–780.
32. Fischer K, Messner K: Reducing troublesome pitch in pulp mills by lipolytic enzymes. Tappi J 1992; 75(2): 130–134.
33. Fischer K, Messner K: Biological pitch reduction of sulfite pulp on pilot scale. Proceedings of 5th International Conference on Biotechnology in Pulp and Paper Industry, Kyoto, Japan 1992; 169–174.
34. Fischer K, Puchinger L, Schloffer K, Kreiner W, Messner K: Enzymatic pitch reduction of sulfite pulp on pilot scale. J Biotechnol 1993; 27: 341–348.
35. Mustranta A, Fagernas L, Viikari L: Effects of lipases on birch extractives. Tappi J 1995; 78(2): 140–146.
36. Sarkar JM, Tseng AM, Hartig ET: Application of enzymes and polymers for controlling pitch in papermaking, Proceedings Papermakers Conference, Chicago, Illinois, Tappi Press 1995; 175–182.
37. Sarkar JM, Finck MR: Method for controlling pitch deposits using lipase and cationic polymer, US Patent 1993; 5, 256, 252: 25 pages.

3 Biopulping: a Less Polluting Alternative to CTMP

At present, pulp is produced from wood either by chemical delignification, mechanical separation of the fibers, or combinations of chemical and mechanical methods. Mechanical pulping methods are being used increasingly because they give much higher yields (>90%) than do chemical methods (40–50%) and require less capital investment. The main disadvantages of mechanical pulping are the high energy requirement, the low strength and the low brightness stability of mechanical pulps (Table 3.1). Addition of chemical pulp is often required to produce papers with adequate strength.

The use of white rot fungi for the treatment of wood chips prior to mechanical or chemical pulping is called biopulping. In biomechanical pulping, the aim in using fungi is to replace chemicals in pretreating wood for mechanical pulping, reduce energy demand, and increase paper strength. For chemical pulping, biopulping is intended to reduce the amount of cooking chemicals, to increase the cooking capacity, or to enable extended cooking, resulting in lower consumption of bleaching chemicals. Increased delignification efficiency results in an indirect energy saving for pulping, and reduces pollution.[1]

Mechanical Pulping Processes

The most commonly used mechanical pulping processes are stone groundwood (SGW) pulping, refiner mechanical pulping (RMP), thermomechanical pulping (TMP), and chemi-thermomechanical pulping (CTMP).[2] Softwoods, especially spruce, pine, balsam, fir, and hemlock, are most often used in mechanical pulping, but hardwood species such as aspen, poplar, birch, beech, and eucalyptus are also used.

The groundwood pulping process, developed in 1844, was the first process used for the production of paper from wood. This process involves pressing a log against a grindstone to pull off fibers, which are continuously washed away by a water stream. High temperatures in the refining zone caused by friction soften the wood and ease fiberization. The yield is high (about 95%) but because most of the lignin remains, the fibers are stiff and bulky. Paper produced from groundwood pulp has low strength and high color reversion, but the opacity is excellent. Groundwood pulping is an energy-intensive process.

Table 3.1. Comparison of wood pulps

Process	Yield (%)	Paper properties	Environmental impact	Net energy requirement (kWh/ton)	Bleaching
Kraft	45	High strength	Moderate		3–5 stages using –
Sulfite		Strength lower than kraft	High		Cl_2, ClO_2, O_2, O_3, HClO, H_2O_2 and/or enzyme
Refiner mechanical pulping (RMP)	95	Strength better than SGW; good opacity; high color reversion	Low	1975–2275	1–2 stage using alkaline H_2O_2
Groundwood pulping (SGW)	95	Low strength; excellent opacity; high color reversion	Low	1300–1675	Same as RMP
Thermomechanical pulping (TMP)	93	Strength better than RMP	Low	2175–2900	Same as RMP
Chemi-thermomechanical pulping (CTMP)	85–90	Strength better than TMP	Moderate	1375–1775	Same as RMP

In the RMP process, chips are processed through a rotating disk refiner. The refiner plate is made up of three zones to first break the chips, then to produce intermediate size fragments, and finally to produce single fibers. This setup produces fibers with better bonding properties, and thus better paper strength than SGW pulp. However, opacity is reduced, color reversion is similar, and the energy expenditure is increased compared to SGW pulping. RMP, as well as TMP and CTMP, is usually performed as a two-stage process. The first stage separates wood into individual fibers; the second stage loosens the structure of the fiber walls to increase fiber flexibility and fibrillation, improving fiber bonding and thus paper strength.

The TMP process is a modification of the RMP process involving a steam pretreatment at 110–150 °C to soften the wood followed by refining. In the first stage, the refiners are at elevated temperature and pressure to promote fiber liberation; in the second stage, the refiners are at ambient temperature to treat the fibers for paper making. The higher temperature during refining in the first step, 110–130 °C, softens the fibers and allows their recovery with minimal cutting and fines generation. The refining is performed just below the glass transition temperature of lignin, approximately 140 °C, so that separation of fibers occurs at the S-1 cell wall layer. This improves fibrillation and access to hydroxyl groups for hydrogen bonding. The high strength of this pulp relative to the other mechanical pulps has made it the most important mechanical pulping process. Energy requirements are 2175–2900 kWh/ton. Over two-thirds of this is used in the primary pressurized refining step, and less than one-third is used in the secondary atmospheric pressure refining step. The pulp yield is >93%. Solubilization of wood components results in relatively high BOD in mill effluents.

The chemithermomechanical pulping (CTMP) process is a further refinement of TMP which involves pretreatment of wood chips with sodium sulfite (about 2% on dry wood) at pH 9–10 or sodium hydroxide (with hydrogen peroxide in the alkaline peroxide method), then steaming at 130–170 °C, and finally refining. Liquor penetration is often achieved by a system that compresses the wood chips into a liquid-tight plug that is fed into the impregnator vessel where the chips expand and absorb the liquor. Unlike TMP, CTMP is effective with most hardwoods, particularly with the cold soda process. Chemical pretreatments of wood chips are used to enhance the strength properties of mechanical pulps. The addition of CTMP to pulp blends may reduce or eliminate the requirement for kraft pulp. Capital expenditures for a CTMP plant are one-fifth those of a kraft mill of comparable size.[3] The energy expenditures are decreased but the yield is also decreased to 85–91% by removing wood substance. The CTMP process generates more pollutants than other mechanical pulping processes and thus increases waste-treatment costs.

Early Development of Biomechanical Pulping

Recently, it was reported that pulps pretreated with fungi appeared quite similar to CTMP in terms of strength and properties[4] and it is expected that the energy requirement and generation of pollution will be much less, as compared to CTMP. White rot fungi have great potential for biotechnological applications. They not only produce the whole set of enzymes necessary for lignin degradation, but can also act as a transport system for these enzymes by bringing them into the depth of the wood chips and create the physiological conditions necessary for the enzymatic reactions.

Fresh wood chips stored for pulp production are rapidly colonized by a variety of microorganisms, including many species of fungi. These organisms compete vigorously while easily assimilable foodstuffs last, and then their population decreases. They are replaced by fungi that can degrade and gain nourishment from the cell wall structure polymers: cellulose, hemicelluloses, and lignin. Left unchecked, these last colonizers, mostly white rot fungi, eventually decompose the wood to carbon dioxide and water. Some white rot fungi selectively degrade the lignin component, which is what chemical pulping processes accomplish. It is these fungi which are useful for biopulping.

Fungal delignification of wood for biopulping was first seriously considered by industrial researchers of the West Virginia Pulp and Paper Company (now Westvaco Corporation) in the 1950s.[5] The researchers wondered whether wood chips could be inoculated with a lignin-degrading fungus during transport and storage, and thereby become partially pulped. They published a survey of 72 lignin-degrading fungi, summarizing knowledge about fungal degradation of lignin. In the 1970s, Eriksson's group at the Swedish Forest Products Laboratory (STFI) launched a more intensive investigation. A fungus isolated in Sweden, *Phanerochaete chrysosporium* (conidial state *Sporotrichum pulverulentum*), was characterized by a high optimum temperature for growth, rapid growth, and selective lignin degradation in incipient stages of birch wood decay. This fungus was proposed to be a useful wood defibrator in the pulping process. A US patent was obtained by STFI for the process.[6] Considerable efforts at STFI were directed towards developing cellulase-less mutants of selected white rot fungi to improve the selectivity of lignin degradation and thus the specificity of biopulping.[7] In one study, using spruce and pine wood, up to 23% energy savings and an increase in tensile index was noticed. On a large scale, success was achieved on bagasse,[8] while the results using wood chips were less encouraging. An energy requirement of 4800 kWh/ton for producing chemi-mechanical pulp (CMP) of 70°SR according to the Cuba-9 process (6% NaOH treatment at 90 °C for 10–20 min) was decreased to 1700 kWh/ton by pretreating the bagasse with fungi as shown in Table 3.2.[8] The strength properties of biochemi-mechanical pulp (BCMP) were better than those of CMP, but there was a small drop in the yield of BCMP due to fungal degradation of bagasse. STFI's work has been summarized in a number of publications on biomechanical pulping and related aspects.[6–14]

Preliminary research on biopulping was conducted at Forest Products Laboratory (FPL-USDA) at Madison, Wisconsin in the 1970s.[15] Kirk et al. at FPL showed that aspen wood chips treated with *Rigidoporus ulmaris* consumed less energy during pulping and produced stronger paper (FPL internal report 1972). Bar-Lev et al.[16] showed that treatment of a coarse mechanical pulp with *P. chrysosporium* decreased the energy requirement (25–30%) for further fiberization and improved the paper strength properties. Akamatsu and coworkers found that treatment of wood chips with any of ten white rot fungi decreased the mechanical pulping energy, and with three of the fungi (*Trametes sanguinea*, *Trametes coccinea* and *Coriolus hirsutus*), increased paper strength.[17]

Table 3.2. Energy requirement for chemimechanical pulp (CMP) and biochemimechanical pulp (BCMP) from bagasse

Refining equipment	Energy input (kWh/ton)	
	CMP	BCMP
Defibrator and PFI mill	4800	1700
Disk refiner	3100	2100

Based on data from Ref. 8.

Biopulping Consortia

A comprehensive evaluation of biomechanical pulping was launched in 1987 at the Forest Products Laboratory at Madison, Wisconsin after the establishment of Biopulping Consortium I, which involved the Forest Products Laboratory, the Universities of Wisconsin and Minnesota and 20 pulp and paper and related companies. The overall goal was to establish the technical feasibility of using fungal treatment with mechanical pulping to save energy and/or improve paper strength. It was assumed that fungal pretreatment would have less environmental impact than chemical pretreatments, which proved to be the case. The consortium research was conducted by seven closely coordinated research teams: fungal, pulp and paper, enzyme, molecular genetics, economics, engineering and scaleup, and information. The scientists of the consortium investigated all fields of research relating to biopulping.[1,5,18–32] The first report of Biopulping Consortium I, a 5-year research and information program, was published in 1993.[15] Biopulping Consortium II was established in 1992 and extended until June 1996, mainly for the scaleup of the process and other important aspects.

Several white rot fungi were screened for their biopulping performance using aspen wood chips.[20,22,25,33] Based on energy savings and improvements in paper strength properties, six fungi – *P. chrysosporium*, *Hypodontia setulosa*, *Phlebia brevispora*, *Phlebia subserialis*, *Phlebia tremellosa*, and *Ceriporiopsis subvermispora* – were selected. The energy-saving potentials of these fungi on biomechanical pulping of loblolly pine are given in Table 3.3.[32,33] Out of about 200 strains, two fungi exhibiting a great deal of intraspecific variation[27,30] seem to be especially useful for biopulping: *Phanerochaete chrysosporium* for hardwoods and *C. subvermispora* for hardwoods and softwoods. Various reactor types, including rotary drums,[20] stationary trays,[27,33] and a static bed bioreactor,[28] were tested on a 2–5 kg scale. The best results were obtained with strains of *C. subvermispora* on aspen and loblolly pine.[28] On aspen, energy savings of 48% were accompanied by increases in burst and tear indices of 40 and 162% respectively. The effects on loblolly pine amounted to 37% energy savings and

Table 3.3. Energy savings from biomechanical pulping of loblolly pine chips with different white rot fungi (4-week incubation)

Fungus	Energy savings (%)
Phanerochaete chrysosporium	14
Hyphodontia setulosa	26
Phlebia brevispora	28
Phlebia subserialis	32
Phlebia tremellosa	36
Ceriporiopsis subvermispora	42

Based on data from Ref. 32.

41 and 54% increase in burst and tear indices, respectively. The optical properties deteriorated with both types of wood. After 4 weeks of treatment, a weight loss of 6% for aspen and 5% for loblolly pine was measured. *C. subvermispora* proved to be superior to other selective white rotters.[32] However, *P. chrysosporium* has the advantage of competitiveness at temperature between 35 and 40 °C. When different strains of *C. subvermispora* were tried on pine, the energy savings ranged from 21 to 37%.[15] Adding nutrient nitrogen to the chips as a defined source (L-glutamate or ammonium tartrate) increased energy savings and improved strength properties but led to a high weight loss. Addition of a chemically undefined N source to aspen chips gave large biopulping benefits with low weight loss, using both *P. chrysosporium* and *C. subvermispora*. Wood batch was found to have little influence on the outcome of biopulping and chip storage method (fresh, air-dried, or frozen) and inoculum age and form (spore, mycelial suspension, or colonized chips) were without significant influence.

Other Developments

Another American group has also reported that aspen chips treated with *C. subvermispora* for 17 days required 20% less energy for pulping, while the refining energy of Norway spruce was reduced by 13%.[14] Strength properties were increased with aspen and spruce, but no increase was found with eucalyptus.

In Japan, biopulping research has been conducted mainly in industrial labs. Kobe steel and Oji paper seem to be the major industrial players. Kobe steel has obtained a broad US patent on the use of white rot fungi, particularly NK-1148, for the treatment of primary mechanical pulp to save energy.[34] Applications for four Japanese patents have been filed: (1) an inoculum method,[35] (2) two improved biopulping strains,[34] (3) a silo-type bioreactor,[36] and (4) treatment of chips during transport in ships with a white rot fungus to enhance pulping.[37] Nishida et al. have developed a screening method for selective lignin degradation which was used to identify the strains used in two of the above patents.[38,39]

Biomechanical pulping of nonwood fibers – straw,[40] kenaf[41] and jute[42] – was also successful. The energy consumption in refining was substantially lower and the strength properties higher for the fungal-treated bast strands.[41,42] The opacity and drainage properties were also superior for biomechanical bast pulps, but the brightness level was lower. Scanning electron microscopy of fungus-treated bast strands after refining showed that fibers appeared to separate more readily from adjacent fibers than in noninoculated treatments. Italian researchers studied treatment of nonwoody raw materials with a mixture of various type of enzymes for saving energy and reducing chemical consumption while maintaining good properties of CTM pulp.[43] The level of energy savings was found to depend on the type of raw material, ranging from

21% for rice straw up to 40% for kenaf bast. Enzyme treatment significantly improved tear index regardless of the cellulose source, whereas the tensile index decreased in wheat straw and kenaf bast samples. Burst index was slightly improved in all the biotreated samples except kenaf. Pulp yields of the biotreated samples were, without exception, significantly higher than those of the corresponding control samples. This was apparently due to the lower chemical charge needed for biotreated samples.

Scaleup and Implementation

One of the major costs foreseen during the scaleup of biopulping was for inoculum production. Akhtar and coworkers discovered that the amount of inoculum could be lowered to 5 kg/ton wood chips (dry weight basis) or less by adding an inexpensive and commercially available nutrient source, corn steep liquor (CSL), to the mycelial suspension.[33,44,45] Subsequent studies have also identified a better strain of *C. subvermispora* that gave up to 38% energy savings and improved tear index by 51% compared to the control in the presence of CSL.[33] After the practical and economical feasibility of biopulping was proved on the laboratory scale, accurate kinetic data were needed to determine the potential for the biopulping process on a large scale. Techniques for monitoring dry weight loss and growth rate as functions of time using carbon dioxide production data have been developed.[46] Other aspects of biomechanical pulping, like prediction of energy saving and brightness stability, were also studied.[47–49] Based on the practical and economical feasibility study of biopulping, a chip pile-based system was proposed (Fig. 3.1).[44]

Recently, the Biopulping Consortium conducted successful 5- and 100-ton trials in outdoor chip piles at the Forest Products Laboratory-USDA, Madison, Wisconsin. The results obtained were similar to those obtained in the laboratory-scale bioreactors: the fungal treatment saved 32% electrical energy.[44] Equipment and techniques for the pilot-scale treatments and scaleup to mill scale are discussed by Scott et al.[50] Contaminating microorganisms on the chip surfaces are controlled by brief exposure of the chips to steam, before addition of the fungal inoculum fortified with corn steep liquor. Metabolic heat from the growth of the fungus on the wood chips is removed by forced ventilation. The fungus *Phlebia subserialis* showed operational advantages over *C. subvermispora*, because less mycelial growth outside the chips and lower compressibility of the treated chips resulted in lower resistance to air flow.[50] Economic analysis of a mill-scale design suggested net savings of about $10/ton and a 25% annual return on investment.[50] The technology is now available for licensing through Biopulping International Inc., USA. At least one paper mill, Consolidated Papers, in Wisconsin Rapids, has already experimented with biomechanical pulping. The Biopulping Consortium has obtained three patents on the process [51–53] and three applications are pending.

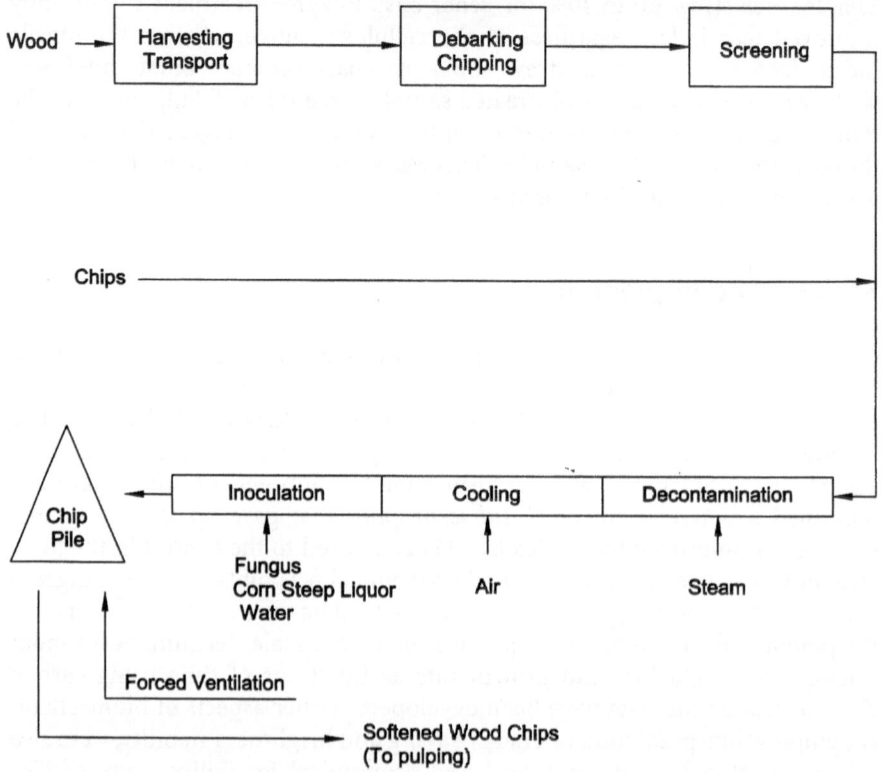

Fig. 3.1. Schematic of a proposed biopulping process. Reproduced from Ref. 44

Effluent Characteristics and Treatability

CTMP. Mechanical pulping processes release organic materials (sugars, low molecular weight lignins, extractives) from wood, and these materials appear in the effluent stream.[13] Mill effluents could contain as many as 100 potentially toxic constituents. Moreover, even low levels of several benign compounds could interact synergistically to produce a toxic effluent.[54] CTMP processes produce effluents of high color and BOD which may be difficult to treat; environmental considerations have kept this process from being used in many locations. Canadian CTMP mills have been reported to generate effluents having pollution loads of 35-65 kg BOD/ton pulp and 70-200 kg COD/ton pulp, depending on the pulping conditions and process as well as wood species used.[55] Lo et al.,[56] based on the characterization of effluents from 17 different sources in a Canadian CTMP pulp and paper mill, showed that the concentrations and loadings of BOD, COD, resin, and fatty acids (RFA), and other polluting constituents in the effluent from CTMP washing were very much higher

Table 3.4. Characteristics of bleached CTMP wastewater

Parameter	Value
pH	6.5
Acetic acid (mg/l)	1360
COD total (mg/l)	9300
COD soluble (mg/l)	5030
TSS (g/kg)	2.45
VSS (g/kg)	1.98
Resin acids (mg/l)	36–40

Based on data from Ref. 58.

Table 3.5. Composition of resin acids in bleached CTMP wastewater

Resin acid	Percentage
Pimaric	7
Sandarocopimaric	6
Isopimaric	12
Levopimaric	21
Dehydroabietic	22
Abietic	26
Neoabietic	6
	100

Based on data from Ref. 58.

than those in other effluents. The approximate pollutant amounts sent to the receiving water from the primary clarifier are BOD 63.3 kg, COD 90 kg, and RFA 0.43 kg per ton pulp. Resin and fatty acid constituents of wood are extracted during mechanical pulping/refining, and are major contributors to effluent toxicity.[57] CTMP wastewater characteristics are given in Table 3.4 and the composition of resin acids are represented in Table 3.5.[58] It has also been observed that the CTMP wastewater can be toxic to anaerobic bacteria (anaerobic treatment is sometimes given to CTMP effluents to reduce their pollution load) because of the presence of resin acids.[59,60] However, Kennedy et al.[58] reported that resin acids were toxic to anaerobic bacteria but were not responsible for all the toxicity in bleached CTMP effluent. Furthermore, resin acid shocks were found to be inhibitory to the batch anaerobic system. However, short-term shocks of resin acids up to concentrations of 400 mg/l (38 mg resin acid/g VSS/day) had little effect on UASB (upflow anaerobic sludge blanket) reactor efficiency. Long-term continuous exposure to abietic acid up to a concentration of 600 mg/l (60 mg resin acid/gVSS/day) did not significantly affect UASB reactor performance. In other words, anaerobic bacterial biomass can be acclimatized, to some extent, to the resin acid-containing wastewater from CTMP plants.

Biomechanical Pulping

The environmental impact of a new pulping process is a critical factor in assessing its viability. Ideally, a new process should be more environmentally compatible than existing processes. One question about biopulping is whether its effluents are more detrimental to the environment than are effluents from traditional mechanical pulping and refining. Specifically, do fungal metabolites or fungal degradation products of wood introduce additional toxicity to the effluent or significantly increase the BOD and COD load?

The microtox method of toxicity test has been used satisfactorily as a rapid screening method in the evaluation of acute toxicity of pulp mill effluents.[61] The microtox assay uses luminescent bacteria to measure acute toxicity. Samples of wastewater from the first refiner passes of aspen chips pretreated with either *P. chrysosporium* or *C. subvermispora* were analyzed for BOD, COD, and microtox toxicity.[15,62] Effluents from pulping fungus-treated chips were substantially less toxic than effluents from pulping raw chips (Table 3.6 and Fig. 3.2).[62] BOD values for effluents from fungus-treated pulps were slightly higher than for refiner mechanical pulps (RMP) of raw chips (Figs. 3.3 and 3.4). The COD for effluents from fungus-treated pulps were considerably higher than for RMP of raw chips (Figs. 3.5 and 3.6), probably because of the release of lignin degradation products.[62] Addition of nutrients to the aspen chips also affected the BOD and COD loads of effluents. BOD decreased after 4-week incubation with nutrient enriched *P. chrysosporium*, while BOD remained unchanged following incubation with nutrient-enriched *C. subvermispora.* (Fig. 3.3). COD remained unchanged after a 4-week incubation with nutrient-enriched *P. chrysosporium*, while COD increased in effluents from chips pretreated for 4

Table 3.6. BOD, COD, and toxicity of nonsterile aspen chips after treatment with *C. subvermispora*

Pulp[a]	BOD (g/kg pulp)	COD (g/kg pulp)	EPA toxicity[b] (100/EC_{50})
Raw chips	18	40	33
Control			
No nutrients	10	30	5
Nutrients added	12	33	9
Fungus treated			
No nutrients	10	33	6
Nutrients enriched	11	35	4

Based on data from Ref. 62.

[a] All pulps incubated for 4 weeks at 27 °C, except raw chips.

[b] EC_{50} is a measure of toxicity.

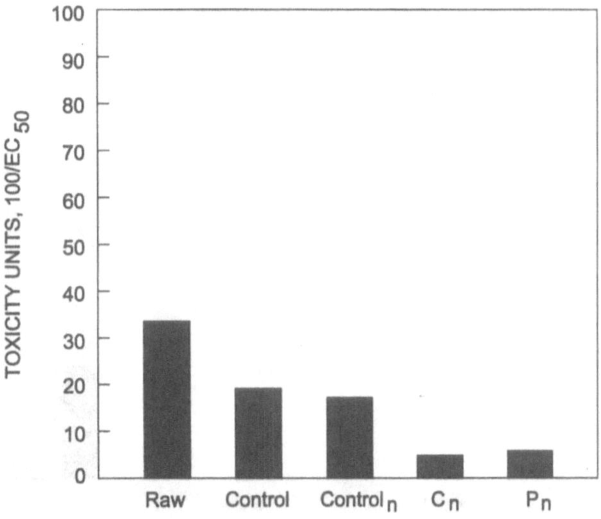

Fig. 3.2. Toxicity of aspen mechanical-pulp effluents with and without fungal pretreatments. All samples were sterilized and incubated for 4 weeks, except the raw chips, and effluent volume was made to 70 l/kg o.d. chips. *P* and *C* refer to *P. chrysosporium* and *C. subvermispora*, respectively, and subscript *N* refers to nutrient-enriched. Reproduced with permission from Tappi Press, Ref. 62

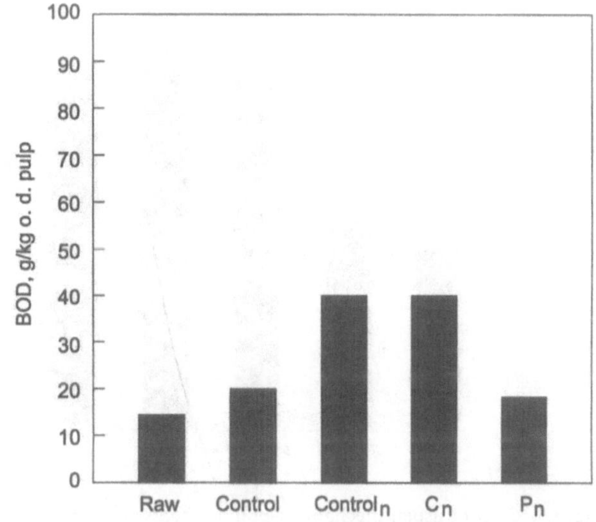

Fig. 3.3. BOD of aspen mechanical-pulp effluents with and without pretreatment with *C. subvermispora* or *P. chrysosporium* (all samples were sterilized and incubated for 4 weeks, except the raw chips). Reproduced with permission from Tappi Press, Ref. 62

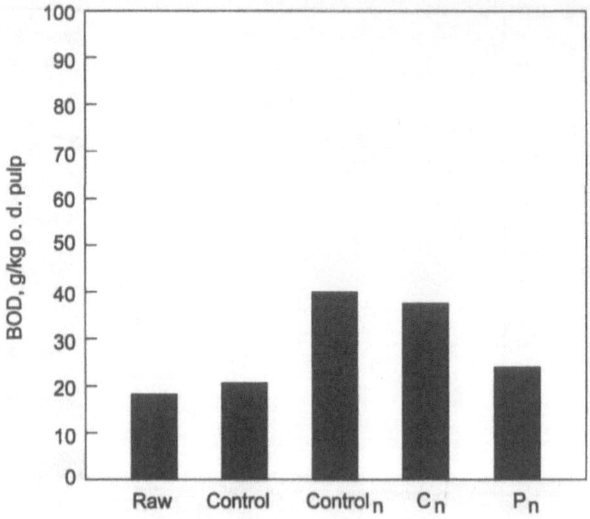

Fig. 3.4. Effect of nutrient enrichment on BOD of aspen mechanical-pulp effluents with and without *C. subvermispora* pretreatment (all samples were sterilized and incubated for 4 weeks, except the raw chips). Reproduced with permission from Tappi Press, Ref. 62

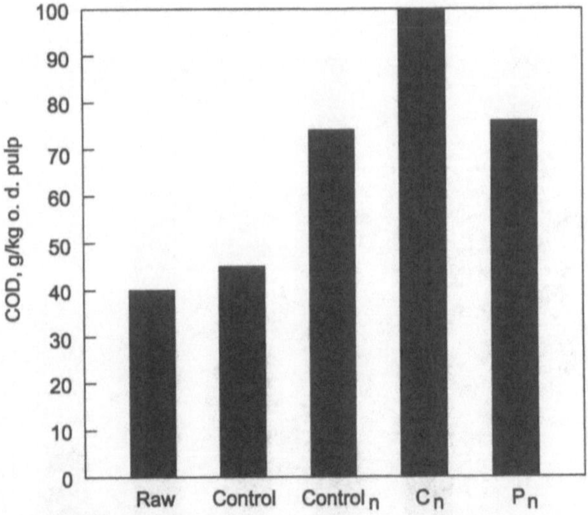

Fig. 3.5. COD of aspen mechanical-pulp effluents with and without pretreatment with *C. subvermispora* or *P. chrysosporium* (all samples were sterilized and incubated for 4 weeks, except the raw chips). Reproduced with permission from Tappi Press, Ref. 62

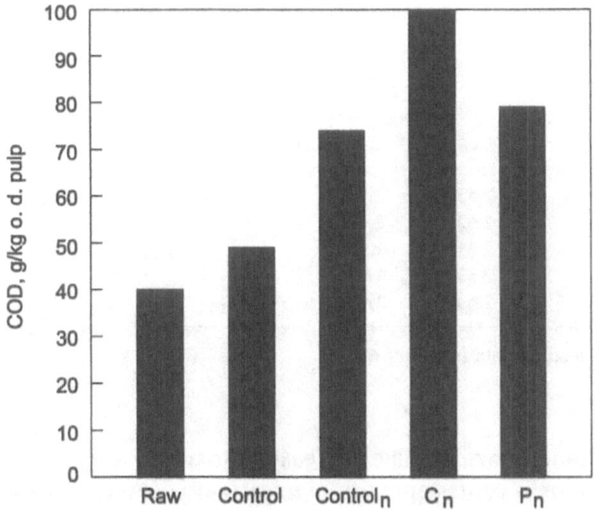

Fig. 3.6. Effect of nutrient enrichment of COD of aspen mechanical-pulp effluents with and without *C. subvermispora* pretreatment (all samples were sterilized and incubated for 4 weeks, except the raw chips). Reproduced with permission from Tappi Press, Ref. 62

weeks with nutrient-enriched *C. subvermispora* (Fig. 3.4). As different white rot fungi used for pulp treatment have differing nutrient requirements, an important factor in optimizing the biopulping process is to establish the minimum amount of nutrient required to assure fungal growth. The BOD and COD data indicate that *C. subvermispora* does not require much added nutrients. It has been concluded that the aspen biopulp effluents were less toxic, and probably contained considerably lower BOD and COD levels, than aspen TMP or CTMP at comparable yields[62] although no data on CTMP were available for comparison. It is likely that the environmental impact of CTMP effluent will be more than that of RMP or TMP effluents because of more extractives in the CTMP effluent. As the resin content of the wood chips decreases on fungal treatment (Table 3.7),[63] it is expected that the amount of these extractives will be less in the effluent after mechanical pulping/refining of these chips, resulting in reduced toxicity.

Published values for BOD load, based on the EPA survey for establishing effluent guidelines, are 20 kg/ton of o.d. pulp for TMP and 95 kg/ton of o.d. pulp for CTMP.[64] A commercial CTMP mill reported a BOD load of 45 kg/air dry ton, or approximately 52 kg/ton of o.d. pulp, for aspen CTMP.[65] *C. subvermispora* biopulping effluents contained approximately 24 kg/ton pulp and *P. chrysosporium* biopulping effluents contained 20 kg/ton pulp, at comparable yield,[62] less than half the commercial CTMP values.

P. chrysosporium has been used to remove color in the MyCoR (mycelial color removal) system,[66,67] and it was later discovered that this also decreased

Table 3.7. Effect of fungal treatment on resin content (% dry wood) of loblolly pine and spruce chips

Time (weeks)	Loblolly pine		Spruce		
	Control	*C. subvermispora*	Control	*C. subvermispora*	*P. chrysosporium*
0	2.55	2.55	1.2	1.2	1.2
1	2.62	2.04	–	–	–
2	2.64	1.93	1.2	0.8	0.9
3	2.55	1.75	–	–	–
4	2.63	1.75			

Based on data from Ref. 63.

effluent toxicity. The decrease of toxicity observed for the fungus-treated biopulps is consistent with biotreatment of mill effluents by the MyCoR system.

Biochemical Pulping

Fungal pretreatment of wood before chemical pulping has received relatively little attention. However, the fungi that are effective in biomechanical pulping have been tested as pretreatments for both kraft and sulfite pulping in a few studies.[68]

Kraft

Kraft pulps prepared from chips of aspen or red oak pretreated with *P.chrysosporium* for 10–30 days cooked faster and gave higher yields at a given kappa number (residual lignin content) than pulps from untreated chips.[69,70] The improved cooking properties of the fungus-treated chips were attributed to enhanced penetration of cooking liquor and a lower lignin content. The fungus-pretreated pulps were more responsive to beating and gave higher tensile strength than the control pulps.[69,70] The environmental consequences of the fungal treatments were not addressed in these studies, but the substantial (up to 17%) wood weight loss during the fungal pretreatment would negate the reported yield improvement, and the darkening of the wood by the fungus would probably require application of more bleaching chemicals.

Pretreatment of mixed hardwood chips with the Cartapip 97 fungus, *Ophiostoma piliferum*, for 21 days improved kraft pulping efficiency so that kappa number could be decreased by 29% or the active alkali concentration in the pulping liquor could be decreased by 20%.[71] Pulp yield was unaffected, and viscosity, an indicator of pulp strength, was increased. The improvements were attributed to enhanced liquor penetration resulting from removal of ray parenchyma cells, resin deposits, and pit membranes. The Cartapip pretreatment could be used to reduce the environmental impact of kraft pulp production by decreasing bleach chemical usage or by decreasing cooking chemical usage.[71]

Screening of 283 Basidiomycetes for ability to improve the efficiency of kraft pulping of pine chips revealed *Coriolus versicolor, Pycnoporus sanguineus,* and *Stereum hirsutum* as the most promising species.[72] Pretreatment of the pine chips with *S. hirsutum* for 9 weeks reduced the cooking time required to reach a kappa number of 28, but increased alkali consumption and lowered pulp yield and viscosity. The fungal pretreatment did not seem to give economical or environmental benefits.

Treatment of birch, maple, oak, seetgum, sycamore, or pine chips with an enzyme mixture containing cellulase and hemicellulase for 24-h disrupted the pit membranes blocking pores between cells and increased the longitudinal and transverse diffusion rates of sodium hydroxide in the chips.[73] Kraft pulping experiments with enzyme-treated sycamore chips confirmed the expected improvement in delignification; addition of pectinase to the enzyme mixture further lowered kappa numbers without reducing yield.[74] The pulps produced from enzyme-treated chips were easier to bleach with chlorine dioxide than control pulps, and had comparable strength properties. Reduced bleaching chemical use should result in lower effluent BOD, COD, and chloroorganic loadings.

Sulfite

Pretreatment of chips with white rot fungi increases the rate of their delignification in sulfite pulping.[68] In magnesium-based sulfite pulping, treatment of birch and spruce chips with *Phlebia tremellosa, P.brevispora, Dichomitus squalens,* and especially *Ceriporiopsis subvermispora* for 2–4 weeks significantly reduced pulp kappa number. However, the fungal treatments also reduced pulp strength and brightness. Pretreatment of loblolly pine chips with *C.subvermispora* for 2 weeks increased the rates of lignin and yield loss to the same extent in sodium bisulfite pulping, but preferentially enhanced delignification in calcium-acid sulfite pulping, and decreased shives production. The fungal treatment darkened the chips, so that equal amounts of bleaching chemicals were needed to brighten the treated and control pulps. BOD and COD levels were the same in effluents from the fungus-treated and control pulps, but the Microtox toxicity in the effluent from the fungus-treated chips was less than half that of the control. The reduced toxicity was attributed to biodegradation of resin and fatty acids by the fungus.[68]

In biochemical pulping, delignification during the fungal treatment partially replaces delignification by the pulping chemicals, without obvious economic or environmental benefit. Pretreatments with nonligninolytic fungi or enzymes which can remove toxic extractives and/or impediments to pulping liquor penetration offer more promise. Fungal delignification is more appropriate to pulp bleaching, where the amounts of lignin are smaller and chemical delignification is more problematic (see Chap. 5).

Advantages of Biomechanical Pulping

It is well established that waste loads increase as pulp yield decreases, temperature increases, or pulping chemicals are added. Consequently, the waste load produced by biopulping should be considerably lower and more benign than effluents currently produced by commercial CTMP mills. In fact, the effluents from fungus-treated mechanical pulps have been found to be less toxic,[62] although sometimes they may contain slightly higher BOD and COD than effluents from untreated pulp. These findings suggest that biopulping is environmentally compatible.

Biomechanical pulping saves a substantial amount of electrical energy or increases mill throughput significantly. It also improves paper strength compared to conventional refiner mechanical pulping. Recent studies suggest that fungal treatment is also effective for depitching wood chips. It decreases dichloromethane-extractable resin by about 30%,[63] including a 60% reduction in triglycerides, which are responsible for sticky deposits on the paper machine.[75] The cost of incorporating the fungal treatment process into existing mills is minimal. It is a relatively simple process that can be carried out in any woodyard.

Limitations and Future Prospects of Biomechanical Pulping

A biopulping process would require inoculum on a regular basis for commercial-scale applications, which would involve additional work and expense. Large-scale production of basidiomycetes is usually difficult. The fungal treatment is lengthy; a minimum 2-week incubation is required to obtain the desired benefits. At first glance, the long reaction time needed for the fungal process seems to be a great disadvantage. However, considering that wood chips are often stored at the mill for at least 2 weeks, time and space should be available in the pulp mill to introduce this process. In fact, this type of bioprocess is already familiar to the pulp and paper industry, as a similar process, based on Cartapip, has been used commercially in many US mechanical pulp mills since 1990.[76–78] The success of the Cartapip process shows that mills are able and willing to insert a biological step into their existing operations. Nevertheless, it is desirable to apply classical or molecular genetic methods to improve the effectiveness of the biopulping fungi, leading to shorter reaction treatment times.

Although a chip pile-based biopulping system has been designed and evaluated on a pilot scale, the process requires demonstrated long-term operation at mill scale.

Fungal treatment reduces the brightness of resulting mechanical pulps by as much as 15–20 Elrepho brightness points in 4 weeks and 8–10 points in 2 weeks. However, aspen biorefiner mechanical pulp (BRMP) could be readily bleached to 60% Elrepho brightness with 1% sodium hydrosulfite, and 80% brightness with a two-step bleach sequence using sodium hydrosulfite and alkaline hydrogen peroxide. Based on the accelerated thermal and photo-aging tests, the brightness stability of BRMP was found to be slightly lower than that of refiner mechanical pulp, but slightly higher than that of CTMP.[49]

Although biomechanical pulping seems to have a high potential to reduce pollution problems, very few data on the environmental performance of biomechanical pulping vis-à-vis chemithermomechanical pulping and other high-yield pulping processes are available. These data need to be generated from systematic studies comparing biomechanical pulping to other pulping processes, for equivalent quality and yield of pulp.

Biomechanical pulping is now at the stage of commercial introduction. It has a good chance of success due to the following four developments: (1) the discovery of a white rot fungus, *C. subvermispora*, which is effective on both hardwood and softwood species, (2) the finding that brief atmospheric steaming can decontaminate the surface of wood chips so that the fungus can take over, (3) the use of unsterilized corn steep liquor to dramatically reduce the inoculum required (from 3 kg to less than 1 g fungus/ton of wood (dry basis), and (4) the demonstration of a successful 40-ton outdoor chip pile where the fungal pretreatment saved 32% electrical energy.

References

1. Kirk TK, Akhtar M, Blanchette RA: Biopulping: seven years of consortia research. Proc Tappi Biol Sci Symp Atlanta: Tappi Press, 1994: 57–66.
2. Leask RA, Kocurek MJ: Mechanical pulping. Montreal: Joint Textbook Committee of the Paper Industry, 1987.
3. Karl W: The 1990s could be the decade for CTMP. Tappi J 1990; 73 (Supplement 2000 and Beyond): 90–92.
4. Wegner TH, Myers GC, Leatham GF: Biological treatments as an alternative to chemical pretreatment in high-yield wood pulping. Tappi J 1991; 74(3): 189–193.
5. Lawson LR, Jr., Still CN: The biological decomposition of lignin – literature survey. Tappi J 1957; 40(9): 56A–80A.
6. Eriksson KE, Ander P, Henningsson B et al: Method for producing cellulose pulp. US Patent 1976; 3,962,033.
7. Johnsrud SC, Eriksson KE: Cross-breeding of selected and mutated homokaryotic strains of *Phanerochaete chrysosporium* K-3: new cellulase-deficient strains with increased ability to degrade lignin. Appl Microbiol Biotechnol 1985; 21: 320–327.
8. Johnsrud SC, Fernandez N, Lopez P et al: Properties of fungal pretreated high yield bagasse. Nord Pulp Pap Res J 1987; Spec Issue 2: 47–52.
9. Ander P, Eriksson KE: Influence of carbohydrates on lignin degradation by the white-rot fungus *Sporotrichum pulverulentum*. Sven Papperstidn 1975; 78: 643–652.
10. Eriksson KE, Grünewald A, Vallander L: Studies of growth conditions in wood for three white-rot fungi and their cellulase-less mutants. Biotechnol Bioeng 1980; 22: 363–376.

11. Eriksson KE, Vallander L: Biomechanical pulping. In: Kirk TK, Higuchi T, Chang H-m, (eds) Lignin biodegradation: microbiology, chemistry, and potential applications. CRC Press, Boca Raton 1980; vol II: 213–233.
12. Eriksson KE, Vallander L: Properties of pulps from thermomechanical pulping of chips pretreated with fungi. Sven Papperstidn 1982; 85: R33-R38.
13. Eriksson KE: Swedish developments in biotechnology related to the pulp and paper industry. Tappi J 1985; 68(7): 46–55.
14. Setliff EC, Marton R, Granzow SG et al: Biomechanical pulping with white-rot fungi. Tappi J 1990; 73(8): 141–147.
15. Kirk TK et al: Biopulping: a glimpse of the future? Madison: Forest Products Laboratory. Res Rep FPL-RP-523 1993.
16. Bar-Lev SS, Kirk TK, Chang H-m: Fungal treatment can reduce energy requirements for secondary refining of TMP. Tappi J 1982; 65(10): 111–113.
17. Akamatsu I, Yoshihara K, Kamishima H et al: Influence of white-rot fungi on poplar chips and thermo-mechanical pulping of fungi-treated chips. Mokuzai Gakkaishi 1984; 30: 697–702.
18. Otjen L, Blanchette R, Effland M et al: Assessment of 30 white rot basidiomycetes for selective lignin degradation. Holzforschung 1987; 41: 343–349.
19. Blanchette RA, Burnes TA, Leatham GF et al: Selection of white-rot fungi for biopulping. Biomass 1988; 15: 93–101.
20. Myers GC, Leatham GF, Wegner TH et al: Fungal pretreatment of aspen chips improves strength of refiner mechanical pulp. Tappi J 1988; 71(5): 105–108.
21. Sachs IB, Leatham GF, Myers GC: Biomechanical pulping of aspen chips by *Phanerochaete chrysosporium*: fungal growth pattern and effects on wood cell walls. Wood Fiber Sci 1989; 21: 331–342.
22. Leatham GF, Myers GC: A PFI mill can be used to predict biomechanical pulp strength properties. Tappi J 1990; 73(4): 192–197.
23. Sachs IB, Leatham GF, Myers GC et al: Distinguishing characteristics of biomechanical pulp. Tappi J 1990; 73(9): 249–254.
24. Akhtar M, Blanchette RA, Myers G et al: An overview of biomechanical pulping research. In: Young RA, Akhtar M (eds) Environmentally friendly technologies for the pulp and paper industry. New York: John Wiley, 1998; 309–340.
25. Leatham GF, Myers GC, Wegner TH et al: Biomechanical pulping of aspen chips: paper strength and optical properties resulting from different fungal treatments. Tappi J 1990; 73(3): 249–255.
26. Sachs IB, Blanchette RA, Cease KR et al: Effect of wood particle size on fungal growth in a model biomechanical pulping process. Wood Fiber Sci 1991; 23: 363–375.
27. Akhtar M, Attridge NC, Myers GC et al: Biomechanical pulping of loblolly pine with different strains of the white-rot fungus *Ceriporiopsis subvermispora*. Tappi J 1992; 75(2): 105–109.
28. Akhtar M, Attridge, MC Blanchette RA et al: The white-rot fungus *Ceriporiopsis subvermispora* saves electrical energy and improves strength properties during biomechanical pulping of both hardwood and softwood chips. In: Kuwahara M, Shimada M (eds) Biotechnology in the pulp and paper industry. Kyoto: UNI Publishers, 1992; 3–8.
29. Blanchette RA, Akhtar M, Attridge MC: Using Simons stain to evaluate fiber characteristics of biomechanical pulps. Tappi J 1992; 75(11): 121–124.
30. Blanchette RA, Burnes TA, Eerdmans MM: Evaluating isolates of *Phanerochaete chrysosporium* and *Ceriporiopsis subvermispora* for use in biological pulping processes. Holzforschung 1992; 46: 109–115.
31. Akhtar M, Attridge MC, Myers GC et al: Biomechanical pulping of loblolly pine chips with selected white-rot fungi. Holzforschung 1993; 47: 36–40.
32. Leatham GF, Myers GC, Wegner TH: Biomechanical pulping of aspen chips: energy savings resulting from different fungal treatments. Tappi J 1990; 73(5): 197–200.
33. Akhtar M, Blanchette RA, Kirk TK: Biopulping: an overview of consortia research. In: Srebotnik E, Messner K (eds) Biotechnology in the pulp and paper industry. Recent advances in applied and fundamental research. Vienna: Facultas-Universitatsverlag, 1996; 187–192.

34. Kobe Steel: Lignin-degrading microorganisms having high activity and selectivity (for lignin). Japanese Patent 1988; EP295063.
35. Kojima Y: Inoculation of lignocellulosic materials with microbes in delignification. Japanese Patent Application 1988; 63/91077.
36. Akamatsu IH, Ueshima KY, Umeda TA et al: Biological pulping apparatus for wood chips. Japanese Patent Application 1988; 63/83537.
37. Heden CG, Eriksson KE, Johnsrud K: Japanese Patent Application 1988; 152/380.
38. Nishida T: Lignin biodegradation by wood-rotting fungi. V. A new method for evaluation of the ligninolytic activity of lignin-degrading fungi. Mokuzai Gakkaishi 1989; 35(7): 675–677.
39. Nishida T, Kashino Y, Mimura A et al: Lignin biodegradation by wood-rotting fungi. I. Screening of lignin-degrading fungi. Mokuzai Gakkaishi 1988; 34: 530–536.
40. Martinez AT, Camarero S, Guillén F et al: Progress in biopulping of non-woody materials: chemical, enzymatic and ultrastructural aspects of wheat straw delignification with ligninolytic fungi from the genus *Pleurotus*. FEMS Microbiol Rev 1994; 13:265–274.
41. Sabharwal HS, Akhtar M, Blanchette RA et al: Biomechanical pulping of kenaf. Tappi J 1994; 77(12): 105–112.
42. Sabharwal HS, Akhtar M, Blanchette RA et al: Refiner mechanical and biomechanical pulping of jute. Holzforschung 1995; 49: 537–544.
43. Giovannozzi-Sermanni G, Cappelletto PL, D'Annibale A et al: Enzymatic pretreatment of non-woody plants for pulp and paper production. Tappi J 1997; 80(6): 139–144.
44. Akhtar M, Blanchette RA, Kirk TK: Fungal delignification and biomechanical pulping of wood. Adv Biochem Eng Biotechnol 1997; 57: 159–195.
45. Akhtar M, Lentz MJ, Blanchette RA et al: Corn steep liquor lowers the amount of inoculum for biopulping. Tappi J 1997; 80(6): 161–164.
46. Wall MB, Cameron DC, Lightfoot EN: Biopulping process design and kinetics. Biotechnol Adv 1993; 11: 645–662.
47. Akhtar M: Biomechanical pulping of aspen wood chips with three strains of *Ceriporiopsis subvermispora*. Holzforschung 1994; 48: 199–202
48. Akhtar M, Blanchette RA, Burnes T: Using Simons stain to predict energy savings during biomechanical pulping. Wood Fiber Sci 1995; 27(3):258–264.
49. Sykes M: Bleaching and brightness stability of aspen biomechanical pulps. Tappi J 1993; 76(11): 121–126.
50. Scott GM, Akhtar M, Lentz MJ et al: Engineering, scale-up, and economic aspects of fungal pretreatment of wood chips. In: Young RA, Akhtar M (eds) Environmentally friendly technologies for the pulp and paper industry. New York: John Wiley, 1997;341–383.
51. Blanchette RA, Leatham GF, Attridge M et al: Biomechanical pulping with *C. subvermispora*. US Patent 1991; 5,055,159.
52. Akhtar M, Attridge MC, Koning JW et al: Method of pulping wood chips with a fungi using sulfite salt-treated wood chips. US Patent 1995; 5,460,697.
53. Akhtar M: Method of enhancing biopulping efficiency. US Patent 1997; 5,620,564.
54. Johnson I, Butler R: Paper mill effluents: a move to toxicity-based consents. Pap Technol 1991; 32(6): 21–25.
55. Environment Canada. Report WTC Bio-07-1988 1988.
56. Lo SN, Liu HW, Rousseau S et al: Characterization of pollutants at source and biological treatment of a CTMP effluent. Appita 1991; 44(2): 133–138.
57. Leach JM, Thakore AN: Toxic constituents in mechanical pulping effluents. Tappi J 1976; 59(2): 129–132.
58. Kennedy KJ, McCarthy PJ, Droste RL: Batch and continuous anaerobic toxicity of resin acids from chemithermomechanical pulp wastewater.J Ferm Bioeng 1992; 73(3): 206–212.
59. Welander T: An anaerobic process for treatment of CTMP effluent. Wat Sci Technol 1988; 20: 143–147.
60. Hall E, Cornacchio LA: Anaerobic treatability of Canadian pulp and paper mill wastewaters. Pulp Pap Can 1988; 89: T188–192.

61. Firth B, Backman C: Comparison of Microtox testing with rainbow trout (acute) and *Ceriodaphnia* (chronic) bioassays in mill wastewaters. Tappi J 1990; 73(12): 169–174.
62. Sykes M: Environmental compatibility of effluents of aspen biomechanical pulps. Tappi J 1994; 77(1): 160–166.
63. Fischer K, Akhtar M, Blanchette RA et al: Reduction of resin content in wood chips during experimental biological pulping processes. Holzforschung 1994; 48: 285–290.
64. Springer AM: Industrial environmental control. Pulp and Paper Industry. New York: Wiley-Interscience, 1986.
65. Jakko Poyry Inc. Multiclient report. Lindingo, Sweden : 2 1985.
66. Eaton DC, Chang H-m, Joyce TW et al: Method obtains fungal reduction of the color of extraction stage kraft bleach effluent. Tappi J 1982; 65(6): 89–92.
67. Joyce TW, Pellinen J: White rot fungi for the treatment of pulp and paper industry wastewater. Proc Tappi Environment Conference. Seattle, 1995.
68. Messner K, Koller K, Wall MB et al: Fungal treatment of wood chips for chemical pulping. In: Young RA, Akhtar M (eds) Environmentally friendly technologies for the pulp and paper industry. New York: John Wiley, 1997; 385–419.
69. Oriaran TP, Labosky P Jr, Blankenhorn PR: Kraft pulp and papermaking properties of *Phanerochaete chrysosporium*-degraded aspen. Tappi J 1990; 73(7):147–152.
70. Oriaran TP, Labosky P, Jr., Blankenhorn PR: Kraft Pulp and papermaking properties of *Phanerochaete chrysosporium*-degraded red oak. Wood Fiber Sci 1991; 23: 316–327.
71. Wall MB, Stafford G, Noel Y et al: Treatment with *Ophiostoma piliferum* improves chemical pulping efficiency. In: Srebotnik E, Messner K (eds) Biotechnology in the pulp and paper industry. Recent advances in applied and fundamental research. Vienna: Facultas-Universitatsverlag, 1996; 205–210.
72. Wolfaardt JF, Bosman JL, Jacobs A et al: Bio-kraft pulping of softwood. In: Srebotnik E, Messner K (eds) Biotechnology in the pulp and paper industry. Recent advances in applied and fundamental research. Vienna: Facultas-Universitatsverlag, 1996; 211–216.
73. Jacobs-Young CJ, Venditti RA, Joyce TW: Effect of enzymatic pretreatment on the diffusion of sodium hydroxide in wood. Tappi J 1998; 81(1):260–266.
74. Jacobs CJ, Vendetti RA, Joyce TW: Effect of enzyme pretreatments on conventional kraft pulping. Tappi J 1998; 81(2):143–147.
75. Fischer K, Messner K: Reducing troublesome pitch in pulp mills by lipolytic enzymes. Tappi J 1992; 75(2): 130–134.
76. Farrell RA, Blanchette RA, Brush TH et al: In: Kuwahara M, Shimada M, eds. Biotechnology in the pulp and paper industry. Kyoto: UNI Publishers, 1992; 27–32.
77. Brush TS, Farrell RL, Ho C: Biodegradation of wood extractives form southern yellow pine by *Ophiostoma piliferum*. Tappi J 1994; 77(1):155–159.
78. Wall MB, Brecker J, Noel Y et al: Cartapip 97 treatment of wood chips to improve chemical pulping efficiency. Proc Tappi Biological Sciences Symp Madison 1994; 67.

4 Pulp Bleaching with Xylanases

The kraft process accounts for 85% of the total pulp production in the United States. Bleached kraft pulp is a relatively high-value component of the total production of kraft paper. Kraft pulping removes most of the lignin and dissolves and degrades hemicelluloses without severely damaging cellulose. The kraft process results in excellent pulp from a wide variety of wood species. Unfortunately, kraft pulping also generates large quantities of chromophores. Chromophores are composed of residual lignin and carbohydrate degradation products. They are hard to extract because they are physically entrapped in and covalently bound to the carbohydrate moieties in the pulp matrix. Manufacturers use elemental chlorine and chlorine dioxide to bleach the chromophores and then they extract them, along with the residual lignin to make white pulp. Because of consumer resistance and environmental regulation of chlorine in bleaching, pulp makers are turning to oxygen, ozone, and peroxide bleaching, even though these may be more expensive and less effective than chlorine. One new technology has evolved to decrease the use of chlorine for bleaching, and that is the treatment of the pulp with xylanase enzyme. The use of xylanase enzymes to enhance the bleaching of the pulp was first reported in 1986.[1] The Finnish forest companies were the first in the world to start mill-scale trials in 1988.

In North America, the first mill trials of xylanase were carried out at Port Alberni in 1991 and ongoing usage in some mills started in 1992. By 1994, more than 8% of Canada's bleached pulp was produced by using xylanase enzyme.[2] Mill applications of xylanases have been widely reported.[3–6] Several reviews on different aspects of xylanase bleaching have been published.[7–12]

Effect of Xylanase Treatment on Pulp Bleaching

The use of xylanases in different bleaching sequences of kraft pulp consistently leads to a reduction in chemical consumption. However, the benefits obtained by enzymes are dependent not only on the type of pulp but also on the chemical bleaching sequence used, as well as the final target brightness and environmental goals of the mill. Originally, xylanases were applied in order to reduce the consumption of chlorine chemicals, especially elemental chlorine. Later, enzymes were combined with various ECF and TCF bleaching sequences to improve the otherwise lower final brightness value of pulp or to decrease

Table 4.1. Effect of xylanase treatment on conventional bleaching[a] for various kappa factor and chlorine dioxide substitutions

Chlorine dioxide substitution (%)	Control		Treated	
	Kappa factor	Final brightness (% ISO)	Kappa factor	Final brightness (% ISO)
10	0.200	90.0	0.05	90.0
	0.233	90.2	0.10	90.8
	0.266	90.3	0.15	91.8
40	0.150	89.2	0.05	90.0
	0.173	90.0	0.10	90.8
	0.200	91.0	0.15	91.8
70	0.200	90.0	0.05	88.7
	0.233	90.6	0.10	90.0
	0.250	90.9	0.15	90.3
100	0.150	83.5	0.10	87.4
	0.173	85.3	0.15	87.4
	0.200	87.0		

Based on data from Ref. 14.
[a] Bleaching sequence C/DEDED.

Table 4.2. Total chlorine and chlorine dioxide charges needed to achieve 90% ISO brightness for xylanase-pretreated and untreated hardwood pulps

Chlorine dioxide substitution(%)	Total chlorine charge on pulp (%)	
	Control	Treated
10	6.75	4.55
40	6.25	4.70
70	7.00	5.05

Based on data from Ref. 13.

the bleaching cost. Results from laboratory studies and mill trials show about 35–41% reduction in active chlorine at the chlorination stage for hardwoods and 10–20% for softwoods, whereas savings in total active chlorine were found to be 20–25% for hardwoods and 10–15% for softwoods.[13–26] The effect of xylanase on conventional bleaching is shown in Tables 4.1 to 4.4. The ability of the xylanase enzyme to reduce bleaching chemical consumption makes it possible to consider significant modifications in the bleaching sequence. It is possible to completely exclude the first chlorination stage in a C/D $E_{OP}DE_{P}D$ and replace it with an enzyme stage(X) to become $XE_{OP}DE_{P}D$. This has been verified

Table 4.3. Effect of xylanase treatment on chlorine requirement in conventional bleaching[a] with 50% chlorine dioxide substitution in the chlorination stage

Total active chlorine (%)	Final brightness (%)	
	Control	Treated
7.0	–	84.5
7.5	–	90.7
7.8	–	91.6
8.0	84.0	92.0
8.6	89.0	–
9.2	91.5	–

Based on data from Ref. 15.
[a] Bleaching sequence-D/CE$_O$D$_E$D.

Table 4.4. Effect of xylanase treatment on the increase in brightness improvement of eucalyptus kraft pulp in a conventional bleaching sequence

Enzyme	Bleaching sequence	Brightness (% ISO)		Increase in brightness due to enzyme treatment (Points)
		Control	Enzyme	
Cartazyme HS-10	CEHH	80.4	84.2	3.8
	CEH	78.1	83.0	4.9
Novozyme 473	CEHH	80.6	83.6	3.0
VAI xylanase	CEHH	80.4	82.5	2.1

Based on data from Ref. 23.
Kappa number of unbleached pulp = 25.5.

in a mill trial in Spain.[3] The advantage of this modified sequence is that filtrates from the E$_{OP}$ stage can be recirculated to the recovery system without risk of chloride-initiated corrosion. This approach helps contribute to closing the water circulation system of the pulp mill and minimize effluent discharge. The following sequence options were tested on hardwood kraft pulp with an incoming kappa number of approximately 14 and viscosity of 850–950 dm^3/kg:[27]

1. XE$_{OP}$DE$_P$D (enzyme / no chlorine)
2. E$_{OP}$DE$_P$D (no enzyme / no chlorine)

The advantage of options 1 and 2 is that filtrate from prebleaching can be kept separate from the filtrates of final bleaching and will therefore not contain components that could not be evaporated and burnt in the recovery boiler. The effluent load will, in these cases, be lower than for the conventional sequence, even if the final bleaching in options 1 and 2 will require slightly higher chlorine dioxide charges. The mill had previously tested the $E_{OP}DE_PD$ sequence without enzyme and could not achieve higher brightness values than 83% ISO. With Cartazyme HS-10, a brightness level of 88% was achieved (Table 4.5).

The potential for cost savings in TCF bleaching is particularly high. It has been shown in both laboratory-scale experiments and mill trials that hydrogen peroxide can be saved.[27] Compared to a QPP sequence, the XQPP sequence can save 5–10 kg H_2O_2/ton of pulp (Table 4.6). In 1992, in a joint venture between Korsnas AB and Dutch company Gist Brocades, Korsnas started a full-scale test on TCF bleaching with enzyme Korsnas T6 xylanase. The enzyme, which was completely free of cellulase side activity, was isolated from *Bacillus stearothermophillus*. It worked well at high pH and temperature and showed a good storage stability. With this enzyme, it was possible to produce TCF pulp of high brightness.[28] The Aanekoski mill in Finland used an enzyme with O_2 delignification and hydrogen peroxide bleaching to produce over 50000 tons of totally chlorine-free pulp.[29] A patent has been filed on this process. Dr. Eriksson's group in the University of Georgia, USA, has developed an Enzone process for bleaching kraft pulps of different raw materials. This process features an oxygen-xylanase-ozone-hydrogen peroxide sequence for hardwood pulp and an additional alkaline extraction stage between ozone and hydrogen peroxide stages for softwood pulps. The main advantage of this process is the

Table 4.5. Summary of plant-scale trial results with xylanase

	No enzyme/no chlorine	Enzyme/nochlorine
Pulp production, ADT/d	380	380
Chemical addition, kg/ADT		
ClO_2	12.7	13.8
H_2O_2	12.7	12.5
Final brightness, % ISO	81.5	88.2
Unbleached kappa number	13.8	14.0
E_{OP} Viscosity, dm^3/kg	799	809

Based on data from Ref. 19.

Table 4.6. Effect of xylanase treatment on chemical usage in different bleaching sequences

Pulp	Bleaching sequence	Chemical consumption (kg/t pulp)				Final brightness (% ISO)
		aCl	NaOH	O_2	H_2O_2	
Hardwood	$DE_{OP}DE_PD$	46.0	18.0	5.0	0.5	89.5
	$XDE_{OP}DE_PD$	37.0	16.0	5.0	0.5	91.0
Softwood	$C_{85}D_{15}E_ODE_PD$	94.0	46.0	5.0	1.0	89.1
	$XC_{85}D_{15}E_ODE_PD$	84.0	44.0	5.0	1.0	88.9

Based on data from Ref. 27.

xylanase step, which increases brightness by 3 to 8 points when compared with pulps bleached without it.[30]

Crestbrook Forest Industries, British Columbia, Canada, used the enzyme Albazyme 10 to remove the bottleneck in chlorine dioxide generation and to increase the production of ECF pulp. Munksjo AB, Sweden, combines the xylanase with the lignox process to produce TCF pulps, while Metsa-Sellu OY, Finland, use it to reduce AOX emissions from its kraft pulp mill. Albazyme 10 is presently available from Genencor International, USA, and Ciba-Geigy, Switzerland.[31] Metsa-Botnia's plant in Kasko now produces chlorine-free spruce pulp by using oxygen delignification and xylanase bleaching. The pulp is used in the production of supercalendered light-weight coated paper and tissue paper. Metsa-Botnia hopes to develop the process into a totally closed system and to be able to bleach birch pulp in similar ways.[32] The Weyerhaeurer Saskatchewan Ltd. mill in Prince Albert, Saskatchewan, Canada, has been using xylanase enzyme successfully since October 1992 to enhance the bleaching of kraft pulp.[6] On average, the enzyme-treated pulp uses 12% less chemicals than untreated pulp. The savings in chemicals is achieved without any significant problems with pulp quality or mill operations. The mill has been able to maintain dioxin-free effluent and is well within the AOX guidelines, and the savings in chlorine dioxide have helped to maintain the desired pulp output.

As the TCF market grows, Canadian mills, which are the world's largest exporters of market pulp, have started to investigate chlorine-free bleaching with xylanase enzymes. European paper makers are now requesting TCF pulps or bleaching pulps with extremely low AOX and / or TOX level in their effluents.[33] Many Finnish companies - Enso Gutzeit OY, Kimi Kymmene, Metsa-Sellu OY, Sunila OY, Veitsiluoto OY, and United Paper Mills - are conducting research to develop chlorine-free bleaching which involves the use of xylanase enzymes, oxygen, peroxide, and ozone.[34]

Xylanases have also been tested in the bleaching of sulfite pulps produced by acid Mg, Mg-bisulfite, two-stage sulfite, and alkaline methods.[10] Due to the different location and limited accessibility of hemicellulose in sulfite pulps, as compared to kraft pulp, lower hydrolysis was obtained with xylanase. Furthermore, the enzymatic pretreatment had no effect on the brightness or the kappa number after one-stage peroxide bleaching.

Xylanase pretreatment has led to reductions in effluent AOX and dioxin concentrations due to reduced chlorine requirement to achieve a given brightness.[4,18,24–26] It decreases the AOX in proportion to the decrease in chlorine compound usage, i.e., a 15% decrease in ClO_2 will decrease the bleach plant AOX discharge by 15%. The level of AOX in effluents was found to be significantly lower for xylanase pretreated pulps when compared with conventionally bleached control pulps (Tables 4.7–4.11).[1,17,25–27] when the softwood kraft pulp pretreated with xylanase was bleached to 90% ISO using a (CD)E_PDED sequence, the required kappa factor was reduced from 0.22 to 0.15, which was below the kappa factor of 0.18–0.19 required for the formation of

Table 4.7. Effect of xylanase treatment on effluent quality

	Control	Treated
Pretreatment	No	Yes
pH	–	4.8
Xylanase charge (unit/kg pulp)	–	5000
Time (h)	–	3
Temperature (°C)	–	37
Kappa number	32	31.3
Viscosity (mPa/s)	25.1	30.7
($C_{70}D_{30}$) Stage		
Cl_2 (kg/mt pulp)	19.2	15.0
ClO_2 (kg/mt pulp)	17.0	13.3
Active Cl_2 (kg/mt pulp)	64.0	49.9
(Eo) Stage		
NaOH (kg/mt pulp)	35.2	27.0
O_2 (kg/mt pulp)	5.0	5.0
Eo kappa number	4.8	4.4
Effluent properties		
Color (kg/mt pulp)	155	118
AOX (kg/mt pulp)	2.02	1.68
BOD (kg/mt pulp)	15	11
COD (kg/mt pulp)	74	62

Based on data from Ref. 1.
Unbleached kappa no. 32.0, viscosity 30.7 mPa s.

Table 4.8. Effect of xylanase treatment on AOX content of effluents[a]

Pulp	Chlorine charge (kg/mt pulp)	AOX (kg/mt pulp)	Brightness (% ISO)
Control	52	4.35	
	63	6.00	91.2
Xylanase-treated	43	3.00	
	52	3.75	91.0
	57	4.50	

Based on data from Ref. 25.
[a] Bleaching sequence $C_{80}D_{20}E_PDED$.

chlorinated dioxin and furans.[35] The AOX concentration in a combined effluent stream was reduced by 33% compared to control.[27] The effluent BOD doubled and there were increases in effluent COD and TOC (total organic carbon). The BOD/COD ratio also increased, indicating that the effluent was more amenable to biological degradation in a secondary treatment plant. Effluent toxicity remained essentially the same. In the same study, xylanase pretreatment of a

Table 4.9. Effect of xylanase treatment on effluent properties of softwood pulp

Parameter	Control		Xylanase-treated			W Control stage	X Xylanase stage
	$W(CD)E_P$	$W(CD)E_P$	$X(CD)E_P$	$X(CD)E_P$	$X(CD)E_P$		
Kappa factor	0.15	0.22	0.10	0.15	0.18		
C-stage residual (% on pulp)	-	0.04	-	-	0.32		
(CD) E_PDED brightness (% ISO)	88.1	92.8	85.8	91.6	94.1		
BOD (mg/l)	51	50	102	113	120	<50	78
COD (mg/l)	600	800	756	948	926	58	370
BOD/COD	0.085	0.063	0.14	0.12	0.13	0.86	0.21
TOC (mg/l)	503	668	585	787	752	54	291
Toxicity (% solution)	33.2	24.3	53.9	25.6	22.7	100	100
AOX (mg/l)	42.7	65.1	22.7	43.5	68.4	-	-

Substitution of chlorine dioxide in chlorination stage: 20%.
Kappa no. of control pulp: 28.3; kappa no. of xylanase-treated pulp: 28.3.
Effluent determinations made on combined W or $X(CD)E_P$ stage.
Based on data from Ref. 26.

Table 4.10. Effect of xylanase treatment on effluent properties of hardwood pulp

Parameter	Control		Xylanase-treated		
	W(DC)E	W(DC)E	X(DC)E	X(C + D)E	X(CD)E
Kappa factor	0.15	0.20	0.10	0.10	0.10
$(DC)E_1DE_2D$ brightness (% ISO)	89.3	90.0	89.5	89.1	89.3
BOD (mg/l)	261	259	212	283	280
COD (mg/l)	1100	1160	956	906	988
BOD/COD	0.24	0.22	0.22	0.31	0.28
TOC (mg/l)	429	464	385	379	402
Toxicity (% solution)	8.9	8.8	4.9	6.6	4.8
AOX (mg/l)	17.6	25.3	13.3	7.5	7.5

Substitution of chlorine dioxide in chlorination stage: 40%.
Kappa no. of control pulp: 12.2; kappa no. of xylanase-treated pulp: 11.6.
Based on data from Ref. 26.

hardwood kraft pulp under the same conditions led to a reduction of 35–40% in required chlorine in the chlorination charge.[27] AOX level in E_P stage effluent was 24% less than in the control, and the BOD/COD ratio was increased. In another study, use of xylanase reduced the amount of chlorophenols and other forms of organically bound chlorine in the spent bleach liquor and COD (Table 4.11).[17] Senior and Hamilton found that organochlorine content of the pulp

Table 4.11. Effect of xylanase treatment on effluent characteristics

Bleaching sequence	COD (kg/TP)	AOX (kg/TP)
No enzyme treatment, Cl_2 bleaching	58.1	2.6
No enzyme treatment, bleaching with mixture of 90% ClO_2 and 10% Cl_2	55.0	1.0
Enzyme treatment, bleaching with mixture of 90% ClO_2 and 10% Cl_2	40.0	0.6
Enzyme treatment, washing and bleaching with mixture of 90% ClO_2 and 10% Cl_2	40.0	0.6

Based on data from Ref. 17.

Table 4.12. Effect of xylanase treatment on strength properties of hardwood pulp[a]

Parameter	Reference	Enzyme
Brightness (% ISO)	88.4	89.2
Tensile index (Nm/g)	68.5	77.1
Burst index (kPa/m²/g)	4.5	5.0
Tear index (mN/m²/g)	2.4	2.9
Porosity (ml/min)	5.3	3.0
Bulk (cm³/g)	1.6	1.6
Viscosity (dm³/kg)	1040	1070

Bleaching sequence $DE_{OP}DE_PD$.
Based on data from Ref. 27.
[a] Oxygen-delignified hardwood pulp was used.

reduced by 41% at chlorine dioxide substitution level of 40%.[27] Bajpai et al. reported that TOCl content in extraction-stage effluent was reduced by 30% when the pulp was first treated with xylanase and then subjected to CEH bleaching.[24] In sequences employing 90% chlorine dioxide substitution and an alkaline extraction after the enzymatic treatment, AOX was reduced to 0.6 kg/ton of pulp and COD to 40 kg/ton of pulp, as compared with the reference treatments in which the AOX and COD loads were 1 and 55 kg/ton, respectively.[13]

Because the bleach plant effluent color is dependent on the chlorine usage, a decrease in chlorine compounds also decreases the effluent color.[11] There is an increase in the bleach plant effluent BOD of 2 to 5 kg/ton resulting from the release of low molecular weight xylan from the pulp. This BOD is of an easily degraded nature in secondary treatment systems and no change in the treated effluent BOD has been observed.

The enzyme-treated pulp show unchanged or improved strength properties (Tables 4.12–4.14).[20,23,27] Also, these pulps are easier to refine than the reference

Table 4.13. Effect of xylanase treatment on pulp properties

Pulp	Bleaching sequence	Brightness (% ISO)	Viscosity (mPa/s)
Hardwood	OXDP	87.7	17.5
kraft pulp	ODP	85.7	15.8
Softwood	OXPDP	88.6	16.3
kraft pulp	OPDP	84.5	14.9

Based on data from Ref. 20.

Table 4.14. Effect of xylanase treatment on strength properties of kraft pulp

Parameter	CEHH		XCEHH	
No. of revolutions (PFI)	1000	1800	1100	1900
Freeness (°SR)	30	40	30	40
Bulk (cm^3/g)	1.55	1.46	1.65	1.40
Tensile index (Nm/g)	47.1	52.5	48.9	52.32
Breaking length (m)	4798	5450	5018	5502
Burst index (kN/g)	3.10	3.81	3.35	3.78
Tear index ($mN/m^2/g$)	5.71	5.30	5.81	5.24
Double fold (no.)	15	42	20	40

Same chemical dose used in CEHH and XCEHH.
Based on data from Ref. 23.

pulps.[27] Improved viscosity of the pulp has been noted as a result of xylanase pretreatment.[13,20,36] This is probably caused by the selective removal of xylan as determined by the pentosan values. Xylan with lower DP than cellulose can be expected to lower the average viscosity of kraft hemicellulose. However, the viscosity of the pulp was adversely affected when cellulase activity was present.[24] Therefore, the presence of cellulase activity in the enzyme preparation is not desirable. The endoglucanases have been found to be most detrimental due to their action on the amorphous regions of pulp cellulose.[10] The cellobiohydrolases, on the other hand, do not affect pulp viscosity significantly even when used in relatively high enzyme dosages.[10] However, when both the activities are present in the enzyme preparation, rapid depolymerization of cellulose occurs due to their synergistic action.

Mechanism of Xylanase Action in Kraft Pulp Bleaching

Xylanases such as endoxylanases are xylan-specific enzymes. They catalyze the hydrolysis of xylose-xylose bonds within the xylan chain and solubilize only a fraction of the total xylan present. However, the actual enzymatic mechanism in bleaching is not yet well understood. One hypothesis suggests that precipitated xylan blocks or occludes extraction and that xylanase increases the accessibility.[37] This model is based on reports that xylan reprecipitates on the fiber surfaces.[38] More recently, it has been reported that no extensive relocation of xylan to the outer surface occurs during pulping, so the occlusion model might not be a sound premise.[10] Another possible explanation for xylanase action in bleaching is that the disruption of xylan chain by xylanase interrupts lignin-carbohydrate bonds improves the accessibility of the bleaching chemicals to the pulps, and facilitates easier removal of solubilized lignin in bleaching.[39] Skjold-Jorgensen et al. found that xylanase treatment decreased the demand for active chlorine for a batch kraft pulp by 15% but decreased active chlorine of pulp from a continuous process by only 6 to 7%.[40] They also showed that DMSO extraction of residual xylan does not lead to an increase in bleachability but that xylanase treatment does. This shows that DMSO-extractable xylan is not involved in bleach-boosting. Paice et al. have shown that the prebleaching effect on black spruce pulp is associated with a drop in the degree of polymerization, even though the xylan content decreases slightly.[39] Prebleaching thus appears to be associated with xylan depolymerization, even though not necessarily with solubilization of the xylan-derived hemicellulose components. Senior and Hamilton have shown that xylanase treatment and extraction change the reactivity of the pulp by enabling a higher chlorine dioxide substitution to achieve a target brightness and that they raise the brightness ceiling of fully bleached pulps.[16]

Xylanase Treatment in Mills

Typically, xylanase is added as an aqueous solution to the pulp at the final brown stock washer. The brown stock, which (though washed) is highly alkaline (pH 9–12), must be neutralized with acid usually sulfuric acid to be compatible with enzyme treatment. The pulp is pumped to the high density storage tower where the enzyme acts. From the high-density storage tower, the enzyme-treated pulp is then pumped into the first bleaching tower where first contact with the oxidizing chemical destroys the enzyme. Unlike other bleaching chemicals, xylanases do not brighten or delignify the pulp. They modify the pulp to make the lignin more accessible to removal by other bleaching chemicals. The enzyme-treated pulp then passes through the bleach plant with decreased chemical requirement for bleaching.

Detailed laboratory work is generally needed to optimize and adapt the enzymatic treatment to individual existing mill conditions.[11] Interestingly, however, xylanase bleaching has been scaled up directly from laboratory scale to the large industrial scale (1000 TP/day) without intermediate pilot stages. It has also been observed that even higher brightness values can be reached on the full scale than those attainable in the laboratory, which is due to the more efficient mixing systems and higher pulp consistencies. No expensive capital investments have generally been necessary for full-scale runs. The most significant requirement is the addition of pH adjustment facilities. Xylanase pretreatment has been shown to be easily applicable with existing industrial equipment, which is a considerable advantage of this technology.

Bleaching with xylanase requires proper control of pH, temperature, as well as retention time.[7–10] The optimum pH and temperature for enzyme treatment vary among enzymes. Generally, xylanases derived from strains of bacterial origin are most effective between pH 6 to 9 while those derived from strains of fungal origin should be used within the pH range of 4 to 6. The optimum temperature ranges from 35 to 60 °C with different enzymes. To obtain the best results from enzyme use, enzyme dosage must be optimized in each single case. In addition, the pulp consistency must be optimized to obtain effective dispersion of enzyme and improve the efficiency of enzyme treatment. Screw conveyers and static mixers are examples of efficient mixing systems. Most of the bleaching effect is obtained after only 1 h of treatment. Usually, the reaction time is set to 2–3 h. Long reaction time must be avoided if cellulases are present. Commercial xylanase enzyme preparations consist mainly of endoxylanases. Most of the enzymes are active at acidic or neutral pH, although some of them function under alkaline conditions. More than 20 commercial xylanases for bleaching are available.[7–9] Xylanases are sold as concentrated liquids and the amount required per metric ton of pulp is very low, less than a liter. The cost of enzyme per ton of pulp varies and depends on the dosage required and the supplier. In 1995, the approximate cost of enzyme treatment was around $2/TP.[10] Due to the low enzyme price and low capital costs of the enzyme stage, the potential economic benefits of enzyme bleaching are significant.

Effect of Mill Operations on Xylanase Performance

Mill operations also affect the performance of the xylanase enzyme. The effect of raw material, pulping process, brown stock washing, and bleaching sequence should be assessed by laboratory testing prior to mill usage of enzymes.[11]

Among raw materials, the important distinction is between hardwood and softwood. The percentage of the bleaching chemicals saved by xylanase treatment is thus greater on hardwood than on softwood. Under good treatment conditions, the decrease in chlorine chemicals is about 20% on hardwood and 15% on softwood. The digestor operation affects the xylan content of the pulp significantly. For example, sulfite pulping destroys most of the xylan and thus sulfite pulp is not suitable for enhanced bleaching by enzyme treatment. In conventional kraft pulping, the xylan content depends strongly on the effective alkalinity. The lower the alkalinity, the higher the xylan content and the benefits from using xylanase enzymes. At high alkalinity (19–22%), much of the xylan is solubilized, which decreases the benefit of xylanase treatment. At low alkalinity (less than about 18%), the xylan structure is more stable, and the bleaching enhancement by xylanase is greater by up to two- to threefold over the high alkalinity pulp. This is often the case for pulp cooked to higher kappa number. Kraft pulping under severe conditions, such as conventional cooking of softwood to kappa number less than 23, also destroys much of the hemicellulose that is accessible to the enzyme. On the other hand, MCC or oxygen-delignified pulps with low unbleached kappa number respond well to enzyme treatment.[41] Tolan reported a much smaller enzyme benefit for batch-cooked pulp at kappa number 21 than for MCC and oxygen-delignified pulp at the same kappa number.[41] The MCC and oxygen-delignified pulps have hemicellulose structures that are similar to that for conventional, high kappa number pulps. Enzyme benefits have been achieved in mills with conventional, MCC, and O_2 delignification systems.[41] The brownstock black liquor properties vary greatly from mill to mill.[11] Some mills' black liquor can inhibit enzyme performance due to the presence of highly oxidizing compounds. This effect differs significantly among enzymes and should always be checked before proceeding with full-scale enzyme use. It is important to note that it is not necessary to wash the pulp after enzyme treatment (before chlorination) to achieve the enhanced bleaching. Identical enzyme benefits with and without a post-enzyme washing have been obtained. The bleaching sequence and brightness target influences the enzymes' benefit to the mill.[11] As a basis of comparison, about 15% of the ClO_2 is saved by using enzyme treatment on softwood, with ClO_2 bleaching to 90% brightness, with the chemical usages optimized both with and without enzyme treatment. The enzyme benefit is greater at higher brightness targets, especially near the brightness ceiling, and lower at lower brightness targets.

Benefits of Xylanase Treatment

Xylanase pretreatment of pulps prior to bleach plant reduces bleach chemical requirements and permits higher brightness to be reached. The reduction in chemical charges can translate into significant cost savings when high levels of chlorine dioxide and hydrogen peroxide are being used. A reduction in the use of chlorine chemicals clearly reduces the formation and release of chlorinated organic compounds in the effluents and the pulps themselves. The ability of xylanases to activate pulps and increase the effectiveness of the bleaching chemicals may allow new bleaching technologies to become more effective. This means that xylanase pretreatment may eventually permit expensive chlorine-free alternatives such as ozone and hydrogen peroxide to become cost effective. Traditional bleaching technologies also stand to benefit from xylanase treatments. Xylanases are easily applied and require essentially no capital expenditure. Because chlorine dioxide charges can be reduced, xylanase may help eliminate the need for increased chlorine dioxide generation capacity. Similarly, the installation of expensive oxygen delignification facilities may be avoided. The benefit of a xylanase bleach boosting stage can also be taken to shift the degree of substitution towards higher chlorine dioxide levels while maintaining the total dosage of active chlorine. Use of high chlorine dioxide substitution dramatically reduces the formation of AOX.

In totally chlorine-free bleaching sequences the addition of enzymes increases the final brightness value, which is a key parameter in marketing chlorine-free pulp. In addition, savings in TCF bleaching are important with respect both to costs and to the strength properties of the pulp. The production of TCF pulp has increased dramatically during recent years. Several alternative new bleaching techniques based on various chemicals such as oxygen, ozone, peroxide, and peroxyacids have been developed. In addition, an oxygen delignification stage has already been installed at many kraft mills. In the bleaching sequences in which only oxygen-based chemicals are used, xylanase pretreatment is generally applied after oxygen delignification to improve the otherwise lower brightness of the pulp or to decrease bleaching costs. The TCF sequences usually also contain a chelating step in which the amount of interfering metal ions in pulp is decreased. It has been observed that the order of metal removal (Q) and enzymatic (X) stages is important for an optimal result.[10] When aiming at the maximal benefit of enzymatic treatment in pulp bleaching, the enzyme stage must be carried out prior to or simultaneously with the chelating stage. In fact, the neutral pH of enzyme treatment is optimum in many cases for chelation of magnesium, iron, and manganese ions that must be removed before bleaching with hydrogen peroxide. The TCF technologies applied today are usually based on bleaching of oxygen-delignified pulps with enzymes and hydrogen peroxide.

Tolan et al. carried out a survey of mill usage of xylanase.[2] The most widely reported benefit of enzyme treatment is a saving in bleaching chemicals. The chemical savings was 8 to 15% with an average of 11% of the total chemical across the bleach plant. The other widespread benefits were in improved effluent including decreases in AOX of 12 to 25%, decreases in effluent color, and other improvements to the effluent. Other benefits of enzyme treatment reported included increased bleached brightness (1 point gain), tear strength (5% gain) and pulp throughput (10% increase).

Problems with Xylanase Treatment

The most common problems with xylanase treatment cited in a mill survey have been corrosion of equipment and maintaining the brownstock residence time.[2] Sulfuric acid corrosion of mildsteel has been encountered in several mills. The brownstock residence time must be maintained for as long as possible but usually for at least 1 to 2 h to obtain the maximum benefits of enzyme treatment. This sometimes means that the mill must maintain the storage tower nearly full, which curtails its ability to act as a buffer between the pulping mill and the bleach plant. Other problems reported with enzyme treatment included difficulties in application and in bleach plant control. These relate to subtle action of enzymes, which is not easily observed on-line or in rapid testing. A decreased tear strength and pitch formation were also reported in some mills.[2]

Future Developments

The development in xylanase bleaching is focusing on improved enzyme properties and improved enzyme performance. Improved properties include higher pH and temperature tolerance of the enzymes, to make the enzyme treatment operations more compatible with existing mill operations. Improved enzyme performance is being approached by tailoring the enzyme action more closely to the hemicellulose structure of the pulp, to result in a greater bleaching benefit or higher pulp yield.

Conclusions

Xylanase enzymes have proven to be a cost-effective way for mills to realize a variety of bleaching benefits without expensive capital investments, including:

1. Reducing AOX discharges, primarily by decreasing chlorine gas usage.
2. De-bottlenecking mills limited by chlorine dioxide generator capacity.
3. Eliminating chlorine gas usage for mills at high chlorine dioxide substitution levels.
4. Increasing the brightness ceiling, particularly for mills contemplating ECF and TCF bleaching sequences.
5. Decreasing cost of bleaching chemicals, particularly for mills using large amounts of peroxide or chlorine dioxide.

These benefits are achieved over the long term when the enzymes are selected and applied properly in the mill.

References

1. Viikari L, Ranua M, Kantelinen A, Sundquist J, Linko M: Bleaching with enzymes. In: Proc. 3rd Int. Conference on Biotechnology in the Pulp and Paper Industry, Stockholm, Sweden, 1986; 67–69.
2. Tolan JS, Olson D, Dines RE: Survey of mill usage of xylanase. In: Enzymes for pulp and paper processing. Jeffries TW and Viikari L (eds.). ACS Symp Ser 1996; 655: 25–35.
3. Turner JC, Skerker PS, Burns BJ, Howard JC, Alonso MA, Andres JL: Bleaching with enzymes instead of chlorine: Mill trials. Tappi J. 1992; 75(12): 83–89.
4. Jean P, Hamilton J, Senior DJ: Mill trial experiences with xylanase: AOX and chemical reductions. Preprints 80th Annual Meeting Technical Section, Montreal, Canada, Feb. 1–2, 1994; A229–233.
5. Scott BP, Young F, Paice MG: Mill scale enzyme treatment of a softwood kraft pulp prior to bleaching. Pulp Pap Can 1993; 94(3): 57–61.
6. Yee Y, Tolan JS: Three years experience running enzymes continuously to enhance bleaching at Weyerhaeuser Prince Albert. Pulp Pap Can 1997; 98(10): T370–T375.

7. Bajpai P, Bajpai PK: Realities and trends in enzymatic prebleaching of kraft pulp. Adv Biochem Eng / Biotechnol. edited by T. Scheper, Springer, Berlin Heidelberg New York, 1997; 56: 1–31.
8. Farrell RL, Viikari L, Senior DJ: Enzyme treatment of pulp. In: Pulp bleaching (Dence CW, Reeve DW. eds.) Tappi Press, Atlanta, 1996; 363–383.
9. Viikari L, Kantelinen A, Sundquist J, Linko M: Xylanases in bleaching: from an idea to industry. FEMS Microbiol Rev 1994; 13: 335–350.
10. Suurnakki A, Tenkanen M, Buchert J, Viikari L: Hemicellulases in the bleaching of chemical pulps. Adv Biochem Eng / Biotechnol. edited by T. Scheper, Springer, Berlin Heidelberg New York, 1997; 57: 261–287.
11. Tolan JS, Guenette M: Using enzymes in pulp bleaching: mill applications. Adv Biochem Eng / Biotechnol. Springer, Berlin Heidelberg New York, 1997; 57: 288–310.
12. Bajpai P, Bajpai PK: Biobleaching of kraft pulp. Process Biochem 1992; 27: 319–325.
13. Senior DJ, Hamilton J: Reduction in chlorine use during bleaching of kraft pulp following xylanase treatment. Tappi J 1992; 75(11): 125–130.
14. Senior DJ, Hamilton J: Bleaching with xylanases brings biotechnology to reality. Pulp Pap 1992; 66(9): 111–114.
15. Tolan JS, Canovas RV: The use of enzymes to decrease the chlorine requirements in pulp bleaching. Pulp Pap Can 1992; 93(5): 39–42.
16. Senior DJ, Hamilton J: Xylanase treatment for the bleaching of softwood kraft pulps: the effect of chlorine dioxide substitution. Tappi J 1993; 76(8): 200–206.
17. Enso-Gutzeit OY: Procedure for the bleaching of pulp. EP 0383,999, 1990.
18. Werthemann D: Prebleaching of *Pinus radiata* pulp using enzymes – technology to reduce AOX. Jpn J Pap Technol 1993; 10: 15–17.
19. Skerker PS, Labbauf MM, Farrell RL, Beerwan N, McCarthy P: Practical bleaching using xylanases: laboratory and mill experience with Cartazyme HS-10 in reduced and chlorine-free bleach sequences. Presented in Tappi Pulping Conference Boston, MA, 1992.
20. Yang JL, Lou G, Eriksson KE: The impact of xylanases on bleaching of kraft pulps. Tappi J 1992; 75(12): 95–101.
21. Allison RW, Clark TA, Wrathall SH: Pretreatment of radiata pine kraft pulp with a thermophillic enzyme. Part I. Effect on conventional bleaching. Appita 1993; 46(4): 269–273.
22. Allison RW, Clark TA, Wrathall SH: Pretreatment of radiata pine kraft pulp with a thermophillic enzyme. Part II. Effect on oxygen, ozone and chlorine dioxide bleaching. Appita 1993; 46(5): 349–353.
23. Bajpai P, Bhardwaj NK, Bajpai PK, Jauhari MB: The impact of xylanases in bleaching of eucalyptus kraft pulp. J Biotechnol 1994; 36(1): 1–6.
24. Bajpai P, Bhardwaj NK, Maheshwari S, Bajpai PK: Use of xylanase in bleaching of eucalypt kraft pulp. Appita 1993; 46(4): 274–276.
25. Senior DJ, Hamilton J: Use of xylanase to decrease the formation of AOX in kraft pulp bleaching. Proc of the Environmental Conference 1991 of the Technical Section, Canadian Pulp and Paper Association, Quebec, Canada, October 8–10, 1991; pp 63–67.
26. Senior DJ, Hamilton J: Use of xylanase to decrease the formation of AOX in kraft pulp bleaching. J Pulp Pap Sci 1992; 18(5): J165–168.
27. Dunlop N, Gronberg V: Recent developments in the application of xylanase enzymes in ECF and TCF bleaching. Preprints 80th Annual Meeting Technical Section, Montreal, Canada, Feb 1–2, 1994; A191–A196.
28. Sandstrom AS: Total chlorine-free bleaching using enzymes. Sven Papperstidn 1993; 7: 40–42.
29. Anonymous: No matter what you call it, chlorine-free bleaching is here to stay. Pulp Paper Can 1992; 93(5): 22–27.
30. Young J: Enzone bleaching, enzynk deinking advance to pilot plant trial. Pulp Pap., Nov 1994; 81–82.
31. Lavielle P: Three large-scale uses of xylanases in kraft pulp bleaching. Proceedings of Pan-Pacific Pulp and Paper Technology, Tokyo, Japan, Sept 8–10, 1992, Part A, pp 59–64.

32. Anonymous: Breakthrough in Finland – chlorine-free bleaching of sulphate pulp. Skogindustri 1991; 45(10): 10.
33. Worster H: Canadian mills begin chlorine-free bleaching as TCF pulp market grows. Pulp Paper 1993; 67(1): 117.
34. Anonymous: Development of bleaching technology in Finland. Paper Puu 1992; 74(2): 102–106.
35. Berry RM, Fleming BI, Voss RH, Luthe CE, Wrist PE: Towards preventing formation of dioxins during chemical pulp bleaching. Pulp Pap Can 1989; 90(8): T279–289.
36. Paice MG, Bernier R Jr, Jurasek L: Viscosity enhancing bleaching of hardwood kraft pulp with xylanase from a cloned gene. Biotechnol Bioeng 1988; 32: 235–239.
37. Kantelinen A, Hortling B, Sundquist J, Linko M, Viikari L: Proposed mechanism of the enzymatic bleaching of kraft pulp with xylanases. Holzforschung 1993; 47: 318–324.
38. Yllner S, Ostberg K, Stockmann L: A study of the removal of the constituents of pine wood in the sulphate process using a continuous liquor flow method. Sven Papperstidn 1957; 60: 795–802.
39. Paice MG, Gurnagul N, Page DH, Jurasek L: Mechanism of hemicellulose directed prebleaching of kraft pulp. Enzyme Microb Technol 1992; 14: 272–276.
40. Skjold-Jorgensen S, Munk N, Pederson LS: Recent progress within the application of xylanases for boosting the bleachability of kraft pulp. In: Kuwahara M, Shimada M (eds) Biotechnology in the pulp and paper industry. Uni Publishers, Tokyo, Japan, 1992; pp 93–99.
41. Tolan JS: Mill implementation of enzyme treatment to enhance bleaching. Proc 78th CPPA Annual Meeting, Montreal Canada, Jan 28–29, 1992; pp A163–168.

5 Pulp Bleaching with White Rot Fungi and Their Enzymes

Lignin is one of the major components in plants. Selective removal of lignin from plant cell walls has significant implications on the pulp and paper industry. Chlorine-based chemicals traditionally have been used to bleach pulps because they selectively and efficiently remove lignin. However, chlorinated organic compounds released in this bleaching process are toxic to the environment. The ever-increasing pressure from environmental protection authorities has forced the pulp and paper industry to seek an environmentally benign bleaching process. Biobleaching is one of the very promising alternatives for eliminating chlorine-based chemicals in the pulp-bleaching process. Since xylanase was found to improve the bleachability of kraft pulp,[1] enormous efforts have been made to develop hemicellulase-aided bleaching. The developments in hemicellulase aided bleaching are reviewed in Chapter 4. Xylanase enzymes are now being produced for the pulp industry by several companies around the world.[2–4] Many mills worldwide have experience in using xylanase prebleaching and a CPPA survey showed that as of early 1995, about 1 million tons/year or 8% of the bleached kraft pulp made in Canada was enzyme-treated.[5] Xylanase applications are often referred to as prebleaching or bleach boosting because the nature of the effect is to enhance the effect of bleaching chemicals rather than to remove lignin directly. The enzyme does not attack the lignin-based chromophores but rather the xylan network by which the residual lignin particles are surrounded and trapped. A limited hydrolysis of the xylan network is often sufficient to facilitate the subsequent chemical attack on lignin with various bleaching chemicals, without sacrificing yield. Mill-scale experience has demonstrated that xylanase pretreatment usually results in up-to 20–25% savings in bleaching chemicals, with simultaneous reduction of pollutant emissions.

The pulp and paper industry responded to environmental challenges not only by modifying the bleaching processes but also by developing new pulping procedures, resulting in pulp with much lower lignin content and therefore decreased demand for chemicals needed to bleach. Some of these developments, especially those using extended delignification, achieve an effect on pulp similar to that of xylanase, thus rendering the enzyme pretreatment less effective. It is well known that many of the fungi that attack wood in nature are able not only to hydrolyze hemicellulose but also to degrade lignin. Thus, there appears to be an opportunity to exploit them for direct attack on lignin in pulps, an attack which can be specific and effective even if the amount of residual lignin is low. Over the past decade, many laboratories have studied the potential of white rot fungi and their enzymes in bleaching kraft pulps, and promising results have been obtained.

White Rot Fungi

White rot fungi which extensively attack lignin in wood are able to bleach kraft pulps. Only a few white rot fungi have been tested for their ability to delignify kraft pulps. Kirk and Yang[6] were the first to attempt to bleach pulp with microorganisms. They observed that *Phanerochaete chrysosporium* and some other white rot fungi could lower the kappa number of unbleached softwood kraft pulp by up to 75%, leading to reduced requirement for chlorine during subsequent chemical bleaching. The pulp was incubated with the fungi in shallow stationary layers for several days, and then extracted with alkali. Kappa number reduction was inhibited by added nutrient nitrogen and enhanced by the oxygen enrichment of the atmosphere, as is lignin degradation by *P. chrysosporium*. Attack on the cellulose of the pulp was severe unless alternative carbohydrate sources were added to the cultures; even in the presence of glucose, the pulp showed a 60% drop in viscosity. Other fungi tested, including *Trametes versicolor*, had lesser effects. In contrast to the results of Kirk and Yang, Pellinen et al.[7] reported that *P. chrysosporium* failed to delignify unbleached softwood kraft pulp in stationary culture, but did so in agitated cultures. Tran and Chambers[8] found that the effects of culture conditions on delignification of unbleached hardwood kraft pulp by *P. chrysosporium* were similar to those observed by others with synthetic lignin or lignin in wood. They did not determine the effects of the fungal treatment on the bleachability or paper-making properties of the pulp. Screening of several white rot fungi at the Pulp and Paper Research Institute of Canada (Paprican) revealed that *Trametes (Coriolus) versicolor* could markedly increase the brightness of hardwood kraft pulp.[9] The fungal treatment was carried out in agitated, aerated cultures for 5 days. Under these conditions, *T. versicolor* performed better than *P. chrysosporium*. The kappa number was decreased from 12 to 8, and the brightness increased from 34 to 48%; it could be further increased to 82% with DED bleaching. In initial experiments, Paice et al. did not observe similar brightening of softwood pulp. Subsequently, *T. versicolor* was found to delignify and brighten softwood kraft pulps.[10] After 14 days of treatment with the fungus followed by alkaline extraction, the pulp had kappa no. 8.5 and could be bleached to 61% ISO brightness with a DED sequence, whereas without fungal treatment, the pulp had kappa number 24 and was bleached to only 33% brightness. Mycelium of *T. versicolor* immobilized in polyurethane foam was also able to delignify the pulp. *T. versicolor* caused a much smaller direct increase in the brightness of softwood kraft pulp than of hardwood kraft pulp. To determine the contributions of higher residual lignin contents (kappa numbers) and structural differences in lignins to recalcitrance of softwood kraft pulps to biobleaching, Reid and Paice tested softwood and hardwood kraft pulps cooked to the same kappa numbers 26 and 12.[11] A low-lignin-content (overcooked) softwood pulp resisted delignification by *T. versicolor*, but a high-lignin-content (lightly cooked) hardwood pulp was delignified at the same rate as a normal softwood pulp. The longer time taken by *T. versicolor* to brighten softwood kraft pulp than hardwood kraft pulp was found to result from the higher residual lignin content of the softwood pulp. Softwood pulps whose lignin contents were decreased by extended modified continuous cooking or oxygen delignification to kappa numbers as low as 15 were delignified by *T. versicolor* at the same rate as normal softwood pulp.[11] More intensive O_2 delignification, like overcooking, decreased the susceptibility of the residual lignin in the pulps to degradation by *T. versicolor*.[11] This fungus was found to delignify and brighten kraft pulp in dilute (1–2% consistency) agitated suspensions. Delignification of both hardwood and softwood pulps with *T. versicolor* was found to be accompanied by a moderate decrease in viscosity, indicating some cellulose depolymerization.[10,12] Addition of excess glucose to repress cellulose biosynthesis did not prevent the loss in viscosity.[13] However, the damage to cellulose was not severe enough to cause important strength losses. Paper sheets made from pulp delignified with *T. versicolor* were slightly stronger than those made from unbleached pulp.[9,10] Delignification by *T. versicolor* was not restricted to conditions of secondary metabolism. The fungus degraded lignin while growing vigorously.[14] Ho et al. reported that when kraft pulp was added to inoculum of *T. versicolor*, the lag before the onset of brightening in subsequent pulp treatment was reduced.[12] The presence of pulp induced production of manganese peroxidase and accumulation of organic acids that can chelate Mn^{3+}.

Addleman and Archibald investigated the ability of 10 dikaryotic and 20 monokaryotic strains of *Trametes versicolor* to bleach and delignify hardwood and softwood kraft pulps.[15] The isolates were found to vary in their bleaching ability. Dikaryotic strains produced brightness changes ranging from −2 to +22 points in hardwood kraft pulp; monokaryons tended to give a higher brightness increase, with a maximum of 26 points. A monokaryon, 52 J, isolated by protoplasting the dikaryon originally used at Paprican, gave slightly higher brightening than its parents and was adopted for further work. In addition to better bleaching, the monokaryon had the advantage of less biomass production, no dark pigment formation, and a simpler genome.[15] In Japan, a 5-day fungal (F) treatment of hardwood kraft pulp with IZU-154 replaced a CE_1DE_2D sequence with an FCED sequence, yielding a target brightness of 88% ISO with 72% less chlorine, 79% less NaOH, and 63% less ClO_2 (Table 5.1).[16] With softwood kraft pulp, similar chemical savings were achieved (Table 5.2).[17] The yield and burst strength of the pulp bleached with a CED sequence after delignification with IZU-154 were equivalent to those of a control bleached with a CEDED sequence. However, the tensile index and tear index of the fungus-treated pulp were decreased by 9 and 2.6%, respectively. To establish an absolutely chlorine-free bleaching process, Murata et al. applied the fungus

Table 5.1. Bleaching conditions and optical properties for conventional five-stage bleaching and biobleaching processes

Bleaching sequence	Dosage (% on pulp)						Brightness (% ISO)		
	C	E_1	D_1	E_2	D_2	As effective chlorine	Before aging	After aging[a]	PC number
CEDED (Conventional process)	5.0	3.6	0.8	0.2	0.3	7.89	88.8	84.2	0.78
FCED (Biobleaching process)	1.4	0.8	0.3	...	...	2.19	88.1	85.3	0.46

Reproduced with permission from Tappi Press, Ref. 16.
[a] At 105 °C for 1 h.

Table 5.2. Bleaching conditions and optical properties of conventionally bleached and biobleached commercial SWKP

Bleaching sequence	Dosage (% on pulp)						Brightness (% ISO)			
	C	E_1	D_1	E_2	D_2	As effective chlorine	Before aging	After aging[a]	PC number	Yield (%)
CEDED (Conventional process)	9.0	6.4	1.0	0.5	0.5	12.95	84.2	82.4	0.38	91.4
FCED (Biobleaching process)[b]	2.4	1.7	0.4	...	...	3.45	84.6	83.4	0.25	91.4

Reproduced with permission from Tappi Press, Ref. 17.
[a] At 105 °C for 1 h.
[b] Commercial SWKP was treated with fungus IZU-154 for 6 days, with kappa number falling from 40.0 to 14.9.

IZU-154 to the delignification and brightening of oxygen-bleached hardwood kraft pulp.[18] The fungus brightened the pulp and simultaneously decreased its kappa number. Brightness was increased by 17 and 22 points by 3-day and 5-day treatments, respectively, and kappa number was decreased from 10.1 to 6.4 by a 5-day treatment.[18] The combination of the 3-day fungal treatment, alkaline extraction (2% NaOH charge) and hydrogen peroxide bleaching with 5% charge of H_2O_2 gave a pulp of 86.3% ISO brightness (Table 5.3). The 5-day-treated pulp was brightened to 87.3% ISO brightness by 4% H_2O_2 bleaching after alkaline extraction. Optical and strength properties of OFEP-bleached pulp were sufficiently comparable to those of conventional OCED-bleached pulp. Tsuchikawa et al. have reported biobleaching of hardwood kraft pulp with

Table 5.3. Optical properties of the bleached pulps

Bleaching sequence	Brightness (% ISO)		Postcolor number	Yield (%)
	Before aging	After aging		
OCED	87.9	85.7	0.36	97.4
OPP	73.8	71.8	0.89	97.4
OFEP, 3 days				
4% H_2O_2	85.0	83.7	0.26	97.7
5% H_2O_2	86.3	84.9	0.26	97.8
OFEP, 5 days				
2% H_2O_2	84.3	83.5	0.17	97.1
4% H_2O_2	87.3	85.6	0.29	97.0

Reproduced with permission from Tappi Press, Ref. 18.

lignin-degrading fungi *Phanerochaete sordida* YK-624.[19] When the hardwood kraft pulp was treated with the fungus for 10 days, the kappa number was decreased from 14.4 to 5.75 and the brightness was increased to 61% ISO. If the fungal incubation was interrupted after 5 days and the pulp extracted with alkali and treated with fungus for another 5 days, the kappa number was lowered to 4.8, and the brightness reached 80% ISO. The intermediate alkaline extraction, which was more effective than water washing, seemed to reactivate the lignin towards degradation as it does with chemical bleaching reagents. The tensile and burst strengths of the fungus-treated pulp were almost as high as those of control pulp, but the tear strength was 34% lower.

Nishida et al. investigated the biobleaching of hardwood unbleached kraft pulp by *P. chrysosporium* and *T. versicolor* in the solid-state and liquid-state fermentation systems with four different culture media (low nitrogen-high carbon, low nitrogen-low carbon, high nitrogen-high carbon and high nitrogen-low carbon).[20] In the solid state fermentation system with low nitrogen and high carbon culture medium, pulp brightness increased by 15 and 30 points after 5 days of treatment with *T. versicolor* and *P. chrysosporium*, respectively. The pulp kappa number decreased with the increasing brightness, and a positive correlation between the kappa number decrease and brightness increase of the fungus-treated pulp was observed.

Wroblewska and Zielinsk examined biodelignification of beech and birch pulp wood by selected white rot fungi.[21] One of the strains, designated DL-Sth-4, was found to be best for selective delignification of beech wood. About 25% of lignin was lost with very little loss in cellulose content. Pazukhina et al. used the culture filtrate of several white rot fungi – *P. sanguinea, C. versicolor, Ganoderma applanatum* and *Trichoptum biforma* – for bleaching hardwood kraft pulp.[22] *P. sanguinea* showed the highest selectivity in lignin degradation. Moreira et al. tested the ability of 25 white rot fungal strains to bleach *Eucalyptus globulus* oxygen delignified kraft pulp.[23] Under nitrogen-limited culture conditions, eight outstanding biobleaching strains were identified that increased the brightness of OKP by more than 10 ISO units compared to pulp incubated in sterile control medium. The highest brightness gain of approximately 13 ISO units was obtained with *Bjerkandera* sp. strain BOS 55, providing a high final brightness of 82% ISO. This strain also caused the greatest level of delignification, decreasing the kappa number of OKP by 29%. When the white rot fungal strains were tested in nitrogen-sufficient medium, the extracellular activities of laccase and peroxidases increased in many strains; nonetheless, the pulp handsheets were either destroyed or brightness gains were lower than those obtained under nitrogen limitation. The titer of ligninolytic enzymes was not found to be indicative of biobleaching potential. However, the best biobleaching strains were generally characterized by a predominance of manganese-dependent peroxidase activity compared to other ligninolytic enzymes and by a high decolorizing activity towards the polyanthraquinone ligninolytic indicator dye Poly R-478.

Iimori and coworkers at Nippon Paper Industries carried extensive screening for pulp-bleaching fungi.[24] By plating samples of decayed wood or fruit bodies directly on agar containing unbleached kraft pulp, they isolated 1758 cultures that produced decolorization zones; 266 of these bleached oxygen-delignified hardwood pulp to 70–81% ISO brightness within 7 days of incubation under solid-state fermentation conditions. The most active isolate, SKB-1152, also brightened agitated suspensions (2% consistency) of O_2-delignified hardwood pulp to 80% ISO. Optimization studies, with the aim of shortening treatment time, showed that dilute pulp suspensions in the range of 0.5–1% consistency were brightened faster than those with higher consistency, inoculum-to-pulp ratio of 6% was found to be adequate. Incubation of the mycelial suspension used as inoculum in a nutrient medium for 24 h before adding it to the pulp eliminated the lag period before the onset of brightening.[25]

Delignification by white rot fungi is strongly influenced by culture conditions.The major culture parameters affecting lignin degradation are: growth substrate, nitrogen availability, other culture conditions, oxygen concentration, and mode of cultivation, etc. *P. chrysosporium* and *C. versicolor* require a carbon source such as glucose or cellulose to metabolize lignin to form carbon dioxide.[26] Glucose supplementation at low concentration (0.47%) was found to stimulate delignification in kraft pulps and lignin in aspen wood.[8,27] However, there was no difference in the degree of delignification for thermomechanical pulp at glucose concentrations ranging from 0 to 35%.[28] Glucose was added in most studies to maintain pulp yield because it is a repressor of endo-1,4,β-glucanase, mannanase, xylanase, aryl-β-glucosidase, pectinase, and cellulase, which attack the carbohydrate fraction of the pulp. The white rot fungus IZU-154 was able to degrade lignin in hardwood kraft pulp without an exogenous energy source and without significantly affecting pulp yield.[16]

Nutrient depletion, particularly nitrogen depletion, appears to trigger the development of the ligninolytic system in most of the white rot fungi.[27,29–32] The effect of nitrogen depletion does not necessarily hold true for all white rot fungi, as the lignin-degrading abilities of *P. sajor-caju* and *L. edodes* do not show strong stimulation on nitrogen depletion.[33] The ligninolytic system of *P. chrysosporium* is also triggered by limitations in carbon, sulfur, Zn^{2+}, Fe^{2+}, and Mo^{4+}, but not phosphorus.[34,35] Other culture conditions such as temperature and pH have not been studied extensively.[34] Tran and Chambers[8] reported that the optimal temperature for delignification by *P. chrysosporium* was 38 °C but Drew and Kadam[36] found that the extent of degradation of ^{14}C-kraft lignin in 15 days at 28 °C was about four times than at 38 °C. Optimum pH values reported for lignin degradation are different for different fungal strains. Tran and Chambers reported that the optimal pH of *P. chrysosporium* was 3.5 for delignification and 4.5 for growth.[8] Kirk et al. reported an optimum pH of 4–4.5 for their strain.[37] Suppression of delignification was noted at pH less than 3.5 and greater than 5.5.[37] Recently, Royle et al. reported increased lignin degradation by *P. chrysosporium* and *L. edodes* at pH 3.[33] O-pthalate was found to inhibit delignification.[38] In one study, lignin degradation rates were doubled by changing the buffer from o-pthalate to 2,2- dimethylsuccinate.[39]

Lignin degradation in white rot fungi is primarily oxidative. It occurred faster in cultures of *P. chrysosporium* which had been flushed with pure oxygen than in those flushed with air.[8,28,40–43] Enhanced lignin degradation due to elevated oxygen partial pressures has also been reported for other white rot fungi including *C. versicolor, Pycnoporus cinnabarinus, Lentinus edodes, Giifola frondosa, Polyporus burmalus*, and *Merulius tremellosus*.[44] However, other species, such as *Gleoporus dichrous, Pleurotus ostreatus, Bondarzewia berkeleyi* and IZU-154, were less responsive to increased oxygen concentations.[16,44] The addition of polydimethylsiloxane oxygen carriers enhanced lignin degradation by *C. versicolor*.[45] However, they alone were not found to appreciably increase the brightness of hardwood kraft pulp.[46] Together with *C. versicolor*, they acted synergistically, resulting in an overall increase of approximately 10 brightness points.[45] Lignin degradation is strongly influenced by the mode of cultivation. Most researchers have studied stationary cultures, although dependence on stationary cultures causes difficulties in scaleup. Also, agitation is an important consideration because of the role of oxygen mass transfer. In a number of studies, agitation in reciprocating and gyratory shakers caused the formation of pellets consisting of mycelia entangled with pulp. The formation of mycelial pellets resulted in suppression of degradation of kraft lignin and lignin model compounds by *P. chrysosporium*.[8,28] Similar delignification rates were noted in agitated and stationary cultures and also with the addition of detergent to the culture.[43,47] Effective delignification has also been reported in agitated cultures of *C. versicolor* in which the formation of mycelial pellets was prevented and in cultures of *P. chrysosporium* in the presence of veratryl alcohol.[12,48,49]

P. chrysosporium and *C. versicolor* have been successfully immobilized on polyurethane foam.[13,45,50] Immobilized and free cultures of *C. versicolor* have been found to bleach hardwood and softwood kraft pulp at a comparable rate and to a similar extent.[10,51] The results showed that intimate contact between the fungal hyphae and pulp fibers was not required as long as the media was renewed through contact with the fungus.[51,52] Immobilization enabled the pulp to be separated from the mycelia. Another advantage of immobilization was that the same fungal biomass could be reused to treat other batches of pulp either immediately or after storage at 4 °C.[51]

Cell-free filtrates from bleaching cultures of *T. versicolor* did not cause detectable pulp brightening.[52] These filtrates contain laccase and manganese peroxidase, which can, under appropriate conditions, delignify and brighten pulp.[53] The failure to detect pulp brightening by culture filtrates can be attributed to lack of H_2O_2 and Tween-80 to support brightening by manganese peroxidase and the absence of mediator for laccase. The ability of intact *T. versicolor* cultures to extensively delignify and brighten kraft pulps in the absence of a mediator for laccase or an analogue of Tween-80 suggests that the fungus produces one or more unknown enzymes that contribute to its pulp-bleaching ability. *Phanerochaete sordida* YK-624 could delignify and brighten hardwood kraft pulp separated from the mycelium by a membrane filter with 0.1 µm pores; a thin polycarbonate membrane supported more bleaching than a thicker cellulose nitrate membrane.[54] *P. chrysosporium* and *T. versicolor* could also be used through the membrane, although to a lesser extent than YK-624.

In order to be adopted by the pulp and paper industry, biological bleaching must not compromise the quality of the pulp. Results from lab-scale fungal bleachings indicate an improvement in strength characteristics (burst factor, tear index, tensile strength, breaking length, stretch, and fold) in both hardwood and softwood pulps. Jurasek and Paice suggested that the lignin may become more flexible and hydrophilic as a result of fungal enzyme action, resulting in a softer pulp with improved bonding and stronger paper characteristics.[55] Reduced color reversion was another benefit noted with the fungus IZU-154.[16] Some viscosity loss, indicating limited cellulose depolymerization, has been reported as a result of fungal bleaching.[10,16,51] However, based upon experiments done with free and immobilized cultures, Kirkpatrick et al. reported that up to 25% of the reduction in the pulp viscosity may be due to the presence of fungal mycelia, rather than cellulose cleavage.[51] Although fungal bleaching is primarily an oxidative process, it appears to be more selective than oxygen bleaching at high pH and at kappa numbers less than 17 because there is a better retention of pulp viscosity.

A serious shortcoming in the fungal bleaching process is the long incubation time required for contact with the biomass. Typical contact periods range from 5 to 14 days for both hardwood and softwood pulps. Softwood incubation periods are likely to be longer than for hardwood pulps because softwood pulps require a longer time lag (6 days) before a kappa number reduction or a brightness increase can occur.[10] Hardwood pulps generally have a lag time of 1–2 days, followed by a rapid and then slower delignification stage.[16,30,51] Unfortunately, the size of the fungal bioreactor would have to be very large considering that daily production could range from 200 to over 1000 air-dried tons

Table 5.4. COD and total color loadings for conventional five-stage bleaching and biobleaching processes

Bleaching sequence	COD (kg/ton of pulp)				Total color (kg/ton of pulp)
	C	E_1	$D_1E_2D_2$	Total	
CEDED (Conventional process)	10.08	17.64	3.07	31.42	38.25
FCED (Biobleaching process)	5.03	8.86	1.78	15.94	8.23

Reproduced with permission from Tappi Press, Ref. 16.

of pulp. Most researchers have performed fungal bleaching experiments at low pulp consistencies of 0.5 to 2% (w/v).[8,30,51] Only Fujita et al. have investigated fungal bleaching at high consistency (16–24%), which would allow for a small reactor.[16] Fungal treatment can be efffected in the unbleached storage itself with minor modifications provided the treatment time is reduced to a practically feasible duration. The reaction time required at its current stage of development makes it economically unattractive.

Only few researchers have measured the impact of fungal bleaching on effluent quality. In a Japanese study, with the FCED bleaching sequence, the COD and color loading in the bleach plant waste water were reduced by 50 and 80%, respectively (Table 5.4), assuming that the filtrate from the fungal bleaching stage was sent to a kraft chemical recovery system.[16] Whether this could occur in practice would depend on the capacity available in the recovery furnace. The authors suggested that higher reductions could be obtained with an FED or FE_1DE_2D sequence, although there may be slight loss in pulp yield.[16] Despite the emphasis on fungal bleaching as a means to reduce the use of chlorine and the associated formation of chlorinated organics, the effect upon chlorinated organic discharges has not been reported. As this is an important factor in the choice of any alternative bleaching sequence, quantitative information in this area is needed.

Enzymes

The most important lignin-degrading enzymes are lignin peroxidases, manganese peroxidases, and laccases. Several reports in the literature suggest that these enzymes could prove useful in bleaching of kraft pulps.[56–72] The lignin peroxidases are hemeproteins which are able to catalyze the oxidation of nonphenolic aromatic rings in lignin to cation radicals in the presence of hydrogen peroxide. Manganese-dependent peroxidases oxidize phenolic units in lignin. These require Mn^{2+}, which is oxidized to Mn^{3+} in presence of chelators and hydrogen peroxide. Mn^{3+} is the real oxidizing agent attacking the lignin molecule (Fig. 5.1). During bleaching by *Trametes versicolor*, oxalate and several chelators are biosynthesized and manganese required for manganese peroxidase as a primary substrate is often present in kraft pulp fibers. Laccase uses molecular oxygen as a cosubstrate. The enzymes oxidizes phenolic subunits in lignin and simultaneously reduces oxygen to water (Fig. 5.2). The substrate range of laccases can be extended to nonphenolic subunits by adding readily oxidized substrates.

Unfortunately, for a long time, any attempt to use the lignin-degrading enzymes of white rot fungi as isolated (single) enzymatic catalysts completely failed. In the absence of the living organism, the various peroxidases and laccases perform only negligible reduction in kappa number. Egan has reported kappa number reductions of 24 and 26% after treatment with ligninase 118 from *P. chrysosporium* followed by alkaline extraction.[73] He reported that pulp viscosity did not change but gave no brightness data. Other researchers have not been able to achieve much bleaching effect from either pure or crude enzyme preparations from *P. chrysosporium*. Viikari et al. were unable to demonstrate any bleaching effects on pine kraft pulp with ligninases and

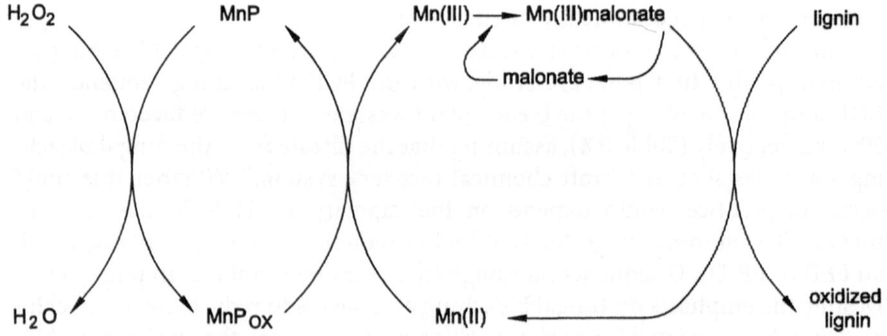

Fig. 5.1. Oxidative pathway for catalytic action of manganese peroxidase on lignin

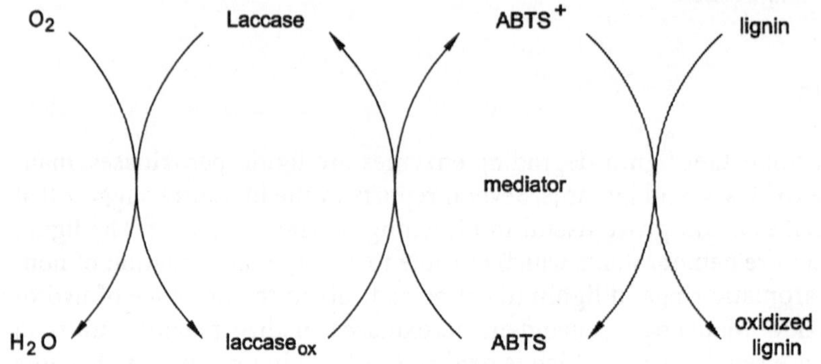

Fig. 5.2. Oxidative pathway for catalytic action of laccase on lignin

oxidase from the white rot fungus *Phlebia radiata* or purified ligninase from *P. chrysosporium*.[74] When the ligninases were applied after treatment with hemicellulases, there were slight reductions in kappa number. However, these were within the accuracy limits of the analytical method. Niku-Paavola et al. treated oxygen delignified pine kraft pulps with lignin-modifying enzymes – laccase, manganese dependent peroxidases, lignin peroxidase – and xylanase in order to improve bleachability.[75] Pulps were treated either with a purified single enzyme or, alternatively, enzymes were successively added in different combinations. The residual pulps were delignified with alkaline hydrogen peroxide and analyzed for kappa number and brightness. Lignin-modifying enzymes did not improve the bleachability when acting alone. Xylanase treatment increased the brightness by 1–2.5 ISO units and xylanase combined with lignin-modifying enzymes increased the brightness by a further 1 unit. By extending the alkaline peroxide step, the final brightness increased in all samples, whereas the relative difference between reference and enzyme-treated samples remained constant.[75] Kantelinen et al. studied the capability of lignin-modifying enzymes of *Phlebia radiata* to improve the bleachability of kraft and peroxyformic acid pulps.[76] The effect of lignin-modifying enzymes alone on kraft pulps was insignificant. The kappa numbers were found to be at the same level as in the control sample with peroxide bleaching alone. The brightness of the pulp was decreased as a result of laccase treatment probably due to oxidation of phenolic hydroxyl groups to the corresponding colored quinones. Bleachability of the kraft pine pulp was found to be improved only when laccase was used after hemicellulase treatment.[76] Hemicellulases apparently increase both lignin extractability and the accessibility of lignin to lignin-modifying enzymes. In the peroxyformic acid pulps, the more oxidized lignin was further oxidized by lignin peroxidases, as indicated by decrease in brightness. Arbeloa et al. used lignin peroxidase enzyme from *P. chrysosporium* to improve the bleachability of hardwood and softwood kraft pulps.[56] Enzyme treatment prior to chemical bleaching increased brightness and decreased lignin content in the pulp. The final brightness of pulp was found to be higher by about 0.8–0.9 points as compared to control.

Milagres et al. studied the effect of number of stages on the treatment of hardwood kraft pulp with xylanase (X) alone and sequentially, with xylanases (X) and laccases (L).[77] They also studied the effect of various orders of sequential treatment with these enzyme preparations. It was noted that a plural-stage process achieved greater delignification than a one-stage process. About 23 and 11.2% delignification was obtained using the sequences X-X-X-L and L-L-L-X, respectively, whereas about 7.9 and 5% delignification was obtained with the sequences X-L and L-X, respectively. Sequence order L-L-X, L-X-L did not have much effect on the lignin content (12.9% delignification). The absolute number of X stages was found to have a greater impact.

Several patents have been filed concerning the use of ligninases for bleaching kraft pulps.[58,59,78–85] One of these makes use of a chemically modified lignin peroxidase.[85] In addition, a European patent application exists for the production of recombinant ligninase.[80] Farrell applied lignin peroxidase followed by alkaline extraction to produce a kraft pulp with higher brightness, but no quantitative data were reported.[79] Vaheri and Miiki described a procedure for bleaching pulp by sequential treatment with laccase- and chlorine-containing chemicals.[83] The procedure produces an effluent with a lower content of chloroorganic compounds, but pulp bleaching is not enhanced. Olsen et al. have developed an enzymatic bleaching procedure using partially purified manganese peroxidase and lignin peroxidase isoenzymes from *P. chrysosporium*.[81] At least one process uses lignin and manganese peroxidases from *P. chrysosporium* sequentially with xylanases from a *Chiana* sp.[82] Intervening alkali extraction steps reduce residual lignin content and result in a brighter pulp without chlorine. Vaheri and Piirainen showed that the oxidizing enzyme laccase can be used in conjunction with manganese ions to reduce the consumption of chlorine chemicals applied in the later stages of bleaching.[84] Although much of the research has focused on lignin peroxidases, these enzymes are not necessarily involved in lignin degradation and may not be secreted by all lignin-degrading fungi. *C. versicolor* produces laccases as well as lignin and manganese-dependent peroxidases. However, Archibald reported that lignin peroxidases secreted by *C. versicolor* did not appear to play an important role in lignin degradation.[40] The enzyme is neither detected during bleaching nor is fungal bleaching enhanced by addition of exogenous enzyme. Furthermore, fungal bleaching is not inhibited by metavanadate ion, while lignin peroxidase activity is. However, Iimori et al. have reported that bleaching by isolate SKB-1152 was better under conditions where lignin peroxidase was detected than in its absence;[25] thus it might contribute to bleaching by this strain. Manganese peroxidase is found to be a key enzyme in fungal bleaching. Several lines of evidence suggest that: (1) Catalase, an enzyme that destroys H_2O_2, inhibits the bleaching. (2) Mutants of *C. versicolor* deficient in manganese peroxidase do not bleach and bleaching activity is partially restored by addition of manganese peroxidase. (3) Isolated manganese peroxidase produces partial delignification of pulps when supplied with H_2O_2, Mn^{2+}, and chelator.[66] *Dichomitus squalens* and *Rigidoporus lignosus* have been reported to produce both laccase and a manganese-dependent peroxidase but do not produce lignin peroxidase.[86,87] Cellobiose quinone oxidoreductase (CBQase), a quinone-reducing enzyme, also plays a number of roles in delignification of kraft pulp. Studies with CBQase have shown that the enzyme can reduce reaction products of laccase or peroxidase such as quinones, radicals, and Mn^{2+}. Concurrently, cellobionate, a potential chelator of Mn^{3+}, is formed.[88–90] However, cellobionate is less efficient than malonate or oxalate as a chelator in the manganese peroxidase catalytic cycle, probably because it has a higher binding constant for Mn^{2+}.[91] Ander[92] has suggested that enzyme mixtures for lignin degradation should probably contain this enzyme. It was shown to decrease the polymerization of kraft lignin by lignin peroxidase. It is now known that lignin peroxidases and laccases play an important role in degrading the lignin in vivo; in vitro the oxidation reactions catalyzed by the enzyme result in further polymerization of the lignin.[93] Hammel and Moen reported depolymerization of a synthetic lignin by a lignin peroxidase in the presence of H_2O_2 and veratryl alcohol, but this effect has not been demonstrated with lignin in wood or pulp.[94] It is likely that fungi possess enzyme systems that prevent polymerization.[92,93]

These results imply that single enzymes are not able to mimic the complete biological system. Small improvements can be achieved by addition of low molecular weight aromatic compounds like veratryl alcohol or other substances such as ABTS and Remazol blue.[81,95]

In spite of this experience with isolated lignin-degrading enzymes, Lignozym GmbH, Germany, continued work with enzymes plus chemical mediators which create a red/ox system throughout the pulp-treatment period.[58–63] Their idea was to find a system which is a good mimic of the natural situation. Starting in 1987 with the enzyme mediator concept, recently Lignozym has improved the performance of the mediator system for the laccase of *Trametes versicolor* by changing and further fine tuning the chemical nature of the component. The mediator is hydroxybenzotriazole (HBT).[63] The treatment of pulp with laccase alone does not result in any degradation of lignin but just in a structural change or repolymerization, the laccase mediator system causes a significant kappa number reduction at reasonable treatment times, even if the enzyme mediator system is applied in several consecutive treatment steps to the same pulp. In contrast to commonly used pulp-bleaching chemicals (oxidizing and reducing), no passivation can be observed in enzymatic delignification, i.e., efficiency does not drop significantly during the consecutive treatment steps. The laccase enzyme contains four atoms of copper per molecule and requires oxygen as a cosubstrate for the oxidation reaction which has to be provided. Oxygen delignification is an existing technology in many mills and, therefore, most of the technical problems related to the introduction of gas into a three-phase system with high consistency have been solved today. According to the present understanding, the laccase, while oxidizing the chemical mediator, is generating a strongly oxidized comediator, which is the real bleaching agent (Fig. 5.3). Table 5.5 illustrates the conceptual difference between the xylanase and the direct enzymatic lignin attack. The general technical conditions for enzymatic bleaching with the laccase mediator system are: temperature 40–65 °C, pH 4–7, consistency 1–20%, pressure 1–14 bar, duration 1–4 h. The results of bleaching of conventional softwood kraft pulp after

Table 5.5. Differences in xylanase and laccase/mediator treatment

	Xylanase	Laccase/mediator
1	No or very poor kappa reduction	Very good kappa reduction
2	Moderate bleaching effect	Good bleaching effect
3	Saving of bleaching chemicals	Saving of bleaching chemicals
4	-	TCF pulp production possible

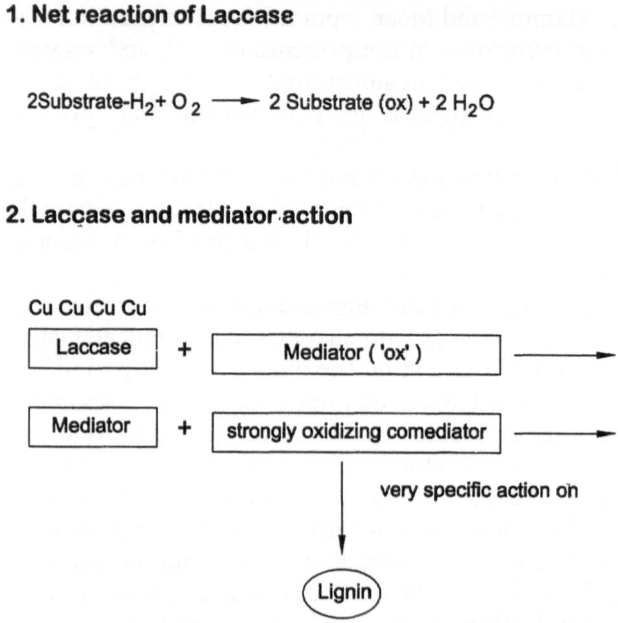

Fig. 5.3. Possible mechanism of laccase and mediator action on lignin

Table 5.6. Bleaching of an extended cooked softwood kraft pulp with laccase mediator system

Treatment	Kappa number	Viscosity (ml/g)	Brightness (% ISO)
Untreated pulp	20	970	28.8
Enzyme-treated pulp	10	950	32.8
TCF sequence	–	850	80.0
Enzyme-treated pulp after TCF sequence	–	850	88.0

Based on data from Refs. 62 and 63.

extended cooking with laccase mediator system are presented in Tables 5.6 and 5.7. The laccase mediator system provides a broad flexibility with respect to the pulp substrate, the technical requirements for application, and the final quality of the pulp. The principal applicability has been demonstrated for softwood and hardwood pulps as well as for annual plant fibers. A repeated enzyme treatment is possible and results in a 50–70% kappa number reduction per treatment step. The laccase mediator system (LMS) is compatible with all other bleaching sequences. Depending on mediator dosage, consumption, and price, a recharging of mediator, an in situ regeneration, and recycling of the mediator are possible. The typical consumption of mediator during the enzymatically catalyzed bleaching process is about 40–50%. The waste after the

Table 5.7. Bleaching of conventional softwood kraft pulp with laccase mediator system

Treatment	Kappa number	Viscosity (cp)	Brightness (% ISO)
Untreated pulp	28.7	27.5	24.7
Enzyme-treated pulp	11.5	31.6	31.6
Enzyme-treated pulp after TCF sequence	–	15.0	84.0

Based on data from Refs. 62 and 63.

Table 5.8. Summary of results from the pilot plant trial with laccase mediator system

Sequence	Pulp	Dosage of enzyme/ mediator (kg/TP)	Degree of delignification (%)	Max. brightness (% ISO)
L-E-Q-P	A	2/13	56.6	76.5
L-E-L-E-Q-P	A	2 × 2/2 × 8	50.6/67.7	82.7
L-E-Q-(P)	B	2/8	44.2	

Conditions:

Parameter	L stage	E stage	Q stage	P stage
Consistency (%)	10	10	5	10
Temperature (°C)	45	60	60	75
pH	4.5	11.5	5	11.2
Residence time (min)	120	60	30	210
Pressure (bar)	2	–	–	–
Dosage	Enzyme: 2 kg/T Mediator: variable	NaOH	0.2% DTPA	3% peroxide

Based on data from Ref. 63.

treatment containing the unreacted mediator can be recharged up to the original mediator level before the next treatment step starts. Under continuous operation, this recharge can be repeated several times. Using an additional reactor in a bypass, the reaction product can be converted into the active form (regeneration step). The performance of the LMS system has been proven in a pilot plant trial. The summary of the results are presented in Table 5.8. A degree of delignification of >50% could be obtained in a single step, even if the mediator dosage is reduced by factor 0.6. The technology is now ready for large-scale application. Recently, Lignozyme has introduced a new mediator, N-hydroxyacetinalide (NHAA), which is biodegradable and has been claimed to be cost-effective.[96] Biobleaching with NHAA also allows the activity of the enzyme to maintain about 80% of its original activity after a 1-h treatment, whereas biobleaching with HBT causes a severe loss of enzyme activity. Since January 1995, Wacker Chemie GmbH has been engaged in the commercialization, taking over the worldwide exclusive licence of the Lignozym technology.

The next steps to be planned are full-scale mill trials and the setup of a production of both the mediator and the enzyme.[97]

Sealey et al. reported that with oxygen-reinforced alkaline extraction, laccase-HBT biobleaching could obtain over 70% delignification in one stage.[98] Further treatment of $L(E_{OP})$ pulp with another laccase HBT stage increased the delignification to about 80%. It seems that laccase HBT biobleaching is capable of reacting with the last vestiges of residual lignin, which are typically very unreactive.

The effect of ABTS on laccase-catalyzed oxidation of kraft pulp was reported by Bourbonnais and Paice.[57] It was found that demethylation (diagnostic for delignification) was enhanced and kappa number was decreased with ABTS. The delignifying effect was confirmed by Klason lignin analysis of the treated pulps and by monitoring the fate of ^{14}C lignin in labeled kraft pulp. However, the brightness of the pulp was decreased because the dye sticks to the pulp. The pulp treatments were found to be quite selective for lignin as pulp viscosities decreased by less than 10% and zero-span breaking lengths were unchanged. Recently, Bourbonnais et al. compared the activities of induced laccases I and II (from *T. versicolor*) with and without ABTS, on hardwood and softwood kraft pulps.[65] It was noted that, in the absence of ABTS, the two purified laccases were not able to reduce the kappa number of either pulp, but both produced small amounts of methanol with the hardwood pulp only. When ABTS was present, the two enzymes were equally effective in delignifying and demethylating either pulp. The amount of methanol produced and the decrease in kappa number were found to be less with hardwood pulp. The enhancement of delignification by the dye ABTS is not fully understood, although it is apparent that laccase initially generates a stable radical cation from the dye. It may be that the ABTS radical cation acts as an electron carrier between residual lignin in the kraft pulp fiber wall and the large laccase molecule. *Trametes versicolor* always produces laccase during bleaching of pulp, but a chemical species equivalent in action to ABTS has not yet been isolated from the fungus.

To make the laccase-ABTS system compatible with industrial bleaching processes, these researchers surveyed several variables, including enzyme and mediator dosage, pulp consistency, oxygen pressure, temperature, reaction time, and pH.[64] A typical reaction was performed for 2h at 60°C, pH 5.0, with 300kPa of O_2, 1% ABTS, and 5 units/g of laccase on 10% w/v pulp. After one-stage treatment followed by alkaline extraction, the extent of delignification varied from 25 to 40% with various kraft pulps and was over 50% with a sulfite pulp. By repeating the treatment with laccase-ABTS and alkaline extraction, the kappa number of a softwood kraft pulp was decreased by 55%. Significant differences in reactivity between various fungal laccases for pulp delignification in the presence of mediators ABTS and HBT were found.[99] A comparison of *T. versicolor* laccase with various mediators, including ABTS, HBT, Remazol blue, nitroso-napthols, and phenothiazines, has shown that HBT gave the most extensive delignification, but deactivated the enzyme and, therefore, required a higher enzyme dosage. Reaction products from lignin dimer oxidation by laccase were found to be different in the presence of ABTS and HBT. From electrochemical studies, it was concluded that mediator species with redox potential around 900 mV are generated by laccase-catalyzed oxidation, and that these species migrate from the enzyme and oxidize lignin. Both the initial lignin content and chemical structure of the pulp significantly influence mediator-aided laccase delignification and subsequent final bleaching with alkaline peroxide with or without a xylanase stage in the bleaching sequence.[100] Three different chemical pulps, i.e., a pine kraft pulp, a two-stage oxygen-delignified pine kraft pulp, and birch formic acid/peroxyformic acid (MILOX) pulps were subjected to HBT- and ABTS-mediated laccase treatments. HBT was more effective than ABTS in the presence of laccase in delignification and gave higher pulp brightness. All the laccase/HBT-treated pulps showed an increased response to alkaline hydrogen peroxide bleaching, and as a consequence, oxygen-delignified pulp and MILOX pulp reached full brightness in one and two stages, respectively. Even the pine kraft pulp reached a final brightness of 83%. Moreover, a xylanase stage before or together with laccase/HBT slightly improved the effect of laccase/HBT and gave a higher final brightness after peroxide bleaching than without the xylanase treatment.

Most of the work on in vitro bleaching of kraft pulp with manganese peroxidase has been done in Japan and Canada. Kondo et al. examined in vitro bleaching of kraft pulp with manganese peroxidase enzyme which was isolated from cultures of *P. sordida* YK-624.[69] When the kraft pulp was treated with partially purified manganese peroxidase in the presence of $MnSO_4$, Tween-80, and sodium malonate with continuous addition of H_2O_2 at 37 °C for 24 h, the pulp brightness increased by about 10 points and the kappa number decreased by about 6 points compared with untreated pulp. The pulp brightness was also increased by 43 points to 75.5% by six treatments with manganese peroxidase combined with alkaline extraction. The brightness was found to be comparable to that caused by whole fungal systems. Although this would be impractical for industrial application, the results demonstrate that under the right conditions, MnP can achieve most of the delignifying effect of the fungus. In another study, Kondo et al. reported that pulp brightness was increased by about 15 points and kappa number decreased by about 8 points, when unbleached kraft pulp was treated with manganese peroxidase from *P. sordida* YK-624 in the presence of $MnSO_4$, Tween-80 and sodium malonate with addition of H_2O_2 at a rate of 3 ml/h at 45 °C for 12 h.[70] To establish an absolutely chlorine-free bleaching process, oxygen-bleached kraft pulp (OKP) was treated with a four-stage biobleaching process consisting of sequential manganese peroxidase treatment, alkaline extraction, manganese peroxidase treatment, and hydrogen peroxide treatment. Full bleached kraft pulp (brightness 91%, yield 97%) was obtained from OKP by combination of enzyme treatment and hydrogen peroxide bleaching. Harazono et al. studied in vitro bleaching of an unbleached hardwood kraft pulp with partially purified manganese peroxidase without the addition of $MnSO_4$ in the presence of oxalate, malonate, or gluconate as manganese chelator.[71] When the pulp was treated without the addition of $MnSO_4$, the pulp brightness increased by about 10 points in the presence of 2 mM oxalate but the brightness did not significantly increase in the presence of 50 mM malonate, a good manganese chelator. Residual manganese peroxidase activity decreased faster during the bleaching with manganese peroxidase without $MnSO_4$ in the presence of malonate than in the presence of oxalate. Oxalate reduced MnO_2 which already existed in the pulp or was produced from Mn^{2+} by oxidation with manganese peroxidase, and thus supplied Mn^{2+} to the manganese peroxidase system. The presence of gluconate, produced by the H_2O_2-generating enzyme glucose oxidase, also improved the pulp brightness, without the addition of $MnSO_4$, although treatment with gluconate was inferior to that with oxalate with regard to increase in brightness. Thus, bleaching of hardwood kraft pulp with manganese peroxidase using manganese originally present in the pulp is possible in presence of oxalate, a good manganese chelator and reducing agent. To improve in vitro bleaching of the kraft pulp with MnP, Kondo et al. characterized various MnPs from white rot fungi.[72] MnP from *Ganoderma* sp. YK-505 was superior to MnPs from *P. sordida* YK-624 and *P. chrysosporium* in stabilities against high temperature and high concentration of H_2O_2. MnP from *Ganoderma* sp. YK-505 also differed in pH activity profile from other MnPs.

At Paprican, manganese peroxidase from *T. versicolor* was examined for bleaching of kraft pulp.[53,67] Studies on the effect of enzyme treatment on delignification of hardwood kraft pulp showed about 14% reduction in kappa number. The delignification was found to be accompanied by the release of methanol from phenolic methoxyl groups in kraft pulp lignin. With softwood kraft pulp, delignification was observed over a wide range of initial lignin contents (kappa number) (Table 5.9). Only the pulp of lowest initial kappa number gave a higher brightness after enzyme treatment. Subsequent treatment with alkaline hydrogen peroxide resulted in pulps upto 7–8 points brighter than those obtained without enzyme (Table 5.10). In the absence of added manganese and malonate buffer, comparable brightness gain of 6 points could be achieved.

Ducka and Pekarovicova used crude ligninases from *P. chrysosporium* for bleaching of softwood kraft pulp.[101] The pulp after oxygen bleaching with kappa number 19.7 and 36.5% MgO brightness was pretreated with ligninases (L) or xylanases (X) and bleached in $QE_{OP}DP$ sequence. The brightness of pulp bleached in this sequence was 87.8%, which is about 3.2 points higher than the control, and is approximately the same as with the use of commercial xylanases.

Limited information is available on the effect of lignin-degrading enzymes on pulp properties, yield, and effluent quality. Due to the specific action on lignin, these enzymes do not affect the viscosity, strength properties, and yield of the pulp. Effluent properties, namely AOX, color, etc., are also expected to improve due to reduced chlorine consumption required to achieve a given

Table 5.9. Effect of manganese peroxidase treatment on softwood kraft pulps[a]

Enzyme treatment	Initial kappa No.	Final kappa No.	Methanol (mg/l)	Final brightness (% ISO)
Control	10.8	9.7	0	35.4
Manganese peroxidase		7.9	8.3	38.6
Control	12.4	10.7	0	37.1
Manganese peroxidase		9.6	9.0	38.1
Control	16.2	15.0	2.5	33.0
Manganese peroxidase		13.1	14.2	28.1
Control	21.2	18.5	2.3	31.2
Manganese peroxidase		17.3	16.1	25.3

Treatment with manganese peroxidase enzyme was run for 24 h with 1 U/ml of enzyme, manganese sulfate (0.5 mM), glucose (10 mM), and glucose oxidase (0.025 U/ml) in sodium malonate buffer (50 mM, pH 4.5). Control contained all components except manganese peroxidase. Based on data from Ref. 67.

[a] Softwood pulp was delignified to the indicated initial lignin contents by oxygen delignification or by alkaline extraction with oxygen (Eo).

Table 5.10. Effect of manganese peroxidase treatment on QP bleaching of softwood kraft pulp[a]

Enzyme treatment	NaOH charge in the peroxide stage (%)	Kappa no. after QP	Brightness (% ISO)	Viscosity (mPa. s)	H_2O_2 consumed (%)
Control	2.5	8.5	71.8	20.7	1.29
Manganese peroxidase		5.2	79.0	20.9	1.60
Control	2.0	8.1	70.7	21.4	1.19
Manganese peroxidase		5.6	78.4	19.5	1.49
Control	1.5	8.3	69.4	22.8	1.04
Manganese peroxidase		5.7	77.3	20.6	1.34

Treatment with manganese peroxidase enzyme was run for 24 h with 1 U/ml of enzyme, manganese sulfate (0.5 mM), glucose (10 mM), and glucose oxidase (0.025 U/ml) in sodium malonate buffer (50 mM, pH 4.5). Control contained all components except manganese peroxidase. Based on data from Ref. 67.

[a] Softwood pulp was used after being oxygen-delignified to kappa no. 15.0.

brightness. In a pilot plant trial with the laccase mediator system, the strength properties and viscosity of the pulp with the L-E-Q-P sequence were not found to be affected.[63] Bourbonnais and Paice have reported that the treatment of kraft pulp with laccase and ABTS did not affect the pulp viscosity and zero-span breaking length.[57] Egan reported that pulp viscosity did not change after treatment of pulp with Ligninase 118.[73] Arbeloa et al. reported that when softwood TMP was treated with lignin peroxidase enzyme, some of the strength properties, viz. burst index, tear index, and breaking length, slightly increased.[56] Bourbonnais and Paice reported the effect of laccase-ABTS treatment on viscosity and strength properties of an oxygen-delignified softwood kraft pulp.[64] Both the control and enzyme-treated pulps had lower viscosity and higher breaking length than untreated pulp, as would be expected from exposure to mechanical action and alkali. However, the tear index was found to be decreased by enzyme-mediator action. In the LA-E-LA-E sequence, the total viscosity loss was 19% compared with the initial pulp. This viscosity loss was found to be comparable with that found following alkaline extraction alone. The viscosity of laccase-ABTS-treated sulfite pulp following the hot alkaline extraction was higher as compared to that of untreated pulp because of hemicellulose removal.

No information is available on the impact of lignin-degrading enzymes on effluent quality and AOX discharges except the work of Vaheri and Mikki, who reported that treatment of pulp with laccase produced an effluent with lower content of chloroorganics.

Advantages, Limitations, and Future Prospects

Pretreatment with fungi has been shown to replace up to 70% of the chemicals needed to bleach kraft pulp. The usual specificity of biological reactions and their mild reaction conditions make biological delignification an interesting alternative to bleaching with chemicals such as pressurized oxygen or ozone. The commercial exploitation of this process is hindered mainly by the slow reaction rate of the living system. The treatment time will have to be reduced in fungal bleaching for it to become an economic alternative.

Treatment with lignin-degrading enzymes – laccase and manganese peroxidase – requires milder conditions and results in more removal of lignin as compared to oxygen delignification, which translates into substantial savings of energy and bleaching chemicals which, in turn, would lead to lower pollution load. Moreover, these enzymes are highly specific towards lignin and there is no damage or loss of cellulose, resulting in better strength and yield of bleached pulp. Commercialization of the bleaching process with these enzymes faces a number of challenges: availability of enzymes, cost of mediators, and enzyme stability. Currently, neither enzyme is available in sufficient quantity for mill trials, and scaleup of enzyme production from fungal cultures may be costly. Cloning of genes for both enzymes has been reported and may provide an alternative production route. The laccase mediators are expensive and alternatives are less effective. There are two ways of improving laccase-mediator system for pulp bleaching. One is to discover new laccases that have extraordinarily high redox potential, the other is to find a very effective laccase mediator. Since the laccase mediators reported so far have very wide structural variations, it should be possible to discover cheaper and more effective laccase mediators than has been done to date. Chelating agents for Mn^{3+} in the manganese peroxidase reaction may also be a signficant cost item. Manganese peroxidase require hydrogen peroxide but are inactivated by concentrations above about 0.1 m mol/l. Even when hydrogen peroxide is kept below 0.1 m mol/l, manganese peroxidase becomes inactive relatively rapidly. Experience with xylanase and with other enzymes such as lipase and amylase has shown that enzymes can be applied successfully in a mill situation. Thus, oxidative enzymes, which can be regarded as catalysts for oxygen and hydrogen peroxide driven delignification, may also find a place in the bleach plant in coming years.

References

1. Viikari L, Ranua M, Kantelinen A, Sundquist J, Linko M: Bleaching with enzymes. In: Proc 3rd Int Conference on Biotechnology in the Pulp and Paper Industry, Stockholm, Sweden, 1986; 67–69.
2. Bajpai P, Bajpai PK: Realities and trends in enzymatic prebleaching of kraft pulp. Adv. Biochem. Eng./Biotechnol. edited by T. Scheper, Springer Heidelberg, Berlin 1997; 56: 1–31.
3. Farrell RL, Viikari L, Senior DJ: Enzyme treatment of pulp. In: Pulp bleaching (Dence CW, Reeve DW. eds), Tappi Press, Atlanta. 1996; 363–383.
4. Viikari L, Kantelinen A, Sundquist J, Linko M: Xylanases in bleaching: from an idea to industry. FEMS Microbiol Rev 1994; 13: 335–350.
5. Tolan JS, Olson D, Dines RE: Survey of mill usage of xylanase. In: Enzymes for pulp and paper processing (Jeffries TW and Viikari L eds). ACS Symp Ser 1996; 655: 25–35.
6. Kirk TK, Yang HH: Partial delignification of unbleached kraft pulp with ligninolytic fungi. Biotechnol Lett 1979; 1: 347–352.
7. Pellinen J, Abuhasan J, Joyce TW, Chang H-m: Biological delignification of pulp by *Phanerochaete chrysosporium.* J Biotechnol 1989; 10: 161–170.
8. Tran AV, Chambers RP: Delignification of an unbleached hardwood pulp by *Phanerochaete chrysosporium.* Appl Microbiol Biotechnol 1987; 25: 484–490.
9. Paice MG, Jurasek L, Ho C, Bournonnais R, Archibald F: Direct biological bleaching of hardwood kraft pulp with the fungus *Coriolus versicolor.* Tappi J 1989; 72(5): 217–221.
10. Reid ID, Paice MG, Ho C, Jurasek L: Biological bleaching of softwood kraft pulp with the fungus *Trametes versicolor.* Tappi J 1990; 73(8): 149–153.
11. Reid ID, Paice MG: Effect of residual lignin type and amount on bleaching of kraft pulp by *Trametes versicolor.* Appl Environ Microbiol 1994; 60(5): 1395–1400.
12. Ho C, Jurasek L, Paice MG: The effect of inoculum on bleaching of hardwood kraft pulp with *Coriolus versicolor.* J Pulp Pap Sci 1990; 16: J78–J83.
13. Kirkpatrick N, Reid ID, Ziomek E, Paice MG: Physiology of hardwood kraft pulp bleaching by *Coriolus versicolor* and use of foam immobilization for the production of mycelium-free bleached pulps. In: Biotechnology in pulp and paper manufacture (Kirk TK, Chang Hm eds.) Butterworth, Heinemann, Stoneham, MA 1990; 131–136.
14. Roy BP, Archibald FS: Effects of kraft pulp and lignin on *Trametes versicolor* carbon metabolism. Appl Environ Microbiol 1993; 59: 1855–1863.
15. Addleman K, Archibald FS: Kraft pulp bleaching and delignification by dikaryons and monokaryons of *Trametes versicolor.* Appl Environ Microbiol 1993; 59: 266–273.
16. Fujita K, Kondo R, Sakai K, Kashino Y, Nishida T, Takahara Y: Biobleaching of kraft pulp using white rot fungus IZU-154. Tappi J 1991; 74(11): 123–127.

17. Fujita K, Kondo R, Sakai K: Biobleaching of softwood kraft pulp with white rot fungus IZU-154. Tappi J 1993; 76(1): 81–84.
18. Murata S, Kondo R, Sakai K, Kashino Y, Nishida T, Takahara Y: Chlorine-free bleaching process of kraft pulp using treatment with the fungus IZU-154. Tappi J 1992; 75(12): 91–94.
19. Tsuchikawa K, Kondo R, Sakai K: Application of ligninolytic enzymes to bleaching of kraft pulp II: Totally chlorine-free bleaching process to bleaching of kraft pulp.II: Totally chlorine-free bleaching process with the introduction of enzyme treatment with crude enzymes secreted from *Phanerochaete sordida* YK-624. Jpn Tappi J 1995; 49: 1332–1337.
20. Nishida T, Katagiri N, Tsutsumi Y: New analysis of lignin-degrading enzymes related to biobleaching of kraft pulp by white rot fungi. 6th Int Conference on Biotechnology in Pulp and Paper Industry. Vienna, Austria, June 11–15, 1995.
21. Wroblewska H, Zielinski MH: Biodelignification of beech and birch pulpwood by selected white rot fungi. 6th Int Conference on Biotechnology in Pulp and Paper Industry, Vienna, Austria, June 11–15, 1995.
22. Pazukhina GA, Soloviev VA, Malysheva ON: Bleaching of kraft pulp with filtrates of white rot fungi. 6th Int Conference on Biotechnology in Pulp and Paper Industry, Vienna, Austria, June 11–15, 1995.
23. Moreira MT, Feijoo G, Sierra-Alvarez R, Lema J, Field JA: Biobleaching of oxygen delignified kraft pulp by several white rot fungal strains. J Biotechnol 1997; 53: 237–251.
24. Iimori T, Kaneko R, Yoshikawa H, Machida M, Yoshioka H, Murakami K: Screening of pulp bleaching fungi and bleaching activity of newly isolated fungus SKB-1152. Mokuzai Gakkaishi 1994; 40(7): 733–737.
25. Iimori T, Yoshikawa H. Kaneko R, Miyawaki S, Machida M, Murakami K: Effects of treatment conditions on treatment times for biobleaching by SKB-1152. Mokuzai Gakkaishi 1996; 42: 313–317.
26. Kirk TK, Connors WJ, Zeikus JG: Requirement for growth substrate during lignin decomposition by two wood-rotting fungi. Appl Environ Microbiol 1976; 32: 192–194.
27. Reid ID: The influence of nutrient balance on lignin degradation by the white rot fungus *Phanerochaete chrysosporium*. Can J Bot 1979; 57: 2050–2058.
28. Yang HH, Effland MJ, Kirk TK: Factors influencing fungal decomposition of lignin in a representative lignocellulosic, thermomechanical pulp. Biotechnol Bioeng 1980; 22: 65–69.
29. Eriksson KE, Goodell EW: Pleiotropic mutants of the wood-rotting fungus *Polyporus adustus* lacking cellulase, mannanase and xylanase. Can J Microbiol 1974; 20: 371–378.
30. Kirkpatrick N, Reid ID, Ziomek E, Ho C, Paice MG: Relationship between fungal biomass production and the brightening of hardwood kraft pulp by *Coriolus versicolor*. Appl Environ Microbiol 1989; 55: 1147–1152.
31. Tien M, Kirk TK: Lignin-degrading enzyme from *Phanerochaete chrysosporium*: purification, characterization and catalytic properties of a unique H_2O_2-requiring oxygenase. Proc Natl Acad Sci USA 1984; 81: 2280–2284.
32. Jeffries TW, Choi S, Kirk TK: Nutritional regulation of lignin degradation by *Phanerochaete chrysosporium*. Appl Environ Microbiol 1981; 42: 290–296.
33. Royle CD, Kropp BR, Reid ID: Solubilization and mineralization of lignin by white rot fungi. Appl Environ Microbiol 1992; 58: 3217–3224.
34. Janshekar H, Fiechter A: Lignin: Biosynthesis, applications and biodegradation. In: Pentoses and lignin. Adv Biochem Eng/Biotechnol 1983; 27: 119–178.
35. Kirk TK, Yang HH, Keyser P: The chemistry and physiology of the fungal degradation of lignin. Dev Ind Microbiol 1978; 19: 51–59.
36. Drew S, Kadam KL: Lignin metabolism by *Aspergillus fumigatus* and white rot fungi. Dev Ind Microbiol 1979; 20: 153–161.
37. Kirk TK, Schultz E, Connors WJ, Lorenz LF, Zeikus JG: Influence of culture parameters on lignin metabolism by *Phanerochaete chrysosporium*. Arch Microbiol 1978; 117: 277–282.
38. Fenn P, Kirk TK: Ligninolytic system of *Phanerochaete chrysosporium:* inhibition by o-pthalate. Arch Microbiol 1979; 123: 307–312.

39. Keyser P, Kirk TK, Zeikus JG: The ligninolytic enzyme system of *Phanerochaete chrysosporium:* synthesized in the absence of lignin in response to nitrogen starvation. J Bacteriol 1978; 135: 790–795.
40. Archibald FS: Lignin peroxidase is not important in biological bleaching and delignification of kraft brownstock by *Trametes verisicolor.* Appl Environ Microbiol 1992; 58: 3101–3109.
41. Reid ID, Paice MG: Biological bleaching of kraft paper pulp. In: Leatham GF (ed) Frontiers in industrial mycology. Chapman and Hall, New York, 1992; 112–126.
42. Bar-lev SS, Kirk TK: Effects of molecular oxygen on lignin degradation by *Phanerochaete chrysosporium.* Biochem Biophys Res Commun 1981; 99: 373–378.
43. Reid ID, Chao EE, Dawson PSS: Lignin degradation by *Phanerochaete chrysosporium* in agitated cultures. Can J Microbiol 1985; 31: 88–90.
44. Reid ID, Seifert KA: Effect on an atmosphere of oxygen on growth, respiration and lignin degradation by white rot fungi. Can J Bot 1982; 60: 252–260.
45. Ziomek E, Kirkpatrick N, Reid ID: Effect of polydimethylsiloxane oxygen carriers on the biological bleaching of hardwood kraft pulp by *Trametes versicolor.* Appl Microbiol Biotechnol 1991; 35: 669–673.
46. Kirkpatrick N, Ziomek E, Reid ID: Effect of increased oxygen availability on the biological bleaching of hardwood kraft pulp by *Coriolus versicolor.* Applications of biotechnology in pulp and paper manufacture (Kirk TK, Chang Hm eds.) Butterworth-Heinemann, Stoneham MA 1990; 137–143.
47. Jager A, Croan S, Kirk TK: Production of ligninases and degradation of lignin in agitated submerged cultures of *Phanerochaete chrysosporium.* Appl Environ Microbiol 1985; 50: 1274–1278.
48. Paice MG, Bernier R Jr, Jurasek L: Viscosity enhancing bleaching of hardwood kraft pulp with xylanase from a cloned gene. Biotechnol Bioeng 1988; 32: 235–239.
49. Leisola MSA, Feichter A: Ligninase production in agitated conditions by *Phanerochaete chrysosporium.* FEMS Microbiol Lett 1985; 29: 33–36.
50. Kirkpatrick N, Palmer JH: Semi-continuous ligninase production using foam-immobilized *Phanerochaete chrysosporium.* Appl Microbiol Biotechnol 1987; 27: 129–133.
51. Kirkpatrick N, Reid ID, Ziomek E, Paice MG: Biological bleaching of hardwood kraft pulp using *Trametes versicolor* immobilized in polyurethane foam. Appl Environ Microbiol 1990; 33: 105–108.
52. Archibald FS: The role of fungus fiber contact in the biobleaching of kraft brownstock by *Trametes versicolor.* Holzforschung 1992; 46: 305–310.
53. Paice MG, Reid ID, Bourbonnais R, Archibald FS, Jurasek L: Manganese peroxidase produced by *Trametes versicolor* during pulp bleaching demethylates and delignifies kraft pulp. Appl Environ Microbiol 1993; 59: 260–265.
54. Kondo R, Kurashiki K, Sakai K: In vitro bleaching of hardwood kraft pulp by extracellular enzymes secreted from white rot fungi in a cultivation system using a membrane filter. Appl Environ Microbiol 1994; 60: 921–926.
55. Jurasek L, Paice MG: Biological treatments of pulps. Biomass 1988; 15: 103–108.
56. Arbeloa M, de Leseleuc J, Goma G, Pommier JC: An evaluation of the potential of lignin peroxidases to improve pulps. Tappi J 1992; 75(3): 215–221.
57. Bourbonnais R, Paice MG: Demethylation and delignification of kraft pulp by *Trametes versicolor* laccase in the presence of 2,2′-azinobis-3-ethylbenzthiazoline-6-sulphonate. Appl Microbiol Biotechnol 1992; 36: 823–827.
58. Call HP: Multicomponent bleaching system. WO 94/29425, 1994.
59. Call HP: Process for modifying, breaking down or bleaching lignin, materials containing lignin or like substances. PCT World patent application; WO 94/29510, 1994.
60. Call HP, Mücke I: State of the art of enzyme bleaching and disclosure of a breakthrough process In: Int Non-Chlorine Bleaching Conf, Amelia Island. F, USA, 1994.
61. Call HP, Mücke I: Enzymatic bleaching of pulps with the laccase-mediator-system In: Pulping Conference AlChE session, San Diego. CA, USA 1994.

62. Call HP, Mücke I: Further improvements of the laccase-mediator system (LMS) for enzymatic delignification and results from large-scale trials. In: Int Non-Chlorine Bleaching Conf, Amelia Island, Florida, USA 1995.
63. Call HP, Mücke I: The laccase-mediator system (LMS) In: Biotechnology in the pulp and paper industry: Recent advances in applied and fundamental research; Proc 6th Int Conf Biotechnol. Pulp and Paper Industry, Srebotnk E and Messner K eds. 1995; Vienna Austria pp 27–32.
64. Bourbonnais R, Paice MG: Enzymatic delignification of kraft pulp using laccase and a mediator. Tappi J 1996; 76(6): 199–204.
65. Bourbonnais R, Paice MG, Reid ID, Lanthier P, Yaguchi M: Lignin oxidation by laccase isozymes from *Trametes versicolor* and role of the mediator 2, 2′-azinobis (3-ethylbenzothiazoline-6-sulfonate) in kraft lignin depolymerization. Appl Environ Microbiol 1995; 61(5): 1876–1880.
66. Paice MG, Bourbonnais R, Reid ID, Archibald FS, Jurasek L: Oxidative bleaching enzymes. J Pulp Pap Sci 1995; 21: J280–284.
67. Paice MG, Bourbonnais R, Reid ID: Bleaching kraft pulps with oxidative enzymes and alkaline hydrogen peroxide. Tappi J 1995; 78(9): 161–170.
68. Reid ID, Paice MG: Biological bleaching of kraft pulps by white rot fungi and their enzymes. FEMS Microbiol Rev 1994; 13: 369–376.
69. Kondo R, Harazono K, Sakai K: Bleaching of hardwood kraft pulp with manganese peroxidase secreted from *Phanerochaete chrysosporium*. Appl Environ Microbiol 1994; 60: 4359–4363.
70. Kondo R, Hirai H, Harazono K, Sakai K: Biobleaching of kraft pulp with lignin-degrading fungi and their enzymes. In: Proc 6th Int Conf on Biotechnol in Pulp and Paper Industry: Recent Advances in Applied and Fundamental Research, Vienna, Austria, 1995: 33–38.
71. Harazono K, Kondo R, Sakai K: Bleaching of hardwood kraft pulp with manganese peroxidase from *Phanerochaete sordida* YK-624 without addition of MnSO4. Appl Environ Microbiol 1996; 62(3): 913–917.
72. Kondo R, Harazono K. Tsuchikawa K. Sakai K: Biological bleaching of kraft pulp with lignin-degrading enzymes. In: Enzymes in pulp and paper processing (Jeffries TW and Viikari L eds.) ACS Symp Ser 1996; 655: 228–240.
73. Egan M: Presented in: 2nd Annual Pulp and Paper Chemical Outlook Conf Nov 12–13, Montreal, Corpus Information Services Ltd, Montreal 1985.
74. Viikari L, Tenkanen M, Buchert J, Ratto M, Bailey M, Siika-aho M, Linko M: Hemicellulases for industrial applications. In: Bioconversion of forest and agricultural plant residues. (Saddler JN eds.) CAB International, Wallingford, 1993: 131–182.
75. Niku-Paavola ML, Ranua M, Suurnakki A, Kantelinen A: Effects of lignin-modifying enzymes on pine kraft pulp. Bioresour Technol 1994; 50: 73–77.
76. Kantelinen A, Hortling BO, Ranua M, Viikari L: Effects of fungal and enzymatic treatments on isolated lignins and pulp bleachability. Holzforschung 1993; 47: 29–35.
77. Milagres AMF, Medeiros MB, Borges LA: Sequential treatment of eucalyptus kraft pulp with *Penicillium janthinellium* xylanase and *Pleurotus ostreatus* laccase. 6th Int Conf on Biotechnol in Pulp and Paper Industry, Vienna, Austria June 11–15, 1995.
78. Farrell R: Use of rldmtm 1–6 and other ligninolytic enzymes. PCT Int Appl WO 87/00, 564, 1987.
79. Farrell RL, Kirk TK, Tien M: Novel enzymes for degradation of lignin WO 87/00550, 1987.
80. Farrell RL, Gelep P, Anillouis A, Javaherian K, Malone TE, Rusche JR, Sadownick BA, Jackson JA: Sequencing and expression of ligninase cDNA of *Phanerochaete chrysosporium*. EP 0216080, 1987.
81. Olsen WL, Slocomb JP, Gallagher HP, Kathleen BA: Enzymatic delignification of lignocellulosic material. EP 0,345,715 A1, 1989.
82. Olsen WL, Gallagher HP, Burris AK, Bhattacharjee SS, Slocomb JP, Dewitt DM: Enzymatic delignification of lignocellulosic material. EP 406, 617, 1991.

83. Vaheri M, Miiki K: Redox enzyme treatment in multistage bleaching of pulp. EP 0,408,803 A1, 1991.
84. Vaheri M, Piirainen O: Bleaching of pulp in presence of oxidizing enzyme and transition metal compound. WO 92/09741, 1992.
85. Gysin B, Griessmann T: Bleaching wood pulp with enzymes. EP 0418201 A2, 1991.
86. Perie FH, Gold MH: Manganese regulation of manganese peroxidase expression and lignin degradation by the white rot fungus *Dichomitus squalens*. Appl Environ Microbiol 1991; 57: 2240–2245.
87. Galliano H, Gas G, Boudet A: Biodegradation of *Hevea brasiliensus* lignocelluloses by *Rigidoporus lignosus*: influence of culture conditions and involvement of oxidizing enzymes. Plant Physiol Biochem 1988; 26: 619–627.
88. Westermark U, Eriksson KE: Cellobiose quinone oxidoreductase, a new wood degrading enzyme from white rot fungi. Acta Chem Scand 1974 Ser B 28: 209–214.
89. Ander P, Mishra C, Farrell RL, Eriksson KE: Redox reactions in lignin degradation: interactions between laccase, different peroxidases and cellobiose: quinone oxidoreductase. J Biotechnol 1990; 13: 189–198.
90. Bao WJ, Usha SN, Renganathan V: Purification and characterization of cellobiose dehydrogenase, a novel extracellular hemoflavoenzyme from the white rot fungus *Phanerochaete chrysosporium*. Arch Biochem Biophys 1993; 300: 705–709.
91. Wariishi H, Valli K, Gold MH: Manganese (II) oxidation by manganese peroxidase from the basidiomycete *Phanerochaete chrysosporium*. J Biol Chem 1992; 267: 23688–23695.
92. Ander P: The cellobiose-oxidizing enzymes CBQ and CBO as related to lignin and cellulose degradation, a review. FEMS Microbiol Rev 1994; 13: 297–311.
93. Haemmerli SD, Leisola MSA, Fiechter A: Polymerization of lignins by ligninases from *Phanerochaete chrysosporium*. FEMS Microbiol Lett 1986; 35: 33–36.
94. Hammel KE, Moen MA: Depolymerization of a synthetic lignin in vitro by lignin peroxidase. Enzyme Microb Technol 1991; 13: 15–18.
95. Bourbonnais R, Paice MG: Oxidation of non-phenolic substrates. An expanded role for laccase in lignin biodegradation. FEBS Lett 1990; 267: 99–102.
96. Amann A: The lignozym process coming closer to the mill. ISWPC Montreal F4-1-F4-5. 1997.
97. Call HP, Mucke I: History, overview and applications of mediated lignolytic systems, especially laccase-mediator-systems (Lignozym process). J Biotechnol 1997; 53: 163–202.
98. Sealey JE, Ragaukas AJ, Runge TM: Biobleaching of kraft pulps with laccase and hydroxybenzotriazole. Proc Tappi Biological Science Symp 1997; 339–342.
99. Bourbonnais R, Leech D, Paice MG, Freiermuth B: Reactivity and mechanism of laccase mediators for pulp delignification. Proc Tappi Biological Science Symp, 1997; 335–338.
100. Pappius-Levlin K, Wang W, Ranua M, Niku-Paavola ML, Viikari L: Biobleaching of chemical pulps by laccase/mediator systems. Proc Tappi Biological Science Symp 1997; 329–333.
101. Ducka I, Pekarovicova A: Ligninases in bleaching of softwood kraft pulp. 6th Int Conf on Biotechnol in Pulp and Paper Industry. Vienna, Austria June 11–15, 1995.

6 Enzymatic Deinking

In recent years, the demand for the use of postconsumer paper products in the production of new paper products has risen dramatically. Some of this demand has been driven by politics and some by consumer demand. The use of secondary fibers in producing different grades of paper has increased greatly over the past two decades particularly as a result of development in deinking processes. The consumption rate of waste paper (secondary fiber) in different countries are given in Table 6.1. The worldwide consumption amounts to 37% (110 million tons) in 1995 and likely to increase to 55% by the year 2000.[1] Currently, 145 deinking facilities are operating, under construction or announced for construction in the USA.[2] Worldwide, deinking capacity is expected to rise to 31 million tons by 2001, with particularly large expansions for newsprint, printing, and tissue grades.[3]

A significant difficulty in dealing with postconsumer (secondary) fiber supplies, a relatively abundant and inexpensive raw material, is the removal of contaminants, particularly ink. The ease, or difficulty, of removing the ink depends primarily on the ink type, printing process, and fiber type. Some paper grades, like newspapers printed with oil-based inks, can be deinked with relative ease by conventional deinking process. Nonimpact printed papers are more difficult to deink[4] and the quantity of such papers continues to grow as a proportion of total recovered paper volume. Similarly, color printing via offset lithography is expanding in the USA at an annual rate of 25%;[5] other countries are also expected to follow. The cross-linking inks used in this process are also difficult to remove. Water-based flexographic newsprint grades of paper can cause problems in recycling systems when using conventional floatation systems.

The most difficult raw material for deinking is mixed office waste (MOW). A large portion of this fiber source has been printed using photocopiers and laser printers which fuse the ink to the fibers, making it difficult to remove by conventional methods. Mixed office waste is a large, virtually untapped source of high-quality fiber which can be used for fine papers and many other products, if the deinking process can be improved upon. Thus, removal of ink remains a major technical obstacle to greater use of recycled paper. Many of the conventional deinking processes require large quantities of chemicals, resulting in high wastewater treatment costs to meet environmental regulations. Deinking processes are substantial sources of solid and liquid waste. Disposal is a problem, and deinking plants would benefit from more effective and

Table 6.1. Consumption of waste paper in different countries

Continent/country	Year	Amount in million tons
Africa	1991	0.55
Australia	1994	0.97
England	1995	3.0
Europe	1995	25.0
Germany	1995	7.8
India	1995	2.06
Indonesia	1992	1.19
Japan	1992	14.8
Korea	1992	3.8
Malaysia	1992	1.24
Philippines	1992	0.32
Singapore	1992	1.01
South Africa	1991	0.21
Sweden	1992	1.2
Taiwan	1992	3.2
Thailand	1992	0.86
USA	1994	3.2
Worldwide	1995	110.0

less polluting processes. Enzymatic deinking seems to be a novel solution to these problems.

Enzymes for Deinking

Different enzymes useful for deinking include lipases, esterases, pectinases, hemicellulases, cellulases, and ligninolytic enzymes. Deinking involves dislodging ink particles from fiber surfaces and separating dispersed ink from fiber suspensions by washing or floatation. Enzymatic approaches involve attacking either the ink or fiber surfaces. Lipases and esterases can degrade vegetable-oil-based inks. Pectinases, hemicellulases, cellulases, and ligninolytic enzymes alter the fiber surface or bonds in the vicinity of the ink particles, thereby freeing ink for removal by washing or floatation. Many patents have been granted or applied for for the use of enzymes in deinking (Table 6.2).[6–17] Several patents specify the use of cellulases, particularly alkaline cellulases for deinking. Few patents claim that esterases can be used, while others specify the use of lipases or pectinases. One patent application mentions the use of laccase enzyme from white rot fungi. Most of the published literature on deinking deals with cellulases and hemicellulases.

Cellulases of fungal and bacterial origin are components of large systems or complexes that hydrolyze cellulose to water-soluble sugars.[18] Hydrolysis of crystalline cellulose requires a three-part system, comprised of endo-1,4-β glu-

Table 6.2. Patents on enzymatic deinking

Title	Enzyme	Reference
Chemical for deinking	Not specified	5
Elimination of ink from reclaimed paper	Alkaline cellulase	6
Deinking chemicals	Not specified	7
Deinking waste-printed paper using enzymes	Cellulases	8
Removal of ink from recycled paper	Lipase	9
Deinking waste paper with the incorporation of lipase	Lipase	10
Elimination of ink from reclaimed paper	Esterase	11
Elimination of ink from reclaimed paper	Cellulase	12
Process for removing printing ink from waste paper	Cellulase or pectinase	13
Biological ink elimination from reclaimed paper	Cellulases and/or pectinase	14
Elimination of ink from reclaimed paper	Esterase	15
Process for waste paper preparation with enzymatic printing ink removal	Ligninolytic enzyme (laccase)	16
Enzymatic deinking process	Alkaline cellulase	17

canases, exo-1,4-β-glucanases, and 1,4-β-glucosidases, with each enzyme performing its particular function. Endoglucanases hydrolyze amorphous cellulose and soluble derivatives by randomly splitting internal β-1,4-glucosidic linkages along cellulose molecules and produces glucose, cellobiose, and other oligomers. Exoglucanases hydrolyze cellulose molecules from the nonreducing end and release glucose or cellobiose units. Glucosidases degrade cellobiose and other oligomers into glucose monomers.[18,19] Overlapping specificities have been noted for endo- and exo-glucanases.[20] In addition, synergistic interaction among them has been reported for crystalline cellulose,[21] with combined activities being greater than the sum of individual activities. Synergistic effects, however, are low with amorphous and extensively hydrated cellulose, and nil with cellulose derivatives. These observations suggest that directed mixtures of cellulases might be more effective in deinking than single enzymes or natural mixtures.

Naturally occurring hemicelluloses are much more variable in composition than cellulose. Due to their complex nature, it requires even more complex enzyme systems for their complete or partial degradation and modification.[18,22] The two main glycanases that depolymerize the hemicellulose backbone are endo-1,4-β-xylanase and endo-1,4-β-mannanase. Small oligosacharides are further hydrolyzed by 1,4-β-D-glucosidase. The side groups are split off by α-L-arabinosidase, α-D-glucuronidase and α-D-galactosidase. Esterified side groups are liberated by acetyl xylan esterase and acetyl galacto-glucomannan esterases. Endo-xylanases are perhaps the best known of xylanolytic enzymes and are noted for initiating endwise attacks on the xylan backbones of common hemicelluloses, β-xylosidases, on the other hand, convert water-soluble dimers and oligomers to xylose.

Mechanisms for enzyme deinking have been studied on different substrates. Nine different possible mechanisms have been reported, from the recent studies, by Welt and Dinus.[23] It has been pointed out by Korean researchers that enzymes partially hydrolyze and depolymerize cellulose between fibers, freeing them from one another.[24] Ink particles are simply dislodged as fibers separate during pulping. Another mechanism proposed by these researchers is that enzymatic treatment weakens the bonds, probably by increasing fibrillation or removing surface layers of individual fibers.[13] The suggestion that enzymatic activity could be sufficient to remove whole surface layers at the low dosages and short reaction times commonly employed, however, is questionable. Woodward et al.[25] suggested that catalytic hydrolysis may not be essential; enzymes can remove ink under nonoptimal conditions. Mere cellulase binding may disrupt the fiber surface in a manner and to an extent sufficient to release ink during pulping. However, Jeffries et al.[26] proposed that deinking may be caused not by enzymes but by additives used to enhance enzyme stability. Residual ink areas obtained with heat-deactivated commercial enzyme preparations, however, did not exceed those observed after pulping in water.[27] Also ink removal varies inversely with enzyme inhibition.[28]

It is also reported that cellulases peel fibrils from fiber surfaces, thereby freeing ink particles for dispersal in suspension.[13] This peeling mechanism has also been suggested in pulp freeness increases after enzymatic treatment of secondary fiber.[29] Enzyme dosages and reaction times, however, seem too low to cause measurable cellulose degradation.

Hemicellulases facilitate deinking by breaking lignin-carbohydrate complexes and releasing lignin from fiber surfaces.[30] Ink particles are dispersed with the lignin. Hemicellulase treatment facilitated ink removal from newsprint and was reported to be accompanied by release of lignin.[31,32]

Enzymatic effects may be indirect, i.e., they remove microfibrils and fines, thereby improving freeness and facilitating washing or floatation.[27] Fines content, however, is not always reduced during enzymatic deinking.[33] Enzymatic treatment of nonimpact printed paper has been reported to remove material from ink particles, thereby increasing particle hydrophobicity and facilitating separation during floatation.[27] However, this promising hypothesis has to be tested with different enzymes, paper grades, and inks.

Mechanical action is supposed to be critical and a prerequisite to enzymatic activity.[28] It was said to distort cellulose chains at or near fiber surfaces, thereby increasing vulnerability to enzymatic attack. Assuming that fiber-fiber friction increases with pulp consistency, such an explanation seems consistent with earlier findings that enzymatic deinking is more effective at medium, as opposed to low, consistency.[27] However, research conducted by Putz et al.[33] disputes the importance of mechanical action. Applying greater shear forces via pulping at higher consistencies or for an extended time did not improve brightness; also, high shear forces can denature enzymes.[34,35] The data of Zeyer et al.[28] also suggest that mechanical action, such as friction on the fiber, is able to alleviate this restriction by opening the outermost layers of the fiber to expose fully the cellulose chains. Only then is the removal of a significant amount of ink possible. In addition, they also provided evidence that only easily accessible cellulose chains are subject to enzymatic cleavage.

It is likely that a particular deinking system would involve more than one of these mechanisms, but the relative importance of each mechanism would be dependent on fiber substrate, ink composition, and enzyme mixture. It should also be noted that none of the above studies used pure enzyme components (e.g., endo-glucanases, exo-glucanases, and cellobiohydrolases) of cellulase or hemicellulase to determine how the different components function in deinking.

Development of the Deinking Process

Realizing that much of the wood-containing paper production occurs at lower pH values, there should be some advantages to be gained from the use of the acidic class of cellulases from *Trichoderma* sp. in this application. The lower pH optimum of these cellulases (~5–5.5) should prevent the alkaline yellowing of the wood-containing papers during the deinking process. This, in turn, could eliminate the need for the normal deinking chemicals used in the pulping stage and improve the efficiency of the postdeinking brightening steps. Several research groups have examined the application of cellulases to the deinking of old newspaper.[24,32,33,36] Korean researchers have also reported that increasing pH from 4.7 to 8.0 decreased the enzyme activity and reduced the brightness of deinked pulp.[24,37] Presoaking with enzymes for 10 min before pulping appeared beneficial.[24] It was speculated that longer presoaking time allowed finely dispersed ink particles to readhere to fiber surfaces or to penetrate into porous parts of fibers, thereby limiting effectiveness of floatation. Soaking after pulping but before floatation adversely affected deinking due to readherence of ink particles to fibers.

Low pH cellulase and hemicellulase mixtures have been evaluated for deinking of letter press and color offset printed newsprint at pH 5.5.[32,36] Highest brightness increase for letterpress paper was obtained with a hemicellulase preparation (xylanase), as shown in Table 6.3, but the lowest residual ink areas as measured via image analysis were achieved with a cellulase preparation (Table 6.4). For colored offset papers, best brightness was obtained with a mixture of cellulases and hemicellulases. These researches also used similar enzymes to deink flexographic-printed newspaper.[31,38] Enzymatic treatment and floatation removed the water-based ink with ease, resulting in brightness levels well above those obtained with conventional deinking (Table 6.3). Inks of this type, however, are so finely divided that floatation is impaired.[39] Such

Table 6.3. Effect of enzymatic deinking on brightness improvement

Old newsprint printed by	Enzyme	Brightness[a] (% ISO)
Black and white letterpress	Cellulase + hemicellulase	58 (53)
Black and white flexo	Cellulase	55 (51)
Colored flexo	Cellulase	51 (46)

Based on data from Ref. 32.
[a] Value in parenthesis is control without enzyme.

Table 6.4. Effect of enzymatic deinking on optical properties of the colored offset newsprint

Enzyme	Enzyme dose/g pulp		Deinked pulp		
	U	ml	ISO brightness (%)	Residual ink area (%)	Scatt coeff. (m^2/kg)
Blank[a]	–	–	41	93	34.5
Control[b]	–	–	49	96	50.2
I	0.2[c]	0.033	52	77	70.5
II	0.2[c]	0.033	52	77	78.0
III	100[d]	0.50	53	65	75.8
IV	19[d]	0.033	54	79	69.9

Enzyme preparation I contained 6 U/ml CMCase, 10.0 U/ml xylanase and 42 U/ml filter paper activity.
Enzyme preparation II contained 6 U/ml CMCase, 6 U/ml xylanase and 1 U/ml filter paper activity.
Enzyme preparation III contained 0.20 U/ml CMCase, 200 U/ml xylanase and 0.005 U/ml filter paper activity.
Enzyme preparation IV contained 26 U/ml CMCase, 580 U/ml xylanase and 50 U/ml filter paper activity.
Based on data from Ref. 36.
[a] Reslushed pulp without enzyme treatment and no floatation.
[b] Reslushed pulp without enzyme treatment but with floatation.
[c] Dosage based on CMC.
[d] Dosage based on xylan.

results suggest that enzyme treatment under acidic conditions would be best for deinking this raw material and emphasize that deinking methods must be tailored to paper ink and printing type.

When the hemicellulases from *Aspergillus niger* and cellulases from *Trichoderma virdie* were evaluated for deinking, brightness was found to increase with increasing enzyme dosage and reaction time.[30] Soaking with enzyme before pulping was found to be beneficial but prolonged soaking reduced ink-particle size, lowered floatation effectiveness, and reduced brightness. An optimal blend of cellulase and hemicellulase gave higher brightness gains than conventional deinking. Regardless of ink type or printing process, enzymatic treatment tends to reduce ink-particle size. It has been reported that reduction in particle size varied with pulping time in the presence of cellulases; overall reduction was greater than that noted in conventional deinking.[24] Prasad et al.[32] and Rushing et al.[40] reported reduction in particle size from 16 to 37% depending on ink type; however, a credible explanation for this was not available.

The requirement of bleach chemicals is usually lower for enzymatic deinking than that for chemical deinking processes. Kim et al.[24] have reported that newspaper pulps bleached after being deinked by enzymatic and conventional means had similar brightness values. In the case of conventional deinking, hydrogen peroxide is used in the pulping as well as the bleaching step, but in the case of enzymatic deinking, hydrogen peroxide is used only in the bleaching process. Enzymatically deinked pulps were thus easier to bleach and required half as much hydrogen peroxide. In a similar study with letterpress, newspaper produced enzymatically deinked pulps with lower initial brightness values than those for conventionally deinked pulps.[40] However, subsequent bleaching with hydrogen peroxide produced similar brightness values with peroxide usage lowest for the enzymatic process. Putz et al.[33] reported that brightness levels obtained from bleaching offset-printed newspaper pulp after enzymatic deinking slightly exceeded those of pulps produced by conventional deinking with the same quantity of hydrogen peroxide applied during pulping.

Increasing amounts of the wood-free paper that comprises mixed office waste (MOW) stream is made under alkaline conditions using calcium carbonate as the filler of choice. The benefits of neutral cellulase for this deinking application was earlier exploited by a French group.[17] This group used a neutral cellulase as a post-treatment to a standard alkaline chemical treatment. An additional brightness and ink-removal benefit were observed. Baret and coworkers evidently did not consider the use of neutral cellulase without any other chemical pretreatment.[41] This other alternative was investigated by researchers at the USDA's Forest Products Laboratory.[27] They claimed enhanced deinking of wood-free, noncontact printed waste with cellulases.[27,42,43] Not surprisingly, the neutral cellulases showed an advantage over the acidic cellulases, even when the pH of the mixed office waste furnish was adjusted to the initial pH region with sulfuric acid.[27] The deinking response observed by the FPL group required a relatively small amount of enzyme to achieve the optimum ink-removal response although the dose response curve reported by this group is unusual and has not been satisfactorily explained to date. Their pilot plant results agreed with the laboratory results for two enzymes.[42] With one of the enzymes, the ink-removal efficiency was 94% in the pilot plant compared to 96% in laboratory trials. Recently, they have reported that in continuous processing of 2300-kg batches of 100% toner-printed office waste paper, cellulases greatly reduce the residual particle count while increasing brightness and freeness.[26] Strength properties and fiber length were essentially unchanged. Novo-Nordisk, Denmark, also performed several lab floatation runs using a neutral cellulase (Novozym 342), produced by them, for

Table 6.5. Effect of enzyme treatment on:

a) Brightness and dirt count

Treatment	Brightness (% ISO)	Dirt count (mm^2/M^2)
Blank	82.1	4200
Control	84.2	1200
Enzyme	88.3	70

b) Freeness

Treatment	Freeness (CSF, ml)
Blank	400
Control	440
Enzyme	490

c) Strength properties

Treatment	Breaking length (km)	Tear index ($mN/m^2/g$)	Burst index ($k Pa/m^2/g$)
Blank	6.5	8.5	4.55
Control	6.2	7.92	4.38
Enzyme	6.8	8.0	4.75

Based on data from Ref. 45.

deinking of copier paper.[44] The response in terms of brightness improvement was similar to that seen by FPL investigators.

Treatment with a pure alkaline cellulase significantly improved brightness levels of photocopied and laser-printed papers relative to pulping in water without enzymes.[45] Brightness improvement of 4 ISO units was observed (Table 6.5a). Residual ink area was reduced by 94%. Enzymatic treatment affected fiber length distributions. Such results might be expected, since these papers typically contain bleached softwood chemical pulp and cellulases are more likely to affect fiber distributions of chemical pulps. Enzyme-treated pulps showed similar increase in freeness (Table 6.5b) and strength properties (Table 6.5c), such as breaking length and burst index relative to control pulps.

Three industrial-scale trial runs were taken by Heise et al. on enzymatic deinking of nonimpact printed toners, at the Voith Sulzer pilot plant in Appleton, Wisconsin, USA.[46] Increased ink removal was achieved at a low level of a commercially available enzyme preparation in combination with a surfactant. The brightness of enzymatically deinked pulp was 2 points higher than that of control pulp (Table 6.6a). The enzyme trials also displayed improved drainage and comparable strength when compared with control (Table 6.6b). No significant differences in the quality and treatability of the process water was

Table 6.6. Effect of enzymatic deinking on:

a) Brightness

Deinking trial	Brightness (% ISO)	
	Floatation feed	Final pulp
Control	74.3	82.2
Enzyme 1	75.8	85.7
Enzyme 2	77.2	86.2

b) Strength properties

Parameter	Control	Enzyme 1	Enzyme 2
CSF (ml)	510	570	565
Kajanni (mm)	1.88	2.01	1.87
Tensile index (kN/m/g)	0.0410	0.0431	0.0412
Burst index (kPa/m^2/g)	2.20	2.42	2.34
Tear index (mN/m^2/g)	4.28	4.39	4.25
Viscosity (mPa/s)	17.2	17.4	16.5

Based on data from Ref. 46.

noted. However, the effluents from these trials had lower oxygen demand and toxicity than the effluents from the control.

The enzymatic deinking of mixed waste papers (laser/ xerographic printed and UV coated waste papers) and old newspapers/old magazines (ONP/OMG) by Yang et al.[47] was found to eliminate or substantially reduce the use of chemicals in the deinking process. The brightness of enzymatically deinked mixed office waste papers containing 90% laser copies, 3% colored paper, and 7% other papers was significantly greater than the chemical method. Enzymatic deinking achieved a 94% lower dirt count (visible) as well as 82% lower total dirt count. The treatment of American old newspaper pulp with blended cellulase in a Korean newsprint mill also resulted in about 2-point improvement in brightness.[48] They also conducted an enzyme mill trial run with white ledger-grade paper pulp at one of the biggest Korean tissue-making mills for several days in August 1994 and the result showed a reduction of residual ink count. The total ink-removal efficiency was found to increase from 93.9 to 98.3% by using the blended cellulase deinking. Zeyer et al.,[49] while studying the performance of enzymes in deinking of old newsprint, demonstrated that the arrangement of unit operations is of importance. No deactivation of enzymes by shear stress could be observed. Statistical investigation of particles on handsheets demonstrated that very likely many ink particles are still on their original location.

Unlike other studies, Novo researchers used monocomponent cellulases SP-476 and SP-613 for deinking of MOW.[44] SP-476 contains three discrete regions,

Table 6.7. Effect of deinking with monocomponent cellulase (SP-613) on brightness

Enzyme dose (U/kg pulp)	Brightness (% ISO)	Ink count (ppm)
0	80.6	>600
200	82.8	400
400	83.5	400
600	84.0	400
800	84.0	300

Based on data from Ref. 44.

a cellulose-binding domain (CBD), a spacer or linker area, which ties the CBD to the active region or core protein. Cellulases like 476, which contain the CBD, have the facility to attack the crystalline region of cellulose structures. SP-613 differs from SP-476 in that it consists exclusively of the core protein region with no CBD or spacer included in its structures. This cellulase is active exclusively on amorphous regions of the cellulose structure and causes less structural change to the cellulose structure. The response obtained using SP-476 was found to be similar to that observed for the multicomponent cellulase preparation. SP-613, on the other hand, demonstrated a dose response closer to what would be expected for a typical enzyme system (Table 6.7). The concomitant increase of brightness and decrease in ink-count values helped confirm that use of brightness as an assessment method could provide a rough measure of response in these systems. These studies show that the monocomponent portion of the multicomponent cellulase systems plays a major role in the enhancement of deinking of the wood-free noncontact-printed waste.

Paper sizing and other additives may prevent or limit contact, but direct physical contact between enzyme and substrate is a prerequisite to activity. Implications of sizing effects have also been studied by many researchers.[28,42,49,50] It has been suggested that sizing physically shields fibers from enzymes. Earlier work with textiles showed that starch sizing must be removed via alpha-amylase or other treatments before cotton fabrics can be altered by cellulase treatment. Paper sizing differs in mode of action and may, therefore, limit contact by various means. Sizing agents may limit enzyme activity by increasing fiber hydrophobicity, physically shielding fiber surfaces from enzyme attachment or preventing access via covalent bonds with cellulose. Alkyl succinic anhydride, for example, increases hydrophobicity but also forms covalent bonds with cellulose. The literature shows that paper sizing reduces enzymatic deinking efficiency and that reductions may vary with sizing agents. For nonimpact printed papers, deinking efficiency was lowest for papers sized with rosin and alum.[42] Such papers have the greatest resistance to wetting and the highest fiber hydrophobicity. Papers sized with alkyl succinic anhydride are less resistant to wetting but are almost as difficult to deink. These findings suggest the need for detailed investigations on the effects of different sizing agents and other additives like coating, dyes, metals, and various fillers, etc.

Alkaline lipases are claimed to facilitate the removal of oil-based inks. Nakano has reported that an alkaline lipase efficiently removed offset printing inks.[51] The effect was attributed to enzymatic hydrolysis of drying oil of thermosetting resin in the inks. Lipases should be effective with the inks carried in vegetable oils, and this approach need to be investigated further, especially since the trends towards greater utilization of such inks is increasing.

Enzymes catalyzing removal of surface lignin may hold promise for deinking of newsprint which contains a proportion of lignin-rich mechanical pulp. This approach has been evaluated using white rot fungi *Phanerochaete crysosporium* and with lignin-degrading enzymes.[52–54] Ink removal by a laccase preparation proved comparable to that of conventional chemical deinking; however, pulps showed high brightness and were easier to bleach. Further studies on such enzymes used alone or in combination with cellulases and/or hemicellulases would lead to better understanding of the interactions between fiber surfaces and ink.

Recently, a novel deinking process, which couples separation technology with cellulase treatment, has been described.[25] According to this, ink particles dislodged from newsprint, presumably by cellulase activity, readhered to smaller fibers originally present or created by enzymatic action. The smaller fibers and adhered ink were then separated from longer deinked fibers. These longer deinked fibers are usable without further treatment. Since ink adhered to the shorter fibers, conventional washing or floatation would be unnecessary and ink would not be released into the effluents. The reason for strong association between ink and short fibers could not be identified. The separation of such fibers is technically feasible.[55,56]

Useful enzymes for deinking are now available in larger quantities and at lower cost than in the past. Commercial use of enzymes for deinking has started recently in the USA.[57] The Enzynk process developed at University of Georgia by Eriksson's group is being commercialized. The process uses a mixture of enzymes in combination with surfactants and a few other chemicals.[58] The enzyme mixture is very much dependent upon the furnish used. This process is claimed to give a higher brightness, lower dirt count, higher freeness, and less sludge compared to chemical deinking.

Effluent Characteristics and Treatability

It seems to be quite obvious that the enzymatic deinking produces a whitewater with a lower COD than conventional alkaline deinking process, as reported by many researchers, thus reducing the load on the wastewater treatment systems.[47]

Some investigators have noted that the wastewater effluent from enzymatic deinking has 20–30% lower COD than wastewater from chemical deinking

Table 6.8. Quality of water entering dissolved air floatation clarifier

Parameter[a]	Control run	Enzyme runs	
		1	2
Total BOD_5	441	327	253
Dissolved BOD_5	234	168	138
Total COD	1080	1190	1455

Based on data from Ref. 46.
[a] All values in mg/l.

Table 6.9. Quality of water exiting dissolved air floatation clarifier

Parameter[a]	Control run	Enzyme runs	
		1	2
Total BOD_5	298	192	115
Dissolved BOD_5	275	165	123
Total COD	510	565	415

Based on data from Ref. 46.
[a] All values in mg/l.

processes.[24] Additional environmental advantages may accrue by avoiding the use of high alkalinity in the pulping stage.[27] However, a recent report indicates that enzymatic treatment produces COD load 50% lower than those for conventional deinking.[33]

The following observations regarding the effluent characteristics have been reported based on the industrial-scale enzyme deinking of nonimpact-printed toners.[46] Process water entering the clarifier from the enzyme runs contained lower total and dissolved BOD and higher COD than that of the comparable control, as shown in Table 6.8. However, the dissolved air floatation (DAF) cell readily clarified the process water, and the water exiting the DAF contained lower BOD and COD than the clarified control water (Table 6.9). The best quality reprocessed water was achieved with Enzyme run-2, which was also the best trial for ink removal. This trial also contained the lowest BOD and COD of the reject effluent stream (Table 6.10). There was no detectable difference in the BOD rate study of different samples from each trial. Toxicity of reject streams was also comparable. However, if the conventional chemical control were used for comparison, the enzyme runs would undoubtedly be less toxic than the conventional run, as has been observed previously on effluents collected from bench experiments.[27,43]

Table 6.10. Quality of reject stream

Trial	Dissolved BOD_5 (mg/l)	Total COD (mg/l)
Control run	219	440
Enzyme run 1	210	427
Enzyme run 2	180	346

Based on data from Ref. 46.

In deinking of offset newsprint, the COD load of the whitewater of the pulp suspension after reaction was lowest, 5 kg/ton of deinked pulp, if no chemicals were added.[33] With the total chemical reference formulation, a COD load of 22 kg/ton is generated due to the alkaline environment in pulping. However, for the enzymatically treated pulp, the COD load depends on the amount of enzymes as well as on the enzyme type. On an average, the following COD were obtained by Putz et al.:[33] 11 kg/ton for 0.2% enzyme addition and 20 kg/ton for 1.0% enzyme addition. Compared to the common alkaline deinking procedure, the COD load of a treatment with 0.2% enzymes was 50% lower. This is an important advantage of the enzymatic deinking process.

Advantages

Conventional deinking is a chemical intensive process and requires extensive wastewater treatment, which is expensive and becoming highly regulated. Enzyme-based deinking offers a potential means for the reduction of chemical use in the deinking process, thus reducing the load on the wastewater treatment systems.

Conventional methods are relatively ineffective in deinking the mixed office waste (MOW). Mixed waste papers present technical and economic challenges to the paper recycler, and of the wide variety of fibers and contaminants present in the paper stock, toners, and other noncontact polymeric inks from laser printing processes are the most difficult to deal with. Toners and laser printing inks are synthetic polymers with embedded carbon black; they do not disperse readily during conventional repulping process. Moreover, they are not readily removed during floatation or washing. Because of these problems, recycled papers contaminated with toners have a relatively low value. Conventional deinking uses surfactants to float toners away from fibers, high temperatures to make toner surfaces form aggregates, and vigorous and high-intensity dispersion for size reduction. Most of the deinking chemicals and high-energy dispersion steps are expensive. The high-energy dispersion step is both capital- and energy-intensive and can also reduce fiber length. Microbial enzymes have been shown to enhance the release of toners from the office waste. Application of cellulases and xylanases to xerographic-printed papers in a medium consistency mixer releases toner particles and facilitates subsequent floatation and washing steps.

Size distribution and shape of the ink removed can be effectively controlled using the enzymatic process to maximize the efficiency of the size-based floatation process. This can be accomplished by selectively varying enzyme composition, its dose and residence time, as well as variation of other additives and pH in the system to effectively dislodge the normally large, flat and rigid ink particles into much finer and nonplatelet forms. Enzymes may also retard redeposition of ink particles onto the fibers.

The most promising implication of high deinking efficiency from enzyme-enhanced deinking is that the dewatering and dispersion steps, as well as subsequent refloatation and washing, may not be essential. This should save capital expenses in constructing a deinking plant as well as reducing electrical energy consumption by the dewatering and dispersion.

The feasibility of enzymatic deinking in acidic environments has been confirmed. Applied commercially, this should reduce overall chemical requirements and minimize yellowing of reclaimed papers after alkaline deinking. Reduced chemical usage means lower waste treatment costs and less impact on the environment. Lower bleaching costs can also be anticipated, as enzymatically deinked pulps have proven to be easier to bleach and require less chemicals than pulps deinked by conventional methods, which will result in reduced environmental pollution.

The enzymatically deinked pulp displays improved drainage and possesses superior physical properties, higher brightness, and lower residual ink compared to chemically deinked recycled pulps. Improved drainage results in faster machine speed, which yields significant savings in energy and thus in overall cost. In addition, more and more recycled fiber will be used in paper making, reducing the requirement of virgin pulp production, thus resulting in great saving in energy required for pulping, bleaching, refining, etc., which will eventually lead to lower environmental pollution problems.

Limitations and Future Prospects

Use of cellulases and hemicellulases with pulps must be done carefully to avoid excessive depolymerization of pulp constituents. Some yield reduction accompanying the release of reducing sugars appears to result from losses of fines and other small particles due to enzyme activity. Otherwise, this might result in fiber loss and high BOD in effluents. More precise control over enzyme dosages and reaction time are expected to minimize such losses. However, there is a lack of information concerning the overall yield loss of an enzyme-deinked pulp after floatation as compared to chemical floatation.

Operating environments are critical to the success of enzyme deinking and many variables must be optimized. These include, among others, temperature, pH, enzyme activity and dosage, time/stage of enzyme addition, reaction time, pulp consistency, and mechanical action. To date, enzyme dosage and treatment time have been determined via trial and error for the enzyme and environment in question. More research is needed in this critical area and also on means for stopping reactions. Adding basic reagents to stop reactions may be counter-productive, i.e., reagents like sodium hydroxide contribute to product yellowing. Enzyme deinking would be especially attractive if surfactants and alkaline chemicals are not added. Operating costs and environmental impacts would be much lower.

Introducing enzymatic deinking technology in a mill-scale operation necessitates extensive customization of the enzyme formulation and process variables to achieve optimal effectiveness. After extensive experience with mill-based trials in the USA,[47] it is clear that the specific enzyme formulations vary widely based on furnish, process water, equipment configuration in various mills, and desired specifications of the deinked pulp. The enzymatic deinking process will naturally lead to a new chemical balance throughout the entire mill-water system. To effectively introduce enzymatic deinking into the pulp and paper industry, the costs and risks of conversion must be minimized.

The ratio of mechanical pulp and chemical pulp as well as hardwood/softwood composition in the mixture might have some effect on the activity and efficiency of the enzyme in the system. Ink composition and printing process also have a direct effect on the deinking, as they might react differently in enzymatic deinking. There are several areas which need to be addressed to further develop enzyme deinking. It is apparent that enzymes can facilitate deinking, but the exact process conditions by which this occurs have not been established. The areas which need further investigation include, (1) interaction of enzyme and fiber surfaces, (2) synergistic effect between various enzyme components, (3) effect of fillers and additives, ion strength, cationic species and pH, (4) effect of ink composition and printing process, (5) methods to evaluate ink redeposition and factors which affect redeposition, and (6) point of enzyme application, consistency, and effects of shear forces.

Useful enzymes for deinking are now available in larger quantities and at lower cost than in the past.[59] Pilot-plant and mill-scale trials have been conducted and promising results have been obtained.[58] Commercial use of enzymes for deinking has started recently in the USA.[57] Producing specific enzymes in quantities sufficient for commercial use involves costly purification. Increased usage and advances in fermentation and purification technology are expected to lower production costs. Alternatively, genetic engineering techniques can be used to identify the gene for a specific enzyme and transfer it to another organism, e.g., *Escherichia coli*, that normally does not produce the enzyme. Transfer and expression of cellulase genes have also been accomplished,[60] and several firms are now producing individual cellulases. The future may also see the development of biomimetic catalysis. These compounds are simpler and have lower molecular weights than enzymes but maintain function and specificity. Such synthetic catalysts would be more stable in commercial deinking environments, thereby permitting wider application.

Although it is quite apparent that enzymatic deinking is a less polluting process than chemical deinking, few data have been reported on the environmental pollution. Efforts should be made to document the pollution loads of different enzymatic deinking processes for various types of papers, in terms of quantity and quality, as compared to conventional deinking processes.

References

1. Dash B, Patel M: Recent advances in deinking technology. Ippta 1997; 9(1): 61–70.
2. Anon: Paper industry hikes its recovery goal to 50%. Recycled Paper News. Jan. 1994; 4(5): 10.
3. Uutela E: The future of recycled fiber for different grades. Pap Technol 1991; 32(10): 44–49.
4. Vidotti RM, Johnson DA, Thompson EV: Comparison of bench scale and pilot plant floatation of photocopied office waste paper. In: Proc. of Tappi Pulping Conf. 1992; 643–652.
5. Urushibata H: Chemicals for deinking. Jap. Pat. 1984; 59-9299.
6. Nomura Y, Shoji S (Honshu Paper Mfg. Co. Ltd): Elimination of ink from reclaimed paper. Jap. Pat. 1988; 59,494.
7. Hgiwara M: Deinking chemicals. Jap. Pat. 1988; 2-80684.
8. Eom TJ, Ow SSK: Deinking waste-printed paper using enzymes. Brit. Pat. 1989; GB 2,231,595A.
9. Guy Vare JA, Lucrelk M, Sharyo M, Sakaguchi H: Removal of ink from recycled paper. Jap. Pat. 1990; 150,984/90.
10. Sharyo M, Sakaguchi H: Deinking used paper with incorporation of lipase. Jap. Pat. 1990; 2,160,984.
11. Fukuda S, Hayashi S, Ochiai H, Ihizumi T, Nakamura K: Elimination of ink from reclaimed paper. Jap. Pat. 1990; 229,290/90.
12. Fukunaga N, Kita Y: Elimination of ink from reclaimed paper. Jap. Pat. 1990; 2-80,683.
13. Eom TJ, Ow SSK: Process for removing printing ink from wastepaper. Ger. Pat. 1990; GB 3,934,772.
14. Gen Y, Go S: Biological ink elimination from reclaimed paper. Jap. Pat. 1991; 882/91.
15. Sugi T, Nakamura K: Elimination of ink from reclaimed paper. Jap. Pat. 1991; 249,291/91.
16. Call HP: Process of treating wastepaper by enzymatic deinking. Ger. pat. 1991; 4,008,894.
17. Baret JL, Leclerc M, Lamort JP: Enzymatic deinking process. Int. Appl. 1991; PCT DK91/00090.
18. Eriksson K-EL, Blanchette RA, Ander P: Microbial and enzymatic degradation of wood and wood components. Springer, Berlin Heidelberg New York 1990; 89–177.
19. Wood TM: Mechanisms of cellulose degradation by enzymes from aerobic and anaerobic fungi. In: Enzyme systems for lignocellulose degradation (Coughtan MP, ed), Elsevier Applied Science, New York 1989; 17–36.
20. Knowles J, Teeri TT, Lehtovaara P, Penttila M, Saloheimo M: The use of gene technology to investigate fungal cellulolytic enzymes. FEMS Symp. Paris, France, September 1987; 43: 153–170.
21. Din N, Gilkes NR, Tekant B, Miller RC, Warren AJ, Kilburn DG: Non-hydrolytic disruption of cellulose fibers by the binding domain of a bacterial cellulose. Bio/Technology 1991; 9: 1096–1099.
22. Eriksson K-EL: Biotechnology in pulp and paper industry. Wood Sci Technol. 1990; 24: 79–101.
23. Welt T, Dinus RJ: Enzymatic deinking – a review. Prog. Paper Recycle 1995; 4(2): 36–47.
24. Kim T-J, Ow S, Eom TJ: Enzymatic deinking methods of waste paper. In: Proc. of Tappi Pulping Conf. 1991; Book 2: 1023–1030.
25. Woodward J, Stephan LM, Koran LJ, Wong KKY, Saddler JN: Enzymatic separation of high-quality uninked pulp fibers from recycled newspaper. Bio/Technology 1994; 12: 905–908.

26. Jeffries TW, Sykes MS, Cropsey KR, Klungness JH, Abubakr S: Enhanced removal of toners from office waste papers by microbial cellulases. In: Proc. of 6th Int. Conf. on Biotechnol. in Pulp and Paper Industry 1995; 141–144.
27. Jeffries TW, Klungness JH, Sykes MS, Rutledge-Cropsey KR: Comparison of enzyme-enhanced with conventional deinking of xerographic and laser-printed paper. Tappi J 1994; 77(4): 173–179.
28. Zeyer Chr, Joyce TW, Heitmann JA, Rucker JW: Factors influencing enzyme deinking of recycled fiber. Tappi J 1994; 77(10): 169–177.
29. Pommier JC, Fuentes JL, Goma G: Using enzymes to improve the process and product quality in the recycled paper industry. Part 1: The basic laboratory work. Tappi J 1989; 72(6): 187–191.
30. Paik KH, Park JY: Enzyme deinking of newsprint waste. I: Effect of cellulase and xylanase on brightness, yield and physical properties of deinked pulps. J Korea Tappi 1993; 25(3): 42–52.
31. Heitmann JA, Joyce TW, Prasad DY: Enzyme deinking of newsprint waste. Proc. Int. Conf. on Biotechnol. in Pulp and Paper Industry 1992; 175–179.
32. Prasad DY, Heitmann JA, Joyce TW: Enzyme deinking of black and white letterpress printed newspaper waste. Prog. Paper Recycl. May 1992; 1(3): 21–30.
33. Putz H-J, Renner K, Gottsching L, Jokinen O: Enzymatic deinking in comparison with conventional deinking of offset news. In: Proc. of Tappi Pulping Conf. San Diego, CA 1994; 877–884.
34. Reese ET, Mandels M: Stability of the cellulases of *Trichoderma reesei* under use conditions. Biotechnol. Bioeng. 1990; 22: 323–335.
35. Kaya F, Heitmann JA, Joyce TW: Cellulase binding to cellulose fibers in high shear fields. J. Biotechnol. 1994; 36: 1–10.
36. Prasad DY, Heitmann JA, Joyce TW: Enzymatic deinking of colored offset newsprint. Nord Pulp Pap Res J 1993; 8(2): 284–286.
37. Ow S, Eom T-J: Enzymatic deinking method of old newspaper. Jpn Tappi J 1991; 45(12): 1377–1382.
38. Prasad DY, Heitman JA, Joyce TW: Enzymatic deinking of flexographic printed newsprint: black and colored inks. Papiripar 1992; 36(4): 122–130.
39. Chabot B, Daneault C, Lapointe M, Marchildon L: Newsprint water-based inks and floatation deinking. Prog. Paper Recycl. 1993; 2(4): 21–29.
40. Rushing W, Joyce TW, Heitmann JA: Hydrogen peroxide bleaching of enzyme deinked old newsprint. 7th Int. Symp. on Wood and Pulping Chemistry, Beijing, China 1993; 233–238.
41. Baret JL, Leclerc M, Lamort JP: US Patent 1994; 5,364,501.
42. Rutledge-Cropsey K, Jeffries T, Klungness JH, Sykes M: Preliminary results of effect of sizings on enzyme-enhanced deinking. In: Proc. of Tappi Recycl. Symp. 1994; 103–105.
43. Sykes M, Klungness J, Abubakr S, Rutledge-Cropsey K: Enzymatic deinking of sorted mixed office waste: recommendations for scale-up. In: Proc. of Tappi Recycl. Symp., New Orleans 1995; 61.
44. Franks NE, Munk N: Alkaline cellulases and the enzymatic deinking of mixed office waste. In: Proc. of Tappi Pulping Conf. 1995; 343–347.
45. Prasad DY: Enzymatic deinking of laser and xerographic office waste. Appita 1993; 46(4): 289–292.
46. Heise OU, Unwin JP, Klungness JH, Fineran WG Jr, Sykes M, Abubakr S: Industrial scale-up of enzyme enhanced deinking of non-impact printed toners. Tappi J 1996; 79(3): 207–212.
47. Yang JL, Ma J, Eriksson K-EL: Enzymatic deinking of recycled fibers – development of the ENZYNK Process. In: Proc. 6th Int. Conf. on Biotechnology in Pulp and Paper Industry 1995; 157–162.
48. Ow SK, Park J-M, Han S-H: Effects of enzyme on ink size and distribution during the enzymatic deinking process of old newsprint. In: Proc. 6th Int. Conf. on Biotechnology in Pulp and Paper Industry 1995; 163–168.

49. Zeyer C, Heitmann JA, Joyce TW, Rucker JW: Performance study of enzymatic deinking using cellulase/hemicellulase blends. In: Proc. 6th Int. Conf. on Biotechnology in Pulp and Paper Industry 1995; 169–172.
50. Zeyer C, Joyce TW, Rucker JW, Heitmann JA: Enzymatic deinking of cellulose fabric: a model study of enzymatic paper deinking. Prog. Paper Recycl. 1993; 3(1): 36–44.
51. Nakano J: Recent research trends of pulping chemistry. J. Korea Tappi 1993; 25(1): 85–91.
52. Call HP, Von Raven A, Leyerer H: Application of ligninolytic enzymes in the production of pulp and paper (biopulping, biobleaching). Papier 1990; 44(10A): V33–V41.
53. Call HP, Strittmatter G: Application of ligninolytic enzymes in the pulp and paper industry - recent results. Papier 1992; 46(10A): V32–V37.
54. Ander P: Biopulping, biobleaching, and the use of enzymes in the pulp and paper industry. J. Korea Tappi 1993; 25(2): 70–76.
55. Floccia L: Fractionation and separate bleaching of wastepaper. In: Proc. Tappi Int. Pulp Bleaching Conf. 1988; 181–197.
56. Eul W, Meier J, Arnold G, Berger M, Suess HU: Fractionation prior to floatation - a new approach for deinking technology. In: Proc. Tappi Pulping Conf. 1990; 757–765.
57. Eriksson K-EL: Commercialization of Enzynk process (personal communication) 1997.
58. Eriksson K-EL, Adolphson RB: Pulp bleaching and deinking plants use chlorine-free process. Tappi J. 1997; 80(6): 80–81.
59. Daniels MJ: Using biological enzymes in papermaking. Pap. Technol. 1992; 33(6): 14–17.
60. Cornet P, Millet J, Beguin P, Aubert JP. Characterization of two CEL (cellulose degradation) genes of *Clostridium thermocellum* coding for endoglucanases. Bio/Technology 1983; 1(7): 589–594.

7 Treatment of Wastewaters with Anaerobic Technology

The forest industry utilizes wood and other lignocellulosic feedstocks as raw materials for the production of paper. The major constituents of wood are cellulose, hemicellulose, lignin, and extractives. Softwoods, hardwoods, and straw have different proportions of chemical components (Table 7.1). The processing of wood in paper mills involves various operations including debarking, pulping, and bleaching that result in the discharge of highly polluted wastewaters. The quantity and types of pollutants in these effluents vary with the type of lignocellulosic feedstock used as raw material, the process conditions applied (pH, temperature, pressure, chemical and mechanical treatments), and the specific water consumption.[1] Chemical additions and, to a lesser extent, high pressures and temperatures result in an increased release of organic matter into the process water and extensive lignin solubilization. Therefore, the pollution loads and the color due to dissolved lignin compounds is very high for chemical as compared to mechanical pulping effluents.[2,3] The COD loads associated with mechanical pulping processes range from 20–50 kg COD/ton of pulp whereas those corresponding to soda pulping processes may be as high as 500–900 kg COD/ton of pulp.[1,4] Nevertheless, the black liquors originating from kraft and soda processes are usually burnt to recover the pulping chemicals and the calorific power from the organic components, diminishing to a great extent, the environmental impact associated with these pulping processes. Conventional recovery processes are not economically viable in small paper mills and in those using nonwoody raw materials with a high silica content.[4,5] Black liquors represent a very important pollution source in several countries where small-scale mills are common.[4–6]

Pulp and paper mill effluents can cause considerable damage to receiving waters if discharged untreated. The environmental impact associated with these wastewaters is not only restricted to the oxygen demand, but also numerous effluents from the forest industry display acute toxicity to fish and other aquatic organisms.[7,8] Furthermore, these wastewater streams often exert inhibitory effects on microorganisms, that can disturb biological treatment systems.[9–11]

Aerobic treatment systems have traditionally been applied for reducing the pollution caused by the pulp and paper industry effluents. However, in the past years, the rising energy prices and the relatively high operation costs of conventional aerobic systems have resulted in an increasing application of the anaerobic wastewater treatment technologies. Recent developments have

Table 7.1. Composition of hardwoods, softwoods, and straw (in %)

Constituent	Hardwood	Softwood	Straw
Lignin	17–26	25–32	17–19
Hemicellulose	22–34	15–18	27–32
Cellulose	58–64	55–61	33–38
Extractives	3–8[a]		1–2[b]
Mineral matter	1		6–8

[a] Alcohol-benzene extract followed by hot water extraction.
[b] Ether extract.

shown that this process is well suited for the treatment of many pulp and paper mill effluents.[12–14]

Comparison Between Anaerobic and Aerobic Treatment

Aerobic biological treatment has been used for a long time to reduce the amount of organic pollutants in pulp and paper mill effluents, and extensive experience of this method is available. The reasons why anaerobic treatment is now in the progress of becoming a viable alternative are its general advantages over aerobic processes.The most important advantages in our opinion are as follows:

1. An effective environmental protection can be accomplished at very low costs. Several specific reasons can be cited to explain the economic advantages. Firstly, the method can be applied in technically simple and relatively inexpensive reactors, including many of the modern high-rate systems. Secondly, anaerobic treatment systems can generally be operated with little (if any) consumptive use of high-grade energy. The method therefore does not depend on the supply of electricity or other energy sources. Finally, instead of consuming energy, a useful energy carrier in the form of biogas is produced.
2. Frequently, very high space-loading rates can be applied in anaerobic wastewater treatment systems, particularly in the modern high-rate systems, so that the space requirements of these systems are relatively small. For specific newly developed anaerobic treatment processes, the space requirements are even exceptionally low, as these systems employ tall reactors instead of the relatively flat systems common for the more conventional high-rate anaerobic (as well as aerobic) systems.
3. The volume of excess sludge produced in anaerobic treatment is significantly lower compared to aerobic treatment method sludge. First of all, because the absolute quantity of excess sludge in organic matter is low, secondly, because the sludge TSS is highly concentrated.

4. Anaerobic organisms can be preserved unfed for long periods of time (exceeding 1 year) without any serious deterioration of their activity. At the same time, other important characteristics of anaerobic sludge also remain almost unaffected, e.g., the settleability of the sludge.

5. The requirement of inorganic nutrients (nitrogen and phosphorus) is lower as a consequence of less biosolids produced. Anaerobically generated biomass is typically about 11% nitrogen and 2% phosphorus.

6. Anaerobic treatment methods can lead to the application of integrated environmental protection systems, e.g., they can be (in principle) combined with posttreatment methods by which useful bulk products like ammonia or sulfur can be recovered, while, in specific cases, effluents and excess sludge could be employed for irrigation and fertilizers or soil conditioning.

7. Unlike aerobic treatment systems, the loading rates of anaerobic reactors are not limited by the supply of any reagent. The potential loadings of the anaerobic system can be increased to the extent to which sludge can be retained in the reactor, provided a sufficient contact between sludge and wastewater can be maintained. Consequently, in the modern high-rate anaerobic wastewater treatment systems, the attempt is made to retain as much active sludge as possible and to achieve maximum contact between incoming wastewater and the bacterial biomass.

In this regard, the upflow anaerobic sludge bed (UASB), expanded granular sludge bed (EGSB), and internal recirculation-granular sludge bed(IC-GSB) reactor systems belong to the category of very promising high-rate anaerobic wastewater treatment processes. In these systems, a high retention of sludge under high organic and hydraulic loading conditions can be achieved as a result of the formation and maintenance of a very well-settleable granular sludge. The requirements for maximum contact between sludge and incoming wastewater are accomplished mainly by the mixing brought about through the gas production.

There are also some general disadvantages attached to the anaerobic treatment processes. These are mainly the following: (1) Lower substrate removal rates per unit of biomass, typically one-fourth to one-tenth those for aerobic treatment of similar substrates. (2)Both the lower substrate removal rates and the lower sludge yields can result in a significantly longer period of time required for initial system startup or for recovery after a process upset (1–6 months). (3) The lower biosolids yields, typically 0.04–0.08 kg TSS/kg COD removed, makes biosolids retention increasingly critical as waste strength declines, in order to maintain adequate biomass in the treatment system. The chemically reduced conditions necessary for anaerobic processes produce compounds, including H_2S, mercaptans, organic acids, and aldehydes, that are both malodorous and corrosive. (4) There is a risk of odor problems especially at high sulfur levels in the effluents. (5) Posttreatment is generally required. The specific limitations of anaerobic treatment are as follows: the effluents from pulp and paper industry contain numerous types of phenolic compounds which range from simple monomers to high molecular weight polyphenolic polymers. Generally, the low molecular weight phenolics are biodegradable, including lignin-derived monomers and chlorinated phenolics.[15,16] As the molecular weight of phenolic compounds increases, a sharp decrease in their anaerobic biodegradability can be expected. High molecular weight lignin and tannins are not biodegradable in anaerobic environments; therefore, the COD corresponding to lignin or high molecular weight tannins cannot be feasibily removed by anaerobic treatment.[17,18] Wastewaters such as black liquors and bleaching effluents, in which lignin can account for 50% of the COD, are generally only 50% biodegradable or less.[19,20] Similarly, semichemical and chemi-thermo-mechanical pulping liquors contain significant fractions of lignin and are thus not fully biodegradable.[21-23] Since color is an important characteristic associated with lignin and other high MW polyphenolics, no significant color removal can be anticipated by anaerobic treatment. Anaerobic treatment is also restricted by the presence of toxic substances which can interfere with the metabolism of readily biodegradable substrates. Common toxic organic substances in pulp and paper industry effluents include: resin compounds, chlorinated phenolics, and tannins.[9,23-26] Resinous components of wood, such as resin acids and volatile terpenes, are important since they are present in many types of effluents generated from industrial processes that involve alkaline treatments of wood. Resin acids and volatile terpenes cause methanogenic inhibition at low concentrations. Low molecular weight chlorinated phenols, which are present in low concentrations in bleaching effluents, are potentially toxic to anaerobic digestion processes. These compounds are highly toxic to methane bacteria at very low concentrations. Generally, the methanogenic toxicity of chlorinated phenols increases with increasing Cl-number as well as with increasing apolarity. Tannic compounds are less toxic than resin compounds and chlorinated phenols; however, they are present at fairly high concentrations in debarking wastewater and in fiber board effluents.[27] The organic toxins do not only present problems for anaerobic digestion processes, they are also known to cause toxicity to the aquatic organisms of the discharge environment. The toxicity to fish has been demonstrated for resin compounds, chlorinated phenols, and tannins.[7,8,28-30] The anaerobic treatment systems have the limited capacity to decrease the aquatic toxicity of forest industry wastewaters.[21] Resin compounds are poorly degraded by anaerobic microorganisms.[31] Low MW tannins of bark are only partially degraded during anaerobic treatment.[32] Monomeric chlorinated phenols, on the other hand, are highly metabolized during anaerobic treatment, which can result in significant aquatic toxicity removal.[33-37]

Pulp and paper manufacturing use sulfur in various forms and processes. Sulfur compounds in wastewaters may inhibit the methane-producing bacteria or act as terminal electron acceptors for sulfate-reducing bacteria which may compete for the available substrates. This may result in low loading-rate potentials and shift the population to SRB instead of MPB.[12]

Anaerobic Effluent Treatment in General

Anaerobic biodegradation is a multistage process involving three basic groups of bacteria. Complex organic compounds are sequentially converted through a series of intermediate compounds to methane and carbon dioxide, as indicated by the four-step process (Fig. 7.1) which involves hydrolysis, fermentation, acetogenesis/dehydrogenation, and methanogenesis.[38] Complex high molecular weight soluble organic compounds (i.e., carbohydrates, lipids, and proteins) and particulates must first be hydrolyzed (stage 1) to simple organics (i.e., simple sugars, amino acids, glycerol, and fatty acids). These simple organics are converted by acid-forming bacteria to higher organic acids (such as propionic and butyric acid), and to acetic acid, hydrogen, and carbon dioxide in a fermentation or acidogenic phase (stage 2). The higher organic acids are subsequently transformed to acetic acid and hydrogen (stage 3) by acetogenic bacteria. The acidogenic and acetogenic bacteria belong to a large, diverse group that includes both facultative and strict anaerobes. Wastewater characteristics determine which bacteria predominate. The final step (stage 4) to produce methane is carried out by three groups of methane bacteria: *Methanobacterium*, *Methanoscarcina*, and *Methanococcus*. These strict anaerobes are capable of metabolizing formic acid, methanol, and carbon monoxide, as well as acetic acid, hydrogen, and carbon dioxide to methane.

In anaerobic processes where inorganic sulfur is a constituent of the wastewater, the sulfate-reducing bacteria, *Desulfovibreo*, are also of importance. Sulfate and/or sulfite is present in most effluents from acid sulfite, neutral sulfite semichemical (NSSC), kraft, chemi-mechanical (CMP), and chemi-thermomechanical pulp mills and where aluminum sulfate is used as a sizing agent for paper production. The sulfur-reducing bacteria use sulfate and sulfite as electron acceptors in the metabolism of organic compounds to produce hydrogen sulfide and carbon dioxde as end products. Sulfur reduction can become a significant factor in the performance and operation of pulp and paper anaerobic treatment systems. The hydrogen sulfide produced can be both toxic and corrosive. As shown in Fig. 7.1, the sulfur-reducing and the methane bacteria use and compete for the same organic compounds, reducing methane yield per unit of substrate removed. The methanogenic step is often the most critical one. Disturbances often result in an inhibition or depression of methane formation followed by an excess formation of fatty acids. A small part of the degraded organic matter is converted into new cellular material. The sludge production rate is low compared with aerobic processes. This means that the sludge retention time, SRT, must be relatively long if a sufficient amount of biomass is to be obtained in the system. A certain amount of biomass is necessary for high treatment efficiencies and a stable process.

Conventional low-rate systems employ long hydraulic retention times, HRT. The SRT value is often equal to the HRT value. These systems are most suitable for concentrated wastes such as sewage sludge. A few recent full-scale installations in the pulp and paper industry are of this type. More recently,

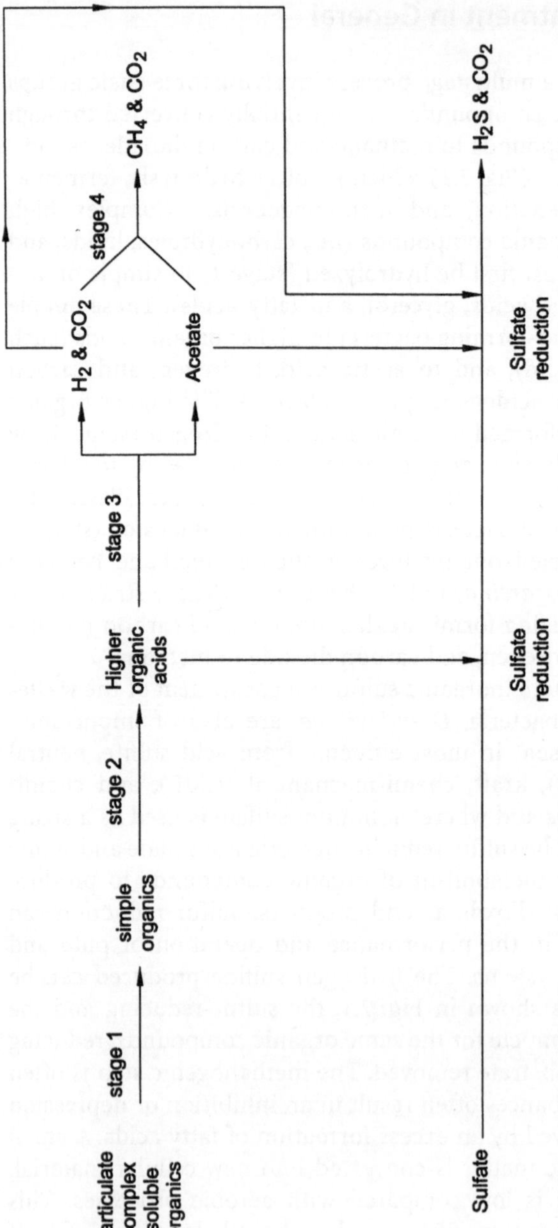

Fig. 7.1. Anaerobic metabolism to methane in competition with sulfate reduction

several high-rate systems have been developed. These systems are characterized by a high SRT/HRT ratio, which means that a large amount of biomass is retained in a relatively small reactor. Most of the recent full-scale applications in the pulp and paper industry are of this type. The following reactor types are applied for full-scale treatment of pulp and paper industry effluents.

Anaerobic Lagoon (Low-Rate). Anaerobic lagoons are the oldest of the engineered anaerobic treatment processes, having been first used in the food-processing industry in Australia in the 1940s.[14] The lagoon is commonly a single cell but can be configured as multiple cells in series or parallel. The lagoon commonly is of earthen construction and is covered with a synthetic membrane to maintain anoxic conditions, to collect biogas generated from anaerobic degradation of organics, and to control odors. The influent is initially mixed with inorganic nutrients and micronutrients and is neutralized as necessary to maintain adequate pH conditions in the lagoon reactor. An anaerobic biomass develops in the lagoon and remains partially suspended from mixing induced by the generated biogas. Mixing and contact with the biomass can be enhanced with intermittent use of low-speed mixers and solids recycle. Solids settling, however, is an important feature that can be designed into anaerobic lagoons to provide the time necessary to hydrolyze and degrade particulate material. Biogas is collected from under the membrane cover normally at multiple points and at slightly negative pressure to hold the membrane against the water surface. Although as not yet widely used in the pulp and paper industry, the anaerobic lagoon concept has several advantages over other anaerobic treatment process configurations, including:
(1) The ability to degrade suspended solids, since particulate that settles can be held in the reactor for months (or years). Anaerobic lagoons are potentially good applications for high-strength process streams from mechanical and chemical pulping, which also have high fiber content and wastewater from secondary fiber mills. (2) For effluent treatment systems that include activated sludge treatment, waste-activated sludge can be combined with the feed to the anaerobic lagoon for anaerobic digestion of these solids; thus, the total quantity of biological sludge requiring disposal from a combined anaerobic/aerobic treatment system is reduced. (3) Simplicity of operation. (4) Equalization of waste characteristics due to the large volume of the reactor. (5) Frequently lower capital and operating costs compared to other alternatives. The disadvantages are: (1) Requirment of large land areas, for many pulp and paper mill applications, minimum hydraulic retention times of 7–10 days would be required to achieve BOD_5 reductions in the range of 75–90%. (2) Solid removal from the lagoon may be required at some time, depending upon the quantity of inorganic solids and the degradability of the suspended material in the influent. (3) Difficulty in collecting gas from larger surfaces.

Anaerobic Contact Reactor. This process is an outgrowth of anaerobic lagoon.[14] This is essentially a closed tank with an agitator followed by a settling tank. Settled sludge is recycled to the reactor tank and thus a high SRT value can be obtained even at short HRT values. After chemical addition to provide essential nutrients and neutralization if needed for pH control, the influent and biomass are brought into intimate contact in a completely mixed reactor. Efficient mixing and solids separation are critical to successful process operation. Mixing induced by biogas generation is usually supplemented by side-mounted or top entering, low-speed mechanical mixers, or mechanically recirculated biogas. A positive pressure in the reactor results in supersaturation of dissolved gases in the effluent. Degassing and reflocculation of the biomass are essential for efficient gravity separation. Because suspended solids in the feed as well as biosolids produced from anaerobic metabolism of soluble organics are separated from the effluent and returned to the reactor, this process can provide the sludge age necessary for hydrolysis of degradable organic particulate material. This is one advantage of anaerobic contact over other high-rate anaerobic processes that makes it particularly well suited to the pulp and paper mill effluents with relatively high TSS concentrations such as mechanical pulping, white-waters, and high-strength secondary fiber effluents. In other high-rate anaerobic processes, biomass produced from the degradation of soluble substrate is held in the reactor principally, as dense granules or biofilms on an inert medium. The retention time of influent-suspended particulate is normally not much longer than the reactor hydraulic residence time and generally is too short for hydrolysis of the cellulosic fibrous material from wood pulping.

Upflow Anaerobic Sludge Blanket (UASB) Reactor. The UASB process was developed by Lettinga and coworkers at the University of Agriculture at Wageningen in The Netherlands. In the UASB reactor, the effluent is distributed into the bottom of the reactor and flows upwards through a layer of microorganisms.[14] Granulated sludge particles with extremely good settling properties are developed in this type of reactor. A separation system is installed at the top of the reactor for separation of gas, sludge particles, and treated effluents in order to minimize the losses of biomass. The SRT value is extremely high for well-adapted systems, and generally this process seems to have the potential to treat more diluted and colder effluents than the contact process. UASB has several advantages compared with other high-rate anaerobic systems. If the reactor is seeded with adapted granular sludge from another full-scale plant treating a similar waste, startup can be very rapid. Depending on the quantity of biosolids and acclimation to the waste, startup time can be as short as a few days. Reactors seeded with granular sludge equivalent to 10–15% of the design biomass typically achieve full load operation within 3 to 6 weeks from initial startup. Startup of anaerobic contact and anaerobic filters may take 3 months or longer to reach full load capability, and anaerobic lagoons can take up to 1 year.

Because of the dense nature and consequent settleability of the granular sludge, the washout rate of biosolids in a UASB can be very low. For this reason, the UASB process can anaerobically treat much lower waste strength than those normally considered feasible as low as 400 mg/l BOD_5. The ability of the high biomass concentration in the reactor to withstand organic or toxic shock loads with minimum adverse effect on performance, as well as with no requirement for mechanical mixing in the reactor, are also advantageous features of the UASB. Granulation of the sludge is the most critical factor in UASB performance. If the characteristics of wastewater are such that granules will not form or that granulation is lost due to toxicity or some other condition adversely affecting growth of the sludge bed, the biogas will be lost and system performance will be greatly affected. Also, high concentration of influent TSS can have an adverse effect on the performance of the sludge bed. This is particularly a problem if biogas production is not adequate to keep the bed sufficiently mixed to allow suspended material to pass through. In general, the suspended solids content of the feed should not exceed 10% of the total COD concentration.

Anaerobic Filter (AF) Reactor. The reactor contains a packing material usually a plastic material with a large surface area.[14] The microorganisms grow on the surface of the material and in void space between the surfaces. Attachment to a fixed surface minimizes or virtually eliminates the risk of washout of the biomass. The biomass film thickness and total quantity in the reactor are influenced largely by process configuration and characteristics of the support media. Anaerobic filters can be operated in an upflow or downflow mode. The main advantages of the upflow mode is that higher biomass concentrations can be developed, because in addition to formation of a biofilm, suspended biomass becomes entrapped in the stucture of the fixed bed media. Entrapment of biosolids in the upflow mode can result in faster startup and greater overall volumetric loading than downflow operation. Entrapment of solids, whether biologically produced or present with the influent, is also the major drawback of the upflow anaerobic reactor, leading to media plugging and hydraulic shortcircuiting. The downflow design largely solves the plugging problem, as biomass is mainly in the form of biofilm. The advantage of the downflow mode over the other high-rate anaerobic configurations, including the upflow anaerobic filter, can be particularly applicable to treatment of pulp and paper mill effluents with a high inorganic sulfur content and a low degradable COD to inorganic sulfur ratio. Downflow operation, in effect, allows for physical separation of the faster-growing, sulfur-reducing bacteria in the upper portion of the reactor-fixed media from the slower-growing methanogenic bacteria in the lower layers. The biogas produced in the lower levels of the reactor tend to strip out the hydrogen sulfide, protecting the toxicity-sensitive methane-forming bacteria. Thus, the downflow anaerobic filter can, in effect, function as a two-stage reactor, and should be able to tolerate much higher inorganic sulfur levels at lower degradable COD to inorganic sulfur ratios than other single-stage anaerobic reactors. Recycling is an essential feature that must be incorporated in both upflow and downflow anaerobic filters, to minimize pH gradients through the reactor bed, to maintain hydraulic conditions that assure good flow distribution and minimize shortcircuiting, and to minimize toxicity effects due to concentration. The severity of a toxic condition in biological treatment systems depends on concentration, exposure time, and adaptation. Biomass can generally stand short duration exposure at high concentrations better than longer-term exposure at low concentration, attached biofilm systems such as anaerobic filters and fluidized bed, are, therefore, much more rêsistant to shock loads than suspended growth processes such as anaerobic contact. The higher biomass density of the fixed-film anaerobic filters allows high volumetric loadings to be achieved.

Anaerobic Fluidized Bed Reactor (High Rate). The effluent is distributed into the bottom of the reactor and flows upwards through a fluidized bed of microorganisms attached on a carrier.[14] A certain amount of water usually has to be recirculated in order to keep the bed fluidized. The reactor bed continues to expand as the biofilm grows on the surface of the media and as soluble organics are metabolized. To control bed expansion and the quantity of biosolids in the reactor, media are withdrawn periodically when the bed reaches a predetermined height and passes through a media/biomass separation device such as an inclined wedge wire static screen. Biosolids separated from the media are wasted and the cleaned media are returned to the reactor.

Bench and pilot-scale studies of the anaerobic fluidized bed have demonstrated its technical viability for efficient treatment of pulp and paper-mill wastewaters and a variety of other industrial effluents. In this reactor, it is possible to obtain the highest volumetric loadings of the high-rate anaerobic process configurations discussed while maintaining similar treatment efficiencies. The ability to operate at very high volumetric loadings with high removal efficiencies, and consequently a smaller reactor volume, is perhaps the primary advantage of the anaerobic fluidized bed process. Its principal disadvantages have been higher power requirements and consequently higher operating costs, perceived (if not actual) increased complexity, and higher overall capital costs.

Hybrid and Two-Stage Anaerobic Configrations. A number of hybrid reactor configurations combining two or more anaerobic process configurations such as the UASB/fixed media system have been developed and evaluated in pilot plants or at full scale.[39] Hybrid systems are designed to take advantage of unique features of two or more process concepts. As an example, where adapted granular sludge is not available, the UASB/fixed-film hybrid may offer a faster startup than UASB alone. Development and entrapment of a flocculant anaerobic biomass, as well as growth of a fixed biofilm, normally proceed more rapidly than development and growth of granular sludge from an initial flocculant seed. Other advantageous features of the UASB/fixed-film hybrid anaerobic reactor include:

(1) high overall reactor biomass concentrations than UASB alone, resulting in a small reactor volume; (2) greater resistance to toxicity shock loads by having both a granular and a fixed film biomass (3) where biomass support media cost is high, the combination of processes may offer a capital cost advantage over an anaerobic filter alone, sized to achieve a similar treatment efficiency.

The primary disadvantage may be eventual plugging of the fixed media operating in an upflow mode and the potential difficulty of optimizing two processes physically housed in a single vessel for a wide range of flow and loading conditions. Separation of the acid-forming and methane-forming phases into two stages, at least in theory, allows the design and operation of each phase to be optimized independently of the other. The facultative acid-forming bacteria in the first stage can provide significant protection to the more sensitive methanogenic strict anaerobes. This is particularly the case when the oxidants such as hydrogen peroxide are present in the wastewater.

Effluent Types Suitable for Anaerobic Treatment

Generally speaking, the anaerobic treatment process requires a certain minimum concentration of degradable organic matter in the effluent if it is to be technically and economically feasible. With the development of new process designs, this limit is gradually decreasing and at present the practical limit is approximately $1000\,g/m^3$ expressed as COD.[14] This means that for many mills today, particularly older mills with high water consumption, anaerobic treatment may not be quite so appealing. However, in many cases, the concentration can be increased by system closures or separations. It should also be noted that laboratory tests indicate that the minimum concentration for anaerobic treatment may be reduced further in the future. Certain chemical compounds, mentioned earlier, may disturb the anaerobic process by toxic action and thus make the effluent less suitable for this type of treatment. However, the present trend indicates some possibilities to eliminate interference from these compounds. A large number of studies with pulp and paper industry effluents have been reported in the literature during the past few years (Table 7.2).[40] These studies cover many effluent types from debarking, mechanical pulping, deinking, chemi-mechanical pulping, semichemical pulping, sulfite pulping, sulfate pulping, peroxide bleaching, chlorine bleaching, and paper-making. Environmental Canada's Wastewater Technology Center screened 42 inplant waste streams from 21 Canadian pulp and paper mills to assess their potential amenability to anaerobic treatment.[41] The screening process consisted of chemical characterization and an anaerobic serum bottle technique to demonstrate biodegradability. Twenty three (55%) of the various effluent streams from kraft, sulfite, mechanical, and semichemical mills were found to be suitable for anaerobic treatment.[41–50] Mechanical and thermomechanical pulping effluents and wastewaters of paper making which are composed predomi-

Table 7.2. Effluent types suitable for anaerobic treatment

1. Kraft mill	Wood room Striper feed Contaminated hot water Evaporator condensate
2. Sulfite mill	Neutral sulfite semichemical Spent liquor Final effluent Combined sewer effluent Acid condensate (hardwood) Washer (softwood)
3. Thermomechanical mills	Final effluent Chip wash Clarifier effluent Chemi-thermomechanical pulp Thermomechanical pulp Thermomechanical pulp liner board
4. Nonsulfur semichemical mills	Controlled effluent Clarifier effluent

Based on data from Ref. 40.

nantly of carbohydrates are easy to treat anaerobically. Likewise, evaporator condensates which are composed mostly of alcohols and volatile fatty acids can be considered as easy for anaerobic treatment. Chemical, semichemical, and chemi-thermomechanical pulping liquors, bleaching and debarking effluents are more difficult for anaerobic treatment. These effluents contain important fractions of recalcitrant organic matter and numerous types of toxic compounds. The chemical process generally contributes to the extraction of lignin in the wastewater and alkaline chemical conditions lead to solubilization of toxic resin compounds. Bleaching operations often result in the formation of highly toxic chlorinated phenolics. Pulping and bleaching chemicals containing sulfur can contribute to the presence of high concentrations of sulfur in the wastewater. The contact of water with bark, i.e, wet debarking of wood, produces effluents in which toxic tannic compounds are extracted. Chemical, biochemical, and physical properties of some mechanical, chemi-mechanical, chemical pulping, and pulp-bleaching effluents that have been successfully anaerobically treated in pilot studies or at full scale are summarized in Tables 7.3–7.6.[41–50] A significant fraction of the organics in these effluents are either organic acids or alcohols. As shown in Fig. 7.1, anaerobic degradation of these compounds bypasses the first two stages (hydrolysis and acidogenesis). Organic acids and alcohols are transferred through acetogenesis and dehydrogenation (stage 3) to the methane precursors (acetate, hydrogen, and CO_2). Formic acid, acetic acid, and methanol are converted directly to methane by methanogenic bacteria.

Table 7.3. Characteristics of effluents from mechanical and chemi-mechanical pulping

Parameter[a]	Thermomechanical pulping	Chemi-thermomechanical pulping
COD	7200	6000–9000
BOD	2800	3000–4000
BOD/COD	0.39	0.44–0.50
Carbohydrates	2700	1000
Acetic acid	235	1500
Methanol	25	–
Total suspended solids	383	500
Total nitrogen (as N)	12	–
Total phosphorus (as P)	2.3	–
Total inorganic sulfur (as S)	167	

Based on data from Refs. 44, 45.
[a] Units of all parameters are mg/ml.

Table 7.4. Characteristics of condensates from kraft mill and sulfite mill

Parameter[a]	Kraft foul condensate	Acid sulfite condensate
COD	16 000	4000–8000
BOD	10 700	2000–4000
BOD_5/COD	0.67	0.5
Total volatile acids	16	3650
Methanol	–	250
Ethanol	–	–
2-Propanol	–	–
Furfural	–	250
Acetone	–	–
Total suspended solids	0	–
Total nitrogen (as N)	306	–
Total phosphorus (as P)	1.0	–
Total inorganic sulfur (as S)	91	800–850
pH	10.2	2.5
Alkalinity (as $CaCO_3$)	1 060	–
Temperature (°C)	–	25–50

Based on data from Refs. 46, 48.
[a] Units of all parameters are mg/ml, except pH and temperature.

Present Status of Anaerobic Treatment in Pulp and Paper Industry

Anaerobic treatment technologies are already in use for many types of pulp and paper industry effluents. In 1993, over 40 full-scale systems were operating and under construction.[39] The first full-scale application of anaerobic treatment in North America was achieved by the Inland Container Corp. Newport, Indiana,

Table 7.5. Characteristics of effluents from semichemical and NSSC pulping

Parameter[a]	Alkaline semichemical and waste paper recycle mill effluent	NSSC effluent		
		Spent liquor	Chip wash	Paper mill effluent
COD	11300–54700	39800	20600	5020
BOD	5300–19500	13300	12000	1600
BOD_5/COD	0.36–0.47	0.33	0.58	0.32
Carbohydrates	–	6210	3210	610
Acetic acid	–	3200	820	54
Methanol	–	90	70	9
Ethanol	–	5	990	–
Total suspended solids	200–18900	253	6095	800
Total nitrogen (as N)	–	55	86	11
Total phosphorus (as P)	–	10	36	0.6
Total sulfur (as S)	500	868	315	97
Temperature (°C)	31.5–39.0	–	–	–

Based on data from Refs. 42, 44.
[a] Units of all parameters are mg/ml, except temperature.

Table 7.6. Characteristics of bleaching effluents

Parameter[a]	Bleached kraft caustic extract
COD	1124–1738
BOD	128–184
BOD_5/COD	0.11
Carbohydrates	–
Acetic acid	0
Methanol	40–76
Ethanol	0
Total suspended solids	37–74
Total nitrogen(as N)	–
Total phosphorus(as P)	–
pH	10.1
Alkalinity(as $CaCO_3$)	275–755

Based on data from Ref. 49.
[a] Units of all parameters are mg/ml, except pH.

in 1978.[51,52] This treatment facility also has an aerobic polishing step following the anaerobic lagoon. The operating data for this system are summarized in Table 7.7. A similar anaerobic lagoon and aerobic polishing system were installed in 1987 at Sonoco Products Company's recycle and paper board mill in Hartsville, South Carolina.[53] The first 18 months of operating data indicated that when the effect of lower winter time temperature is taken into

Table 7.7. Full-scale anaerobic lagoon systems with aerobic polishing

Mill name	Inland Container Corporation	Sonoco Products Company
Mill location	New Port, Indiana	Hartsville, South Carolina
Startup date	1979	1987
Waste paper source	Recycled paper	Recycled paper
Anaerobic volume (MG)	5	24
Influent flow (MG/d)	0.69	4
Retention time (days)	7.2	6
BOD_5 ... Influent (mg/l)	1898	597–901
Anaerobic effluent (mg/l)	297	275–451
Aerobic effluent (mg/l)	89	21–97
Anaerobic removal (%)	84.4	38.1–60.8
Overall removal (%)	95.3	89.2–97.1
Biogas production (m^3/kg BOD_5 removed)	-	0.81
Methane content	-	65–75

Reproduced with permission from Tappi Press, Ref. 39.

account, equilibrium conditions were attained about 10–12 months after startup. In India, two full-scale anaerobic lagoon systems are operating.[54,55] Orient paper mills, Amlai, India, an integrated bleached sulfate pulp and paper mill, has operated a full-scale anaerobic lagoon since 1976.[54] The plant treats washing and screening effluents from the pulp mill and caustic extraction effluents from the bleachery. The treatment system is presedimentation-anaerobic lagoon-aerated lagoon-clarification pond. Biogas is not collected. Gwalior Rayon Mill, Mavoor, India, manufactures rayon-grade pulp by a pre-hydrolysis sulfate process and is treating the prehydrolysis effluent in an anaeobic lagoon.[55] The treatment sequence is neutralization-sedimentation-cooling-anaerobic lagoon treatment-aerated lagoon treatment. Biogas is not collected from the lagoon. About 73% COD removal has been achieved at an influent COD of 80 t/day (flow rate 1700 m^3/day). On a pilot scale, ADI International Fredericton, New Brunswick, Canada, has demonstrated low-rate anaerobic lagoon treatability of a variety of paper-mill effluents, including kraft and sulfite evaporator condensates, soda and neutral sulfite semichemical spent liquor, bleachery effluents, board mill effluent, and paper mill white-waters.[39]

The first full-scale application of anaerobic contact systems in the pulp and paper industry was at Swedish sulfite mills in 1983, a semichemical pulp and waste paper mill in Spain, a sulfite pulping and cellulose derivative manufacturing facility in Sweden in 1984, and a ground wood mill in Wisconsin in 1986.[56–58] Table 7.8 presents the performance data of some of the full-scale anaerobic contact process facilities. Purac AB, a Swedish Company, now licenses the technology developed by the Swedish Sugar Co, Infilco Degremont

Table 7.8. Full-scale anaerobic contact process facilities

Mill location	Wastewater source	Start-up date	Design influent conditions					Reactor loading rate (kg COD/m³/day)	Removals		Performance	
			Flow (MG/d)	BOD_5 (mg/l)	COD (mg/l)	TSS (mg/l)	S (mg/l)		BOD (%)	COD (%)	Gas production (m³/day)	CH_4 content (%)
Hylte Bruks AB, Sweden	TMP, groundwood magnetite deink	1983	2.3	1300	3500	520	–	2.5	71	67	7500	60–70
SAICA Zaragoza, Spain	Waste paper alkaline cooked straw	1984	1.3	10000	30000	–	–	4.8	94	66	40800	70–75
MoDo Cell AB Domsjo, Sweden	Sulfite condensate, CTMP	1984	3.4	4000	10000	–	700	4.3	95	65	50400	55–90
Hannover Paper Alfeld, Germany	SEC	1984	0.4	3000	6000	–	50	4.2	97	85	2900	70–90
Niagara of Wisconsin Wisconsin	CTMP	1986	1.4	2500	4800	3300	–	2.7	96	77	–	–
SCA Ostrand Ostrand, Sweden	CTMP	1987	1.0	3700	7900	–	–	6.0	50	40	4000[a]	–
E. Holtzmann & Cia AG Karlsruhe, Germany	Sulfite	1988[b]	1.0	4000	6400			4.0	–	–	8100[c]	70–80[c]
Modern Karton Istanbul, Turkey	Waste paper alkali wheat straw	1988[b]	0.7	7000	20000			–	–	–	–	–
Alaska Pulp Corporation Sitka, Alaska	Sulfite condensate bleach caustic extract, pulp whitewater	1988[b]	4.3	3500	10000	–	–	3.0	85	49	20100[c]	65

– No data available.
Reproduced with permission from Tappi Press, Ref. 39.
[a] Methane yield.
[b] Under construction.
[c] Expected performance.

Inc., a worldwide supplier of water and wastewater treatment equipment, also currently has licensed anaerobic contact technology operating in the pulp and paper industry.[39]

Reactor volatile solids concentrations reported for anaerobic contact systems operating in the pulp and paper industry have ranged from 3000–5000 mg/l to over 10000 mg/l,[46,58] resulting in volumetric loadings in the range of 1–2 kg BOD removed/m^3/day at BOD removal efficiencies greater than 90% and at optimum temperatures of 35 ± 5 °C. These volumetric loading rates are perhaps 20–50% of those that can be achieved by other high-rate anaerobic treatment configurations.

With commercialization in the late 1970s and early 1980s,[59] the UASB has been used increasingly in pulp and paper industry[59–64] and other industries. A noteworthy increase in the construction of UASB plants has been after 1987 reflecting the recent increased popularity of anaerobic treatment.[12] The loading rates achieved for pulp and paper industry effluents in full-scale UASB plants range from 5 to 27 kg COD/m^3/day. The efficiencies vary from 50 to 80% of the COD, depending mostly on the biodegradability of the particular wastewater being treated. The BOD removal efficiencies are high, in most cases between 75 and 99%, indicating that anaerobic treatment is particularly useful for the elimination of readily biodegradable organic matter. In India, two full-scale UASB plants are operating.[63,64] Harihar Polyfibers and APR Ltd. are using the UASB process to treat the prehydrolysis effluent. Table 7.9 shows the performance of some of the full-scale UASB facilities. UASB technology is offered by several suppliers including Paques B.V., The Netherlands, Gist Brocades, The Netherlands, and Biotin N.V., Belgium.

The full-scale anaerobic filter for treating pulp and paper mill effluent was constructed in 1987 at an integrated mill with CTMP pulping in Lanaken, Belgium.[39] The performance data for the KNPC facility in Lanaken, Belgium, are presented in Table 7.10.

The higher biomass density of the fixed film anaerobic filters has allowed volumetric loadings of 4–15 kg BOD removed/m^3/day to be achieved in pilot studies on pulp and paper mill wastewaters, as well as in full-scale application in other industries.[65,66] Licensed anaerobic filter technology is available from a number of suppliers including Bacardi Corp., Badger, and Infilco Degremont.

Donovan et al., in a pilot study, treated evaporator condensate from an ammonia-based sulfite mill in an anaerobic filter, a UASB, and an anaerobic fluidized bed reactor side by side.[66] Each reactor configuration had similar BOD removal efficiencies of 80–90% but the fluidized bed was able to operate at volumetric loadings of 17–41 kg BOD removed/m^3/day compared to 3.3 kg COD removed/m^3/day for the UASB, and about 10 kg COD removed/m^3/day for the anaerobic filter.

Following a year of pilot studies at a paperboard mill in D'Aubigne Racan, France, a full-scale anaerobic fluidized bed system was reported to have been constructed.[39] Table 7.11 presents the performance of this system. Suppliers of

Table 7.9. Full-scale upflow anaerobic sludge blanket (UASB) facilities

Mill location	Wastewater source	Start-up date	Reactor volume (m^3)	Flow (MG/day)	BOD_5 (mg/l)	COD (mg/l)	Temp (°C)	Reactor loading rate (kg COD m^3/day)	BOD removed (%)	COD removed (%)
Ceres, Holland	Box board	1983	70	0.03	3150	6300	35	9	85	70
Roermond, Holland	Corrugating	1983	750	0.9	2750	5500	35–40	24	85	75
Celtona, Holland	Tissue	1984	700	0.8	600	1200	20–30	5	75	60
Southern Paper Converters, Australia	Waste papers	1984	100	0.026	–	10000	–	10	>80	>80
Industrie-water, Holland	Corrugating box board	1985	2200	3.2	550	1100	29	6	80	70
Davidson, United Kingdom	Linerboard	1986	1600	1.3	1440	2880	35	9	90	75
Mayor Mclnhof, Austria	Folding box board	1987	1500	1.6	900	1800	35–40	7.5	60	50
Emin Leydier, France	Corrugating	1987	1000	0.6	1700	3550	30–35	8.5	85	70
Tillman, West Germany	Corrugating	1987	150	0.4	7500	15000	25–35	15	90	80
Chimicadel Friulli, Italy	Sulfite condensate	1989	3000	0.6	12000	15600	30–40	12.5	90	80
Italcarta, Italy	Corrugating testliner	1988	1900	1.7	1250	2500	35	8.5	85	70
APPM, Australia	Fine Paper	1988	95	0.06	1500	4000	27	7	90	75
Quesnel River pulp, Canada	TMP/CTMP	1988	7000	4.9	3000	7800	35	18	60	50
Lake Utopia Paper, Canada	NSSC	1988	3000	1.0	6000	16000	35	20	80	55
PWA Redenfeldor, West Germany	Waste paper	1988	950	0.95	1400	2800	–	10	90	75

Consolidated Bathurst, New Brunswick	NSSC/CTMP	1988	15600	4.2	5000	11750	–	11.9	85	60
Mac Millan Bloedel, Canada	NSSC/CTMP	1989	7000	1.7	7000	17500	–	15	80	55
Nordcarta, Italy	Corrugating	1989	700	0.2	1400	2800	30–35	10	85	70
Europa Carton, West Germany	Corrugating testliner	1989	1000	1.0	1500	3000	30–40	11.5	85	75
EnsoGutzeit, Finland	Bleached TMP/CTMP	1989	1500	1.3	1800	4000	35	13.5	75	60
Delkaskamp, West Germany	Corrugating	1989	300	0.3	2000	4000	30–40	15	85	75
Model AG, Switzerland	Box board	1989	650	0.6	1733	3000	–	10.2	85	75

Reproduced with permission from Tappi Press, Ref. 39.

Table 7.10. Full-scale anaerobic filter facility

Facility	KNPC
Location	Lanaken, Belgium
Startup date	1988
Wastewater source	CTMP
Influent conditions	
Flow (MG/day)	13
BOD (mg/l)	4000
COD (mg/l)	7900
TSS (mg/l)	-
Sulfur (as S mg/l)	-
Reactor loading (kg COD/m^3/day)	12.7
Removals - COD (%)	70
- BOD_5 (%)	85
Biogas production (m^3/h)	443
Methane content (%)	85

Reproduced with permission from Tappi Press, Ref. 39.

Table 7.11. Full-scale anaerobic fluidized bed facility

Facility	Allard Paper Company
Location	D'Aubigne Racan, France
Anticipated startup	Late 1988
Wastewater source	Paperboard
Influent conditions	
Flow (MG/day)	0.40
COD (mg/l)	3000
BOD_5 (mg/l)	1500
Reactor loading (kg COD/m^3/day)	35
Removals - COD (%)	72.2
- BOD_5 (%)	83.3
Effluent TSS (mg/l)	53
Biogas production (m^3/h)	75
Methane content (%)	65

Reproduced with permission from Tappi Press, Ref. 39.

licensed anaerobic fluidized bed technology include Infilco Degremont, Air products and Chemicals, and Gist Brocades.

At least two hybrid systems have been installed at pulp and paper mills in Finland and New Zealand. Both of these facilities are two-stage systems where the acidogenic and methanogenic stages are physically separated into sequential reactors.[39] The Taman process, developed by Tampella, was piloted in the early 1980s and a full-scale system was constructed in 1985 at the Anjala Paper mill (Anjalinkoski, Finland) to treat mechanical pulping and debarking plant effluents.[39] Little information on operating data from the combined fixed growth packed bed and sludge bed process have been released for publication. Caxton Paper Mills, Ltd, Kaiwerau, New Zealand, has installed a two-stage anaereobic system in conjunction with construction of a new CTMP mill.[39] The methane-producing second stage combines the UASB process with highly porous polyurethane foam, a carrier material developed by Biotin N.V., The Netherlands. The combined UASB/polyurethane carrier reactor is designed to provide a fast startup where granular sludge is not available and to provide high reactor stability with a wastewater known to be difficult to treat.

A full-scale anaerobic-aerobic biological treatment plant treating the total waste effluent from a pulp mill is in operation.[67] All of the condensate and about half of the caustic extraction liquor generated in the sulfite-pulping process are treated in the anaerobic process. The anaerobic plant is designed to treat 60000kg COD/day, of which half comes from sulfite evaporator condensate and half from caustic extraction liquor. The anaerobically treated water is led to the aerobic treatment stage where it is treated with rest of the extraction liquor and other effluent streams. The aerobic stage is designed for the treatment of 102000kg COD/day and 37000kg BOD/day. The excess sludge from the aerobic treatment plant is further treated in an alkaline hydrolysis stage followed by an anaerobic digestion. Results from operation of the aerobic stage with 70000kg COD/day and 30000kg BOD/day show good sludge characteristics and good treatment results: 65% COD reduction and outlet concentrations of TSS <50mg/l, total P <1mg/l and inorganic N <6mg/l.

Anaerobic treatability studies of bagasse wastewater have been conducted in North Carolina.[68,69] An anaerobic downflow fixed-film reactor process has been studied. At the anaerobic stage, 90–95% of the total chemical oxygen demand was removed at loading rates of 3–18kg COD/m^3/day without supplying any nutrients or trace elements. A yield coefficient of 0.156 cells/g COD was calculated at a high COD-loading rate of 18kg COD/m^3/day. It was estimated that with this process, 1 m^3 of bagasse wastewater with a COD content of 13000mg/l can produce about 3–5 m^3 of methane. Intermittent checks on the system alkalinity revealed that feed neutralization to maintain alkalinity would be necessary with sodium bicarbonate at approximately 2500mg/l for achieving steady-state high treatment efficiency.

The effect of intermittent feeding on the performance of an acclimatized anaerobic sludge blanket reactor treating CTMP wastewater has been reported by Kennedy et al.[70] Intermittent feeding did not cause process instability when the anaerobic reactor contained acclimatized anaerobic granular sludge. The reactor could be left idling in a substrate-limited state for up to 1 week without causing biomass inhibition.

Rintala and coworkers studied anaerobic treatment of thermomechanical pulping whitewater in semicontinuously fed batch digesters at 35, 55, and 65 °C and in UASB reactors at 55 and 70 °C.[71,72] Also, they studied posttreatment of the 55 °C UASB effluent by thermophilic activated sludge process and by acid precipitation. The batch digestor studies demonstrated treatability of whitewater at all applied temperatures. In UASB reactors up to 65–75% COD removal was obtained at 55 °C at loading rates of 14–22kg COD/m^3/day. About 60% COD removal was maintained at 55 °C in the UASB reactor at a loading rate as high as 80 kg COD/m^3/day, which corresponded with a hydraulic retention time of 55 min. In the UASB reactor at 70 °C, loading rates of about 13 kg COD/m^3/day were reached with about 60% COD removal. Both the batch and UASB studies showed increased effluent residual COD at increasing treatment temperatures. Compounds other than volatile fatty acids and carbohydrates accounted for most of the COD increment. Activated sludge process and acid precipitation further removed the COD present in anaerobically treated whitewater.

Dr. Lettinga's group studied anaerobic treatability of TMP effluents and soda pulping liquor.[73] Continuous experiments were conducted in laboratory-scale UASB reactors inoculated with anaerobic granular sludge at 30 ± 2 °C. TMP wastewaters were found highly suitable for anaerobic treatment. The application of high organic loadings (31 g COD/l/day) was feasible by the end of the continuous experiment with TMP wastewaters with 68 and 98%, total COD, and biodegradable COD elimination efficiencies, respectively. Unlike TMP effluents, soda pulping wastewaters were highly inhibitory to methanogenic bacteria and they contained important fractions of recalcitrant organic matter. Wood resin constituents were shown to be responsible for most of the methanogenic inhibition in these wastewaters. Nonetheless, anaerobic wastewater treatment was feasible for removing the biodegradable substrate in soda pulping wastewaters if the wastewaters were diluted to subtoxic levels or detoxified by pretreatment with the adsorbent Amberlite XAD-2 prior to biological treatment. Low COD removal efficiencies were observed during the continuous experiment (45 to 50%) with soda pulping liquors, due to high amounts of recalcitrant lignin in these wastewaters.

Future Perspectives for Anaerobic Treatment of Pulp and Paper Industry Effluents

In order to facilitate the anaerobic treatment of difficult pulp and paper industry waste waters, numerous measures can be taken involving either the operation or design of the wastewater treatment system. These measures can be used to overcome the previously discussed limitations confronting the application of anaerobic treatment technologies to pulp and paper industry effluents.

During the operation of the bioreactors, anaerobic bacteria should be acclimatized to toxic organic compounds. The wastewaters should be diluted to subtoxic concentrations during reactor startup and the dilution of the wastewater should be reduced in an incremental fashion in accordance with the degree to which the microorganisms adapt to the toxicity and develop capacities to degrade the organic compounds.[74–78] Additionally, immobilizing anaerobic bacteria and maintaining high concentrations of biomass in the reactor are factors which are known to improve the tolerance to toxic substances by anaerobic treatment systems.[79,80] Several researchers have investigated the use of granular active carbon as a support material for the immobilization of anaerobic bacteria treating phenolic wastewaters.[81–84] The active carbon helps to maintain lower effective concentrations of the phenolics in the reactor, reducing the toxic effect and favoring phenol degradation. The success of these systems depends on the capacity of the anaerobic bacteria to bioregenerate the adsorption sites on the active carbon. However, if the wastewaters contain a significant fraction of nonbiodegradable toxic compounds, then the active carbon systems are not effective. The design of the anaerobic treatment system can include certain pretreatments that eliminate or detoxify aromatic compounds or otherwise modify aromatic compounds to improve their anaerobic biodegradability. These pretreatments include precipitation pretreatments and oxidative pretreatments. Precipitation pretreatments have been applied to anaerobic wastewater treatment systems to remove either toxic compounds or recalcitrant fractions. Examples of toxicity removal include the precipitation of long chain fatty acids and resin acids by calcium.[26,85] Precipitation pretreatments can also be used to remove recalcitrant lignin fractions from the wastewater.[86–88] However, the disadvantage of this method is that the precipitates have to be separated from the wastewater and disposed of. Oxidative pretreatments involve (1) oxidations that polymerize the aromatic compounds and (2) oxidations that destroy the aromatic structure. Polymerization pretreatments detoxify aromatic compounds such as tannins by converting them to high MW humus compounds, permitting the anaerobic treatment of the biodegradable components in the wastewater.[32] Although humus is recalcitrant, it is nontoxic to aquatic organisms in the discharge environment.[30] Strong oxidants such as O_3 and H_2O_2 can be utilized to destructively oxidize recalcitrant xenobiotic aromatic compounds to simple nonaromatic aliphatic carboxylic acids.[32,89] Unfortunately, the high costs of the chemical oxidants renders the application of these pretreatment methods economically unfeasible. The destructive method of oxidation offers good perspectives as a pretreatment for the anaerobic treatment of complex aromatic wastewaters. These oxidations can potentially convert recalcitrant and toxic aromatic compounds to nontoxic biodegradable compounds. Therefore, less expensive methods of obtaining destructive oxidations still need to be developed. Ligninolytic fungi that have the capacity to destructively oxidize aromatic compounds should be evaluated. These fungi are known to depolymerize recalcitrant lignin present in wastewaters and they have a high biodegradative capacity towards numerous xenobiotic compounds.[90] Also, aerobic posttreatment of anaerobic effluents can be considered. These posttreatments are known to be effective in eliminating residual aquatic toxicity in anaerobically treated pulp and paper industry waste water.[21]

Sulfate and sulfite reduction, combined with sulfur recovery, is a promising application for wastewaters from production processes with high sulfur losses. The sulfide produced can be stripped in the gas phase, absorbed in the scrubber, and further chemically recovered as elemental sulfur. Biological posttreatment of anaerobic effluents to convert sulfide to elemental sulfur at short HRT is another alternative.[91,92] Simultaneous methane production, and complete sulfate reduction would serve both for sulfur recovery and organic matter removal. If that is not possible in one reactor due to sulfide inhibition, an additional methane reactor can be used.[93] SRB are inhibited by sulfite and sulfide at concentrations of 1000 mg S/l and 150 mg S/l.[94,95] Nearly complete sulfate reduction is obtained with pulp and paper industry effluents with a sufficiently high biodegradable COD/SO_4 ratio, if acetate is used by SRB. Sulfate reduction should be minimized if sulfide removal or sulfur recovery is not used. Most properly, this is done by selecting for treatment waste streams containing or producing substrates that favor MPB over SRB, and adjusting the environmental conditions to favor MPB. In this regard, the addition of selective inhibitors of SRB has been investigated.[96] The inhibitory effects of free hydrogen sulfide and sulfite to MPB can be reduced by dilution of the influent or by stripping the sulfide from the digester. Considerable stripping might occur naturally with the biogas production if the wastewater has a high biodegradable-COD/SO_4 ratio.[97] Increase of the pH to shift the dissociation towards unionized sulfide is proposed to decrease the inhibition.[98]

At higher temperatures, biochemical reactions can proceed more rapidly, which implicates a decrease of retention times or a smaller reactor volume of the wastewater treatment plants. Many effluents of pulp and paper industry are discharged at high temperatures which make them attractive for thermophilic treatment as no further energy input is required. In spite of this, very few feasibility studies have been performed on the thermophilic treatment of pulp and paper mill wastewater. Pilot-scale studies showed that improved BOD and COD reduction were obtained with anaerobic treatment of thermomechanical pulp mill wastewater at 60 °C in comparison with its mesophilic counterpart.[99] Biodegradability studies on TMP wastewater showed comparative results at 55 as well as 65 °C.[100] Recently, thermophilic treatment has been shown to be suitable for several types of wastewaters. High-strength wastewaters can be treated at very high loading rates with high COD removal efficiency.[101,102] At high temperature, the liquid viscosity is lower, which might benefit the biomass holdup in upflow reactors if low-strength wastewater is treated. To ensure the highest process stability, excess temperature fluctuations should be prevented. The activity of thermophilic sludge decreases with decreasing temperature but is still considerable and comparable with mesophilic sludge at temperatures of 40–45 °C. This is due to the fact that the growth rate and maintenance consumption of thermophilic bacteria is two to three times higher than the mesophilic counterparts.[103] Startup of the thermophilic reactors was demonstrated to be easy and can be done by increasing the reactor temperature directly to the desired level or following a gradual increase with only a few degrees difference during several months.[103,104] Besides temperature, process stability is also dependent on the chosen reactor type. Systems with a high biomass retention tend to be less sensitive than completely mixed reactors because of the higher variety of available selection criteria, and thus variety of biomass, inside the reactor.

Application of Anaerobic Process in Dechlorination of Bleach Plant Effluents

In the pulp and paper industry, in the past few years, considerable attention has been focused on organochlorine compounds released in the effluents from kraft and sulfite chemical pulp mills that bleach pulp with chlorine compounds.[105,106] Regulatory agencies in a number of countries throughout the world have established or proposed guidelines or limitations on either the overall discharge of chlorinated organic compounds, using a broad spectrum measurement parameter such as adsorbable organic halogen (AOX) or total organic chlorine (TOCl) or specific compound groups (i.e., polychlorinated dibenzo-p-dioxins and furans PCDD/Fs), or individual compounds (i.e., 2,3,7,8-tetrachlorodioxin TCDD).[107–111] Anaerobic biological treatment and sequential anaerobic/aerobic two-stage treatment systems have been used for the removal of organochlorine compounds from the bleach plant effluents.[106,112–115] The anaerobic dechlorination of chlorolignins is due to the combination of energy metabolism, growth, chemical hydrolysis, and probably adsorption and /or insolubilization.[116] For details see Chapter 8.

Conclusions

Anaerobic treatment process has proved to be both cost-effective and efficient in many cases. It will probably be one of the most important techniques for the treatment of certain pulp and paper mill effluents in the coming years. Its high effectiveness for waste streams composed mainly of readily biodegradable organic matter such as (thermo-)mechanical pulping effluents and condensates can no longer be doubted. Moreover, the time is approaching rapidly in which the anaerobic treatment method is becoming applicable to the category of "difficult" pulp and paper industry effluents, i.e., from (semi-)chemical pulping and bleaching processes. These complex wastewaters are highly toxic and contain a large fraction of recalcitrant lignin. Dilution to subtoxic levels should be applied where possible as well as careful startup procedures enabling the maximum adaption of the anaerobic microorganisms. If these measures are not feasible, a wide variety of detoxification pretreatments can be considered such as adsorption, precipitation, and oxidation of organic toxins. These physicochemical detoxification treatments are often also effective in the removal of recalcitrant lignin. There still is a great need for developing biological-based detoxification and lignin-removal pretreatments to replace or minimize the chemical requirements of the physicochemical methods.

Anaerobic treatment can also be applied as an essential part of an integrated treatment resource preservation system. For pulp and paper industry wastewaters, sulfate and (sulfite) reduction during anaerobic digestion can be combined with sulfur recovery systems. Particularly, biological recovery of elemental sulfur from hydrogen sulfide in anaerobically treated pulp and paper industry effluents has a promising future.

Reductive dechlorination to remove chlorinated organic compounds from chlorine-bleached chemical pulps is a potential new application for anaerobic treatment in the pulp and paper industry.

Several anaerobic process configurations have been successfully applied to the treatment of pulp and paper mill wastewaters including anaerobic lagoon, anaerobic contact, UASB, anaerobic filter, anaerobic fluidized bed, and hybrids of two or more of these basic configurations. No single process is best for all applications. Each configuration has unique advantages as well as tradeoffs compared with other anaerobic processes. Treatment objectives, waste characteristics, and site-specific factors will influence which systems are technically and economically the most attractive for each application.

References

1. Stemberg E, Norberg G: Effluent from manufacturing of thermomechanical pulps and their treatment. International Mechanical Pulping Conference 1977, Helsinki, Finland, June 6–10.
2. Corson JA, Lloyd JA: Refiner pulp mill effluent. Part I. Generation of suspended and dissolved solid fractions. Paperi ja Puu 1978; 60(6–7): 407–410.
3. Virkola NE, Honkanen K: Wastewater characteristics. Water Sci. Technol. 1985; 17(1): 1–28.
4. Anonymous: In Comprehensive industry document for small pulp and paper industry (Chakrabarti SP and Kumar A eds) Comprehensive industry document series COINDS/22/1986. Central Board for the Prevention and Control of Water Pollution, New Delhi, India.
5. Velasco AA, Frostell B, Greene M: Full scale anaerobic-aerobic biological treatment of a semichemical pulping wastewater. Proc. 40th Industrial Waste Conf., Purdue University, West Cafayette. 1985; pp. 297–304.
6. Gonenc IE, Ilhan R, Orhan D: A rational approach to the design of a rotating disc system for the treatment of black liquor. Water Sci. Technol. 1990; 22(9): 207–214.
7. Roger IH: Isolation and chemicals identification of toxic components of kraft mill wastes. Pulp Pap. Mag. Can. 1973; 74: 111–116.
8. Leach JM, Thakore AN: Toxic constituents of mechanical pulping effluents. Tappi J. 1976; 59(2): 129–132.
9. Benjamin MM, Woods SL, Ferguson JF: Anaerobic toxicity and biodegradability of pulp mill waste constituents. Water Res. 1984; 18: 601–607.
10. Ferguson JF, Benjamin MM: Studies of anaerobic treatment of sulfite process wastes. Water Sci. Technol. 1985; 17(1); 113–122.
11. Welander T: An anaerobic process for treatment of CTMP effluents. Water Sci. Technol. 1988; 20(1): 143–147.
12. Lettinga G, Field JA, Sierra-Alvarez R, vanLier JB, Rintala J: Future perspectives for the anaerobic treatment of forest industry wastewaters. Water Sci. Technol. 1991; 24(3/4): 91–102.
13. Springer AM: Bioprocessing of pulp and paper mill effluents – past, present and future. Paperi Ja Puu-Paper and Timber. 1993; 75(3): 156–161.
14. Simon O, Ullman P: Present state of anaerobic treatment. Paperi Ja Puu-Papper Och Tra. 1987; 6: 510–515.
15. Colberg PJ: Anaerobic microbial degradation of cellulose, lignin, oligolignols and monoaromatic lignin derivatives. In: Biology of anaerobic microorganisms (Zehnder AJB ed.) John Wiley and Sons Inc., New York 1988; 333–342.

16. Schink B: Principles and limits of anaerobic degradation: environmental and technological aspects. In: Biology of anaerobic microorganisms (Zehnder AJB ed.) John Wiley and Sons Inc., New York 1988; 771–846.
17. Zeikus JG, Wellstein AL, Kirk TK: Molecular basis for the biodegradative recalcitrance of lignin in anaerobic environments. FEMS Microbiol. Lett. 1982; 15: 193–197.
18. Colberg PJ, Young LY: Anaerobic degradation of soluble fractions of (^{14}lignin)-lignocellulose. Appl. Environ. Microbiol. 1985; 49: 345–349.
19. Sierra-Alvarez R, Kortekaas S, van Ekert M, Lettinga G: The anaerobic biodegradability and methanogenic toxicity of pulping wasteswaters. Water Sci. Technol. 1991; 24(3): 113–125.
20. Qui R, Ferguson JF, Benjamin MM: Sequential anaerobic and aerobic treatment of kraft pulping wastes. Water Sci. Technol. 1988; 20(1): 107:120.
21. Wilson RW, Murphy KL, Frenette EG: Aerobic and anaerobic treatment of NSSC and CTMP effluent. Pulp Paper Can. 1987; 88(1): T4-8.
22. Welander T, Anderson PE: Anaerobic treatment of wastewater from the production of chemi-thermomechanical pulp. Water Sci. Technol. 1985; 17(1): 103–112.
23. Jurgensen SJ, Benjamin MM, Ferguson JF: Treatability of thermomechanical pulping process effluents with anaerobic biological reactors. Proc. Tappi Environmental Conference, TAPPI, Atlanta, GA, USA 1985; 83–92.
24. Salkinoja-Salonen M, Valo R, Apajalahati J, Hakulinen R, Silăkoski L, Jaakola T: Biodegradation of chlorophenolic compounds in wastes from wood processing industry. In: Current perspective in microbial ecology. (Klug MJ, Reddy CA eds.) American Society for Microbiology, Washington, DC, 1984; 668–676.
25. Guthrie MA, Kirsch EJ, Wukasch RF, Grady Jr CPL: Pentachlorophenol biodegradation-II. Water Res. 1984; 18: 451–461.
26. Field JA, Leyendeckers MJH, Sierra-Alvarez R, Lettinga G, Habets LHA: The methanogenic toxicity of bark tannins and the anaerobic biodegradability of water soluble bark matter. Water Sci. Technol. 1988; 20(1): 219–240.
27. Field JA, Kortekaas S, Lettinga G: The tannin theory of methanogenic toxicity. Biol. Wastes. 1989; 29: 241–262.
28. Kaser LE, Dixon DG, Hodsor PV: QSAR studies on chlorophenols, chlorobenzenes and para-substituted phenols. In: QSAR in environmental toxicology (Kaiser KLE ed.) D. Reidel Publishing Co. Boston, USA 1984; 189–206.
29. Junna J, Lammi R, Miettinen V: Removal of organic and toxic substances from debarking and kraft pulp bleaching effluents by activated sludge treatment. Publ. of Water Research Institute, National Board of Waters, Finland 1982, No. 49.
30. Temmink JHM, Field JA, van Haastrecht JC, Merckelbach RCM: Acute and sub-acute toxicity of bark tannins to carp (*Cyprinus carpio* L). Water Res. 1989; 23: 341–344.
31. Schink B: Degradation of unsaturated hydrocarbons by methanogenic enrichment cultures. FEMS Microbiol. Ecol. 1985; 31: 69–77.
32. Field JA: The effect of tannic compounds on anaerobic wastewater treatment. Doctoral thesis, Dept. Water Pollution Control, Agricultural University of Wageningen, Wageningen, The Netherlands, 1989.
33. Wood SL, Ferguson JF, Benjamin MM: Characterization of chlorophenol and chloromethoxybenzene biodegradation during anaerobic treatment. Environ. Sci. Technol. 1989; 23: 62–68.
34. Mikesell MD, Boyd SA: Complete reductive dechlorination and mineralization of pentachlorophenol by anaerobic microorganisms. Appl. Environ. Microbiol. 1986; 52: 861–865.
35. Tiedje JM, Boyd SA, Fathepure BZ: Anaerobic degradation of chlorinated aromatic hydrocarbons. Dev. Ind. Microbiol. 1987; 27: 117–127.
36. Hakulinen R, Salkinoja-Salonen M: Treatment of kraft bleaching effluents: Comparison of results obtained by Enso-Fenox and alternative methods. In: Int. Pulp Bleaching Conf. 1982; 97–106.
37. Hakulinen R, Salkinoja-Salonen M: Treatment of pulp and paper industry wastewaters in an anaerobic fluidized-bed reactor. Proc. Biochem. 1982; 17(2): 18–22.

38. Edmond-Jacques N: Biomethanation process. Biotechnology Vol 8. (Rehm HJ, Reed G eds.) VCH, Weinheim, Germany 1986; 207–267.
39. Lee JW: Anaerobic treatment of pulp and paper mill wastewaters. In: Industrial environmental control, pulp and paper industry (Springer AM ed.) Tappi Press, Atlanta, GA 1993; 405–446.
40. Hall ER: Treating wastewater from pulp and paper mills with anaerobic technology. Environmental Technology Notes, published by Environment Canada, Burlington, Ontario, 1988.
41. Hall ER, Cornacchio LA: Anaerobic treatability of Canadian pulp and paper mill-wastewaters. Proc. Tappi Environmental Conference, Portland, Oregon, 1987; 235–240.
42. Velasco AA, Bonkoski WA, Sarner E: Full scale anaerobic biological treatment of a semi-chemical pulping wastewater. Proc. Tappi Environmental Conference, Portland, Oregon, 1987; 197–201.
43. Jurgensen SL, Benjamin MM, Ferguson JF: Treatability of thermomechanical pulping process effluents with anaerobic biological reactors. Proc. Tappi Environmental Conference, Mobile, Alabama, 1985; 83–88.
44. Hall ER, Robson PD, Prong CF, Chmelauskas AJ: Evaluation of anaerobic treatment of NSSC wastewater. Proc. Tappi Environmental Conference, New Orleans, Louisiana, 1986; 207–211.
45. Welander T, Anderson PE: Anaerobic treatment of wastewater from production of chemi-thermochemical pulp. Water Sci. Technol. 1985; 17:103–108.
46. Waters JG, Kanow PE, Dalppe HL: A full scale anaerobic contact process treats sulfite evaporator condensate at Hannover. Paper, Alfred, Germany. Proc. Tappi Environmental Conference, Charleston, South Carolina, 1988; 309–313.
47. Frostell B: Anaerobic pilot scale treatment of a sulfite evaporator condensate. Presented at CPPA 69th Annual Meeting, Montreal, Quebec. 1983.
48. Cocci AA, Landine RC, Brown GJ, Tennier AM, Hall ER: Anaerobic treatment of kraft foul condensates. Proc. Tappi. Environmental Conference, Mobile, Alabama, 1985; 67–71.
49. Qui R, Ferguson JF, Benjamin MM: Anaerobic-aerobic treatment of kraft pulping wastewaters. Proc. Tappi Environmental Conference, Portland, Oregon, 1987; 165–170.
50. Pipyn P: Anaerobic treatment of kraft pulp mill condensates. Proc. Tappi Environmental Conference, Portland, Oregon, 1987; 173–177.
51. Priest CJ: Inland Container saves money with anaerobic-aerobic treatment plant. Tappi J. 1981; 64, 11–16.
52. Priest CJ: A change to anaerobic-aerobic treatment made expanded production possible without expansion of wastewater treatment facilities. Proc. 35th Annual Purdue Industrial Waste Conference, Purdue, 1980; 142–146.
53. Winslow FB: Start-up and operation of an anaerobic-aerobic treatment system. Presented at the Southern Regional Meeting, NCASI, New Orleans, June 1988.
54. Dubey RK, Khare A, Kaul SS, Singh MM: Performance of waste treatment plant at Orient Paper Mills, Amlai. International Seminar Management Environmental Problems in Pulp and Paper Industry, New Delhi, India, Feb. 24–25, 1982.
55. Nambisan PNK, Raja KCJ, Mohanchandran TM, Balakrishnan E: Effluent treatment in a rayon grade pulp mill. IPPTA 1980; 17: 2–10.
56. Janson B: Hylte Braks pioneers in New Generation effluent treatment. Pulp and Paper Sweden 1984; 1: 72–76.
57. Sarner E, Hultman B, Berglund A: Anaerobic treatment using new technology for controlling H_2S toxicity. Proc. Tappi Environmental Conference, Portland, Oregon, 1987; 227–232.
58. Schmutzler DW, Eis BJ, Lee JW, Olsen JE: Start up and operation of a full-scale anaerobic treatment system at a groundwood and coated paper mill. Proc. Tappi Environmental Conference, Charleston, South Carolina, 1988; 227–230.
59. Habets LHA, Knelissen JH: Anaerobic wastewater treatment plant at Papierfabriek Roermond – working successfully and saving expenses. Proc. Tappi Environmental Conference, Mobile, Alabama, 1985; 93–97.
60. Habets LHA: Experience with the UASB reactor under optimal and suboptimal loadings. PIRA Conference Cost Effective Treatment of Paper Mill Effluents Using Anaerobic Technologies, Leatherhead, England, Jan. 14–15, 1986.

61. Habets LHA, Knelissen JH: Application of the UASB reactor for anaerobic treatment of paper and board mill effluent. IAWPRC Symposium, Anaerobic Treatment of Forest Industry Wastewaters. Tampere, Finland. June 11–15, 1984.
62. Rekunen S: Taman Process and its applicability to wastewaters from papermaking. PIRA Conference Cost Effective Treatment of Paper Mill Effluents Using Anaerobic Technologies, Leatherhead, England, Jan. 14–15, 1986.
63. Personal Communication: Harihar polyfibers, Grasim Industries Ltd, Kumarapatnam, Karnataka, India 1998.
64. Personal Communication: APR LTD., Kamalapuram, Andhra Pradesh, India, 1998.
65. Szendrey ML: Anaerobic treatment of fermentation wastewater. Environ. Prog. 1984; 3(4): 222–229.
66. Donovan EJ, Drysinski DA, Subbaramu K: Anaerobic treatment of sulfite liquor evaporator condensate. Proc. Tappi Environmental Conference, Savannah, Georgia, 1984; 209–214.
67. Dalentoft E, Jonsson M: Anaerobic-aerobic treatment of total waste effluent from a pulp mill including sludge hydrolysis. Proc. Tappi Environmental Conference, Portland, Oregon, 1994; 145–152.
68. Prasad DY: The anaerobic biodegradation of a bagasse-based paper mill waste in fixed-film reactor. J. Chem. Technol. Biotechnol. 1992; 53: 67–72.
69. Prasad DY: Biogas generation from bagasse based paper mills by anaerobic digestion. J. Chem. Technol. Biotechnol. 1991; 51: 515–524.
70. Kennedy KJ, Andras E, Elliott CK, Methven B: Effect of intermittent feeding on anaerobic chemithermomechanical pulp wastewater treatment. Environ. Technol. 1991; 12: 291–296.
71. Rintala J, Lepisto SS: Anaerobic treatment of thermomechanical pulping wastewater. Water Res. 1992; 26(10): 1297–1305.
72. Rintala J, Vuoriranta P: Anaerobic-aerobic treatment of thermomechanical pulping effluents. Tappi J. 1988; 71(9): 201–207.
73. Sierra-Alvarez R, Harbrecht J, Kortekaas S, Lettinga G: The continuous anaerobic treatment of pulping wastewaters. J. Ferment. Bioeng. 1990; 70(2): 119–127.
74. Chou WL, Speece RE, Siddiqui RH, Mckeon K: The effect of petrochemical structure on methane fermentation toxicity. Prog. Water Technol. 1978; 10(5): 545–548.
75. Chou WL, Speece RE, Siddiqui RH, Mckeon K: Acclimation and degradation of petrochemical wastewater components by methane fermentation. In: Biotechnology and Bioengineering Symp. No. 8 John Wiley and Sons Inc. New York. 1979; 391–414.
76. Fedorak PM, Hrudey SE: Anaerobic treatment of phenolic coal conversion wastewaters in semicontinuous cultures. Water Res. 1986; 20: 113–122.
77. Vogel P, Winter J: Anaerobic degradation of phenol and cresol in petrochemical wastewater. In: Anaerobic digestion (Tilche A, Rozzi A eds.) Bologna, Italy 1988; 829–832.
78. Yang J, Speece RE, Parkin GF, Gossett J, Kocher W: The response of methane fermentation to cyanide and chloroform. Water Sci. Technol. 1981; 13(2): 977–989.
79. Parkin GF, Speece RE: Attached versus suspended growth anaerobic reactors: response to toxic substances. Water Sci. Technol. 1983; 15(8/9): 261–289.
80. Blum JW, Hergenroeder R, Parkin GF, Speece RE: Anaerobic treatment of coal conversion wastewater constituents: biodegradibility and toxicity. JWPCF. 1986; 58(2): 122–131.
81. Kim BR, Chian ESK, Cross WH, Cheng SS: Adsorption, desorption and bioregeneration in an anaerobic granular activated carbon reactor for the removal of phenol. JWPCF. 1986; 58(1): 35–40.
82. Wang YT, Suidan MT, Rittman BE: Anaerobic treatment of phenol by an expanded-bed reactor. JWPCF. 1986; 58(3): 227–233.
83. Fox P, Suidan MT, Pfeffer JT: Anaerobic treatment of a biologically inhibitory wastewater. JWPCF. 1988; 60(1): 86–92.
84. Nakhla GF, Suidan MT, Pfeffer JT: Operational control of an anaerobic GAC reactor treating hazardous wastes. Water Sci. Technol. 1989; 21(4/5): 167–173.
85. Koster IW: Abatement of long chain fatty acid inhibition of methanogenesis by calcium addition. Biol. Wastes. 1987; 22: 295–301.

86. Milstein O, Haars A, Majcherczyk A, Trojanowski J, Tautz D, Zanker H, Huttermann A: Removal of chlorophenols and chlorolignins from bleaching effluent by combined chemical and biological treatment. Water Sci. Technol. 1988; 20(1): 161–170.
87. Hakulinen R: The use of enzymes in the waste water treatment of pulp and paper industry-a new possibility. Water Sci. Technol. 1988; 20(1): 251–262.
88. Schmidt RL, Joyce TW: An enzymatic pretreatment to enhance the lime precipitatability of pulp mill effluents. Tappi J. 1980; 63(12): 63–67.
89. Wang Y, Pai P, Latchaw JL: Effects of preozonation on the methanogenic toxicity of 2,5-dichlorophenol. JWPCF. 1989; 61(3): 320–326.
90. Paszczynski A, Crawford RL: Potential for bioremediation of xenobiotic compounds by the white-rot fungus *Phanerochate chrysosporium*. Biotechnol. Prog. 1995; 11: 368–379.
91. Buisman CJ, Witt B, Lettinga G: A new biotechnoligical process for sulphide removal with sulphur production In: 5th Int. Symposium on Anaerobic Digestion (Tilche A, Rozzi A eds.) Bologna, Italy 1988; 19–22.
92. Buisman CJ, Geraats BG, Ijspeert P, Lettinga G: Optimization of sulphur production in a biotechnological sulphide-removing reactor. Biotechnol. Bioeng. 1990; 35: 50–56.
93. Saerner E: Influence and control of H_2S on full-scale plants and pilot plant experiments. In: Proc. EWPCA Conf. Anaerobic Treatment, a Grown-up Technology, Amsterdam, The Netherlands, September 15–19 1986; 189–204.
94. Puhakka JA, Ferguson JF, Benjamin MM, Salkinoja-Salonen M: Sulfur reduction and inhibition in anaerobic treatment of simulated pulp mill wastewater system. Appl. Microbiol. 1989; 11: 202–206.
95. Hilton BL, Oleszkiewicz JA: Sulfide induced inhibition of anaerobic digestion. J. Env. Eng. Div ASCE. 1988; 114: 1377–1391.
96. Tanimoto Y, Tasaki M, Okamura K, Yamaguchi M, Minami K: Screening growth inhibitors of sulfate reducing bacteria and their effects on methane formation. J. Ferment. Bioeng. 1989; 68: 353–359.
97. Rinzema A: Anaerobic treatment of wastewater with high concentrations of lipids or sulfate. Doctoral thesis, Dept. Water pollution control, Wageningen, Agricultural University, Wageningen, The Netherlands 1988.
98. Koster IW, Rinzema A, de Vegt AL, Lettinga G: Sulfide inhibition of the methanogenic activity of granular sludge at various pH levels. Water Res. 1986; 20: 1561–1567.
99. Salkinoja-Salonen M, Valo R, Apajalahati J: Treatment of pulp and paper wastewaters in an anaerobic reactor: from bench size units to full scale operation. In: 3rd Int. Symp. on Anaerobic Digestion, Boston, USA, Aug. 12–19 1983; 107–121.
100. Rintala J, Sanz JL, Lettinga G: Themophillic anaerobic treatment of sulfur rich pulp and paper industry process water. Water Sci. Technol. 1991; 24(3): 149–160.
101. Weigant WM, Claassen JA, Borghans AJ, Lettinga G: High rate anaerobic thermophilic digestion for the generation of methane from organic wastes. In: Proc. Conference Anaerobic Waste Water Treatment (vander Brink WJ ed.), The Netherlands Nov. 23–25, 1983; 392–396.
102. Weigant WM, Claassen JA, Lettinga G: Themophilic anaerobic digestion of high strength wastewaters. Biotechnol Bioeng. 1985; 27: 1374–1381.
103. Schraa G, Jewell WJ: High rate conversion of soluble organics with the thermophilic attached film expanded bed. JWPCF. 1984; 56: 226–232.
104. Buhr HO, Andrew JF: The thermophilic anaerobic digestion process. Water Res. 1977; 11: 129–143.
105. Bajpai P, Bajpai PK: Organochlorine compounds in bleach plant effluents – genesis and control. PIRA International, UK 1996: 127 pages.
106. Bajpai P, Bajpai PK: Reduction of organochlorine compounds in bleach plant effluents. Adv. Biochemical Eng./Biotechnol. (Eriksson KEL ed.) Springer, Berlin Heidelberg New York 1997; 57: 213–259.
107. Hinsey NW: EPA's proposed cluster rule: the end of end-of-pipe. Tappi J. 1994; 77(9): 65–74.
108. Lewis J: Cluster bombs rock industry. World Pap. 1994; 219(7–8): 18–19.

109. Demel I, Oller HJ: State of the art of AOX avoidance in effluents from German paper mills. Wochenbl. Papierfbr. 1995; 123(21): 975–978.
110. Taguchi T: Environmental topics in 1993–1994 in Japan. Proc. 1995 International Environmental Conference, Atlanta, GA USA, 7–10 May, 1995. Book 2: 665–673.
111. Vice K, Trepte R, Staurt P, Johnson T: The cluster rule – the road to promulgation. Tappi J. 1997; 80(12): 34–36.
112. Raizer-Neto E, Pichon M, Rouger J: Decreasing chlorinated organics in bleaching effluents in an anaerobic fixed bed bioreactor. Proc. of the 4th Int. Biotechnol. Conf. in Pulp and Paper Industry, Raleigh, North Carolina 16–19 May 1989; 279–287.
113. EK M, Eriksson KE: External treatment of bleach plant effluent. 4th Int. Symp. on Wood and Pulping Chemistry, Paris 1987.
114. Armentate PM, Kafkewitz D, Lewandowski G, King CM: Integrated anaerobic-aerobic process for the biodegradation of chlorinated aromatic compounds. Environ. Prog. 1992; 11(2): 113–118.
115. Poggi-Varaldo HM, Campos-Velarde MD, Rios-Leal E, Villagomez-Hernandez G: Degradation and detoxification of effluents contaminated by 2,4,6-trichlorophenol and phenol using an anaerobic treatment. Biotechnol. Apl. 1996; 13(3): 215.
116. Fitzsimons R, Ek M, Eriksson KE: Anaerobic dechloronination/degradation of chlorinated organic compounds of different molecular masses in bleach plant effluents. Environ. Sci. Technol. 1990; 24(11): 1744–1748.

8 Decolorization and Detoxification of Bleached Kraft Effluents

The pulp and paper industry ranks third in terms of water consumption and fifth among the major industries in its contribution to water pollution problems in the USA. Pulping, bleaching, and paper-making operations are the three major wastewater sources of the industry.

In the production of bleached kraft pulp, most of the lignin is removed in the cooking stage in a closed system at high temperature with concentrated pulping liquors. In many modern pulp mills, additional delignification is achieved by applying oxygen prebleaching. The spent liquors from kraft pulping and oxygen prebleaching are recovered in a multistage countercurrent washing system with a minimum amount of wash water and concentrated in an evaporation system to a high solid content and then burnt to recover the cooking chemicals and energy associated with organic matter in the cooking liquor. In doing so, the pollution load associated with the spent liquor is eliminated. The final delignification and brightening of kraft pulp require the use of various chlorine-related species and occurs in a multistage bleaching process (three to seven stages). Spent liquors from the bleach plant are not suitable for recovery and incineration, mainly due to the presence of potentially corrosive chloride ions, and, therefore, require separate treatment and disposal. Pollution loads from a conventional bleach plant account for 40–90% of the dissolved material (COD) and rest is due to carryover from the pulping process.[1] Bleaching with chlorine chemicals usually starts with an acid treatment with elemental chlorine at low temperature (usually ambient), pH 1.5, and consistency 3–5%. During chlorination, wood components - lignin and some carbohydrates - are structurally modified, degraded, and chlorinated. Some of these chlorinated compounds (mostly low molecular weight material) are dissolved into the spent chlorination liquor. This stage is followed by an alkaline extraction stage using high temperature (70 °C), pH 10.5, and consistency 10%. In the extraction stage, chlorinated oxidized lignins, not soluble in the acidic chlorination stage, are solubilized and dissolved into the alkaline spent liquor. Approximately 75–90% of the dissolution of organic matter and 95% of the chlorinated organic typically produced in the bleach plant occur in the chlorination and first extraction stages.[1] The final bleaching is achieved using oxidized chemicals such as chlorine dioxide, hypochlorite, and hydrogen peroxide. The substances dissolved in the later stages are more strongly oxidized, and pollution loads from these stages are minor. Chlorinated organics generated during pulp bleaching not only exert oxygen demand (BOD and COD) but also cause effluent color and toxicity (acute and chronic) and are also responsible for the mutagenicity and carcinogenecity of the effluent.[2–7] The amount of chlorinated organics (AOX) produced during pulp bleaching varies with wood species, kappa number of pulp, bleaching sequence, and the conditions employed. Typically, AOX in the effluent from a bleached softwood kraft pulp by a conventional sequence is 5–8 kg/ton pulp bleached, representing about 10% of the total chlorine charged in the chlorination stage. A physicochemical classification of this chlorinated organic material, present in spent liquors from conventionally pulped and bleached softwood kraft pulp, shows that about 20% of the organically bound chlorine found in the bleaching effluent corresponds to relatively low molecular mass ($M < 1000$) material.[8–10] In recent years, considerable research effort has been directed towards characterizing this fraction with respect to its individual chlorinated compounds, as this fraction is expected to contain those compounds which are potentially toxic to aquatic organisms, because of their ability to penetrate cell membranes or their propensity to bioaccumulate in the fatty tissues of higher organisms.[6] Some of the major components of this low molecular mass fraction were found to consist of relatively water-soluble substances such as chlorinated acetic acids or chlorinated acetone, which are easily broken down[8,9] before or during biotreatment and are thus of minimal environmental significance. The fraction of AOX which is extractable by a nonpolar organic solvent and is referred to as EOX (or extractable organically bound halogen) accounts for about 1–3% of the total organically bound chlorine. This fraction contains relatively lipophilic (i.e., fat-soluble) neutral organic compounds, primarily of low molecular weight, and is therefore of greater environmental significance than the remaining 99% or so of the AOX material. The EOX material can be further fractionated according to its octanol/water partition coefficient (Pow). The fraction having a partition coefficient greater than 1000 (i.e., log Pow >3) makes up only 0.1% (or less) of the AOX and contains those compounds which are most readily bioaccumulable and considered to be potentially the most toxic and persistent. A component of particular concern in this fraction is the polychlorinated aromatic material that has a relatively high level of chlorine substitution, typically three or more chlorine atoms per aromatic ring. The low molecular weight fraction of the chlorolignins is the main contributor to the effluent BOD and acute toxicity. The high molecular weight chlorinated compounds contribute little to BOD and acute toxicity due to their inability to pass through the cell membranes. However, these compounds are major contributors to effluent color, COD, and chronic toxicity.[11] Color is not only esthetically inacceptable but also inhibits the natural process of photosynthesis in the stream due to absorbance of the sunlight.[6] This leads to chains of adverse effects on the aquatic ecosystem, as the growth of the primary consumers as well as secondary and tertiary consumers is adversely affected. Discharge of untreated or partially treated wastewaters results in color persisting in the receiving body for a long distance. Under natural conditions, these compounds are degraded slowly to various chlorinated phenolics, which may be methylated under aerobic conditions. The low molecular weight phenolics and their methylated counterparts (which are more lipophilic) cause toxicity and are bioaccumulable in fish.[11] Because of these problems, it is necessary to decolorize and detoxify the bleach plant effluents before disposal.

Governments in many countries including the United States, Canada, Scandinavia, Germany, Japan, and India have imposed limits on the amount of AOX discharged from pulp mills into the receiving water. At present, the strictest emission limit value for AOX is 1 kg/ton of pulp. This limit applies to sulfite pulp production in Austria, Germany, and Norway. This is because AOX emissions are easier to reduce with sulfite pulp than with the softwood kraft pulp. However, in Sweden, kraft pulp mills have individual limits as low as 0.3 kg/ton pulp.[12] Some decided or planned regulations for AOX are shown in Table 8.1.[12] The permitted levels are likely to decrease to about 0.5–1 kg/ton in the next few years in many countries. In several countries or provinces, emission limits are decided mill by mill, taking into account the local factors. Thus, for individual mills, stricter emission limits may apply to those given in Table 8.1 (e.g., Sweden). According to a decision in 1992 by PARCOM (Paris Convention for the Prevention of Marine Pollution from Land-Based Sources and Rivers) signed by 12 European countries, a general AOX emission limit was set at 1 kg/ton from 1995. The limit applies to all types of bleached chemical pulp and it has been accepted by Belgium, Denmark, France, Germany, Ireland, Luxembourg, The Netherlands, Norway, Portugal, Spain, Sweden, and the UK. In the USA, the so-called Cluster Rule, first proposed in December 1993[13,14] was signed by EPA on 14 November 1997, after some modifications.[15] The compliance period is up to 3 years from the date of publication of the rule in the Federal Register. The best available technology (BAT) for the bleached paper grade kraft and soda subcategory is complete substitution of chlorine dioxide for chlorine. The BAT for calcium-, magnesium-, and sodium-based paper-grade sulfite mills is TCF bleaching. The BAT for ammonium-based and speciality grade paper sulfite mills is ECF bleaching. The specific limit for AOX is

Table 8.1. Standards for discharge of organochlorine compounds, AOX (kg/t pulp), from bleaching of chemical pulps

Country	1995–2000
Australia	1.0
Austria	0.5–1.0
Canada	1.0–2.0
Finland	1.0–2.0
Germany	1.0
India	2.0
Japan	1.0–1.5
USA	0.5
Sweden	0.3–1.0

Based on data from Ref. 12.

0.512 kg/ton (annual average). This new regulation will most likely have a major impact on the industry. A specific AOX level has never been stipulated before in the USA. In Germany, the use of elemental chlorine-free (ECF) and totally chlorine-free (TCF) pulps has increased considerably. It has been reported that AOX levels were reduced by 31 to 93% between 1987 and 1988 and 1992 and 1993.[16] Germany is presently discussing legislation which will ban not only production of pulp using chlorine-containing chemicals but also consumption of pulps other than TCF. In Japan, all kraft mills now operate at an AOX rate of under 1.5 kg/ton of pulp.[17]

Biotechnological Methods for Treatment of Bleached Kraft Effluents

Several methods have been attempted for decolorization and detoxification of bleached kraft effluents. These include physicochemical and biotechnological methods. The problems underlying the physicochemical treatments are those associated with cost and reliability. Coagulation and precipitation produce a voluminous sludge which is very difficult to dewater. Usually, an extreme pH range is used for optimum treatment and the pH needs to be readjusted to neutral before discharge. Oxidation using ozone and hydrogen peroxide are costly, and oxidation using chlorine species generates secondary pollutants such as chlorinated organics. The membrane techniques require pretreatment and a large capital investment. Membrane fouling is also a problem with the membrane technique. Biotechnological methods have the potential to eliminate/reduce the problems associated with physicochemical methods and are described below.

1. Bacterial Treatment

A number of studies on aerobic and anaerobic treatment of bleach plant effluent have been carried out. Biological oxidation is the most widely used technique to remove BOD, COD, and chlorinated organics due to its effectiveness and low cost. Fulthorpe and Allen[18] studied the ability of three bacterial species to reduce AOX in bleached kraft mill effluents. *Ancylobacter aquaticus* A7 exhibited the broadest substrate range, but could only affect significant AOX reduction in softwood effluents. *Methylobacterium* CP13 exhibited a limited substrate range but was capable of removing significant amounts of AOX from both hardwood and softwood effluents. By contrast, *Pseudomonas* sp. Pl exhibited a limited substrate range and poor to negligible reductions in AOX levels from both effluent types. Mixed inocula of all the three species combined and inocula of sludge from mill treatment systems removed as much AOX from softwood effluents as did pure populations of *Methylobacterium* CP13 (Table 8.2). Melcer et al.[19] carried out pilot-scale investigation of activated sludge (AS)

Table 8.2. Adsorbable organic halogen losses from hardwood and softwood effluents

Parameter	Effluent	CP13		A7		P1		Mixture	
		(mg/l)	(%)[c]	(mg/l)	(%)	(mg/l)	(%)	(mg/l)	(%)
Total loss	Soft-1	1.10	6.1*	3.85	21.4***	2.28	12.7*		
	Soft-2	4.74	21.7**	2.94	13.5*	NS	0	4.6[a]	22.4
	Soft-3	2.91	13.9***	-		-		3.7[b]	19.6
Average			14.1 ± 6.2		17.6 ± 4.0		6.4 ± 6.4		
	Hard-1	0.86	13.7**	NS		NS			
Recalcitrant loss	Soft-1	2.75	14.1**	2.78	14.2**	NS			
	Soft-2	3.12	21.8**	2.85	19.6*	0.65	4.6**	3.4[a]	23.8
	Soft-3	3.09	16.6**	-		-		3.7[b]	21.3
Average			17.6 ± 3.3		17.0 ± 2.8		2.3 ± 2.3		
	Hard-1	0.57	14.8*	NS		NS			
	Hard-2	2.02	20.8**	NS		NS			

*, **, *** Statistical significance of difference from, control, one-tailed t-test: * $P < 0.05$, ** $P < 0.01$, *** $P < 0.001$.
NS no statistically significant difference.
Reproduced from Ref. 18.
[a] Mixture of CP13, A7, P1.
[b] Inoculum of nonsterile lagoon fluid.
[c] [(Final AOX of control-final AOX of treated sample)/final AOX of control)] × 100.

treatment of bleached kraft effluent at a Northern Ontario mill site over an 8-month period. The AS system was operated at 1 day HRT, 25–30 days SRT, and 30 °C. Treated effluents were found to pass all acute and chronic toxicity tests as measured by Rainbrow trout LC_{50}, Microtox, and *Ceriodaphnia* LC_{50} and IC_{25} tests. A high level of effluent quality was achieved with low concentrations of AOX (4–13 mg/l), total chlorophenolics (0.3–0.32 mg/l), toxicity equivalents (TEQ-PCP) (0.4–5 mg/l), total resin and fatty acids (0–4 mg/l), BOD (4–12 mg/l), and soluble COD (142–274 mg/l) being recorded over the whole period of investigation. An 8- to 17-fold reduction in hepatic MFO enzyme activity was measured in the treated effluents over the influent wastewaters.

Rogers et al.[20] treated the bleached kraft mill effluent in a bench-scale aerated lagoon for 29, 58, and 99 h and showed that toxicity, BOD, and resin acids were most consistently reduced during the 99-h treatment. Leach et al.[21] reported the biodegradation of seven compounds representing the major categories of toxicants in a laboratory batch aerated lagoon. Resin acids, which are the major source of acute toxicity, were readily biodegradable but only part (less than 30%) of the load of chlorophenolic compounds was removed. However, Gergov et al.[8] have reported that biological treatment, especially the activated sludge process, removed chlorophenolics by 75–95%. Chlorinated neutral organic compounds were removed effectively.[22] Valenzuela et al.[23] studied the degradation of chlorophenols by *Alcaligenes eutrophus* TMP 134 in bleached kraft mill effluent. After 6 days of incubation, 2,4-dichlorophenoxyacetate (400 ppm) or 2,4,6-trichlorophenol (40 to 100 ppm) were extensively degraded (70 to 100%). In short-term batch incubations, indigenous microorganisms were unable to degrade such compounds. Degradation of 2,4,6-trichlorophenol by strain JMP 134 was significantly lower at 200 to 400 ppm of compound. This strain was also able to degrade 2,4-dichlorophenoxyacetate, 2,4,6-trichlorophenol, 4-chlorophenol, and 2,4,5-trichlorophenol when bleached kraft mill effluent was amended with mixtures of these compounds. On the other hand, the chlorophenol concentration and the indigenous microorganisms inhibited the growth and survival of the strain in short-term incubations. In long-term (>1 month) incubations, strain JMP 134 was unable to maintain a large, stable population, but an extensive 2,4,6-trichlorophenol degradation was still observed. When combined effluents of a kraft pulp mill were treated in a lab-scale activated sludge system, the average TOC and AOX removal efficiencies were found to be 83 and 21%, respectively.[24] The highest AOX removal occurred at larger SRTs. Mass balance on the system revealed that the principal AOX removal mechanism was metabolization at long SRT. About 90% of the AOX removed was metabolized. As SRT was lowered, AOX removal efficiency decreased.

When the bleaching effluents from chlorination and extraction stage were treated in an activated sludge process, the AOX reduction was found to be 30–40% in 8 days. About 70–80% of the total AOX reduction was achieved in about 4 days.[25] The presence of high molecular weight (HMW) material in the bleached kraft effluent was found to improve the removal of chlorophenolic compounds. Growth experiments using microorganisms from a lab-scale activated sludge reactor showed that HMW material had a significant role in soluble COD and chlorophenol removal.[26] Large decreases in the soluble COD and increases in the biomass were observed with the addition of HMW materials to the low molecular weight fraction. The addition of mono- and dichlorinated phenolic compounds at concentrations up to 10 mg/l were found to have no effect on the metabolism or growth of the microorganisms in the activated sludge. While 6-chlorovanillin (6-CV) 2,4-dichlorophenol (2,4-DCP) and 4,5-dichloroguaiacol (4,5-DCG) were found to be stable in uninoculated controls and inoculated LMW effluent over a 160-h period, these compounds decreased significantly when LMW with three times the original concentration of HMW material was inoculated with microorganisms. The removal rates of these compounds increased in the order: 6-CV > 4,5-DCP > 2,4-DCP.

The removal efficiency of chlorinated organics in biological treatment is still a matter of debate. While the Swedish researchers found that the biological treatment is ineffective in removing AOX, American mills reported, on an average, 50–90% AOX removal by an aerated lagoon or activated sludge process. Gergov et al.[8] investigated pollutant removal efficiencies in mill-scale biological treatment systems. They found that 48–65% AOX was removed in the activated sludge process. The aerated lagoon was found to be less efficient than the activated sludge. Recently, Deardorff et al.[27] reported that the efficiency of AOX removal through biotreatment of combined bleach plant effluent increases with increasing chlorine dioxide substitution. Biological treatment in an aerated lagoon reduced the concentration of polychlorinated phenolic compounds by 97%. Jokela et al.[28] reported that aerobic lagoon systems removed 58 to 60% of the organochlorine compounds from the water phase and the full-scale activated sludge plants removed 19 to 55%. Both biotreatments removed all sizes and classes of organochlorine molecules and slightly changed the relative size distribution of the compounds remaining in the water phase towards the larger molecular weights. Eriksson and Kolar[29] have shown that the high molecular weight fraction in bleach plant effluents cannot be degraded in an aerated lagoon. In another study, it has been shown that chloroform is stripped during the biological treatment and COD, AOX, and high molecular weight material are reduced to a lesser extent.[30]

The combined effects of oxygen delignification, ClO_2 substitution, and biological treatment on pollutants levels in bleach plant effluents were examined. Biological treatment did not reduce color but reduced COD, BOD, AOX, and toxicity.[31] ClO_2 substitution reduced the discharge of all five pollutants, with a large reduction in AOX. Oxygen delignification reduced discharges of the five pollutants, and effluents from the sequence with oxygen delignification were easier to treat by aerobic methods. Treatment of bleaching effluent in sequential activated sludge and nitrification systems revealed that dechlorination of bleaching effluent took place in both the systems.[32] In the activated sludge system, released inorganic chloride was 4.5–7 mg/l at TOC loading rate of 0.03–0.07 mg/mg VSS/day, respectively; but it was decreased from 10 mg/l to 3 mg/l at TOC loading rate of 0.006–0.06 mg/mg VSS/day, respectively.

Anaerobic biological treatment can efficiently destroy chlorophenolic compounds, mutagenicity, and acute toxicity.[33–35] Dorica and Elliott[36] studied the treatability of bleached kraft effluents using anaerobic and anaerobic plus aerobic processes. BOD removal in the anaerobic stages varied between 31 and 53%, with the hardwood effluent responding slightly better to the treatment. Similarly, the AOX removal from the hardwood effluents was higher (65 and 71%, for single-stage and two-stage treatment, respectively) than that for softwood effluents (34 and 40%). Chlorate was removed easily from both softwood and hardwood effluents (99 and 96%, respectively) with little difference in efficiency between the single-stage and two-stage anaerobic systems (Fig. 8.1). Biogas yield decreased with increasing BOD load for both the softwood and hardwood effluents. Anaerobic plus aerobic treatment removed more than 92% of BOD and chlorate. AOX removal was 72 to 78% with hardwood effluents and 35 to 43% with softwood effluents (Fig. 8.2). The Enso-Fenox process was capable of removing 64–94% of the chlorophenol load, toxicity, mutagenicity, and chloroform in the bleaching effluent.[37]

The sequential treatment of bleached kraft effluent in an anaerobic fluidized bed and aerobic trickling filter was found to be effective in degrading the chlorinated high and low molecular weight material.[38] The treatment significantly reduced the COD, BOD, and AOX of the wastewater. COD and BOD reduction was greatest in the aerobic process, whereas dechlorination was significant in the anaerobic process. With the combined aerobic and anaerobic treatment, over 65% reduction of AOX and over 75% reduction of chlorinated phenolic

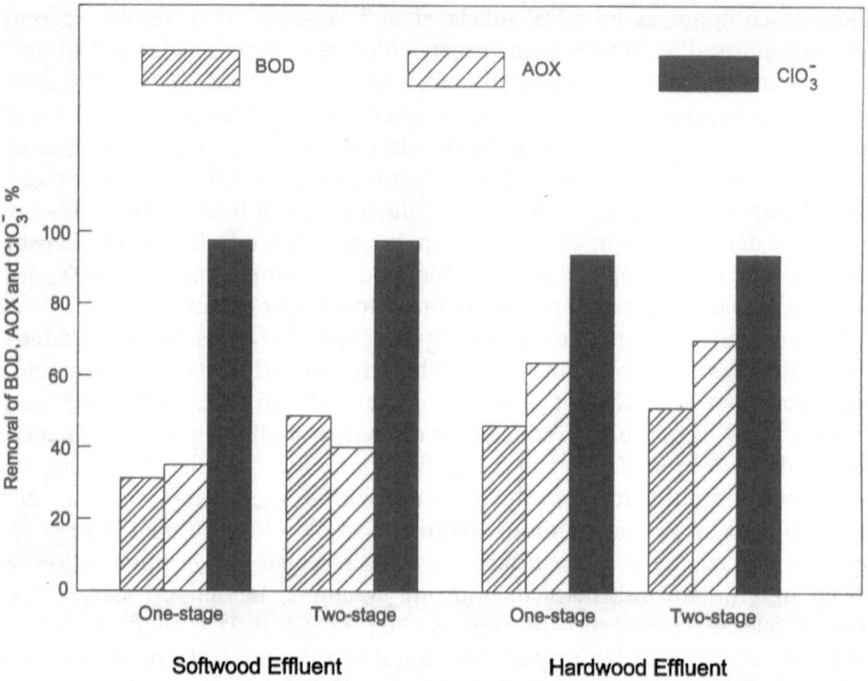

Fig. 8.1. Removal of BOD, AOX, and chlorate from softwood and hardwood bleached kraft effluents by one-stage and two-stage anaerobic treatment. Reproduced with permission from Tappi Press, Ref. 36

compounds were observed (Table 8.3). The COD/AOX ratio of the wastewater was similar before and after treatment, indicating that the chlorinated material was as biodegradable as the nonchlorinated. Microbes capable of mineralizing pentachlorophenol constituted approximately 3% of the total heterotrophic microbial population in the aerobic trickling filter. Two aerobic polychlorophenol-degrading *Rhodococcus* strains were able to degrade polychlorinated phenols, guaiacols, and syringols in the bleaching effluent (Table 8.4).

Hall et al.[39] examined the feasibility of the anaerobic membrane bioreactor concept for the treatment of segregated kraft bleach plant wastewaters by comparing AOX removal achievable by anaerobic biological treatment, ultrafiltration, and coupled anaerobic/ultrafiltration treatment. AOX removal efficiencies were found to improve with the application of both ultrafiltration and anaerobic membrane bioreactor treatments.

Many compounds in the bleaching effluent have been found to be toxic to anaerobic microorganisms. These compounds include chelating agents, extractives, and chlorinated compounds.[40] Therefore, acclimatization of microorganisms to the effluent containing these compounds is necessary for achieving successful treatment. Raizer-Neto et al.[40] studied the efficiency of anaerobic

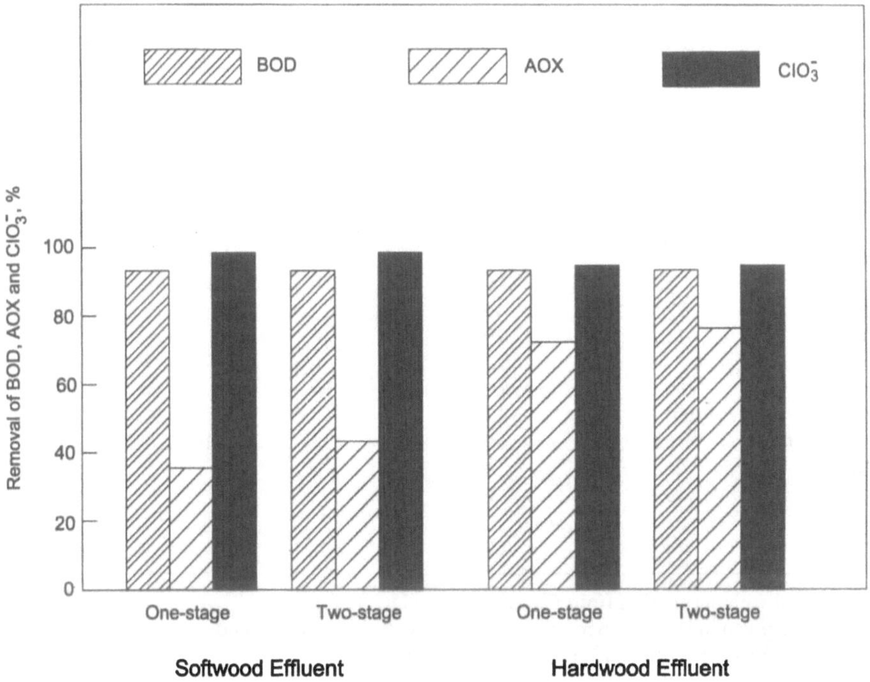

Fig. 8.2. Removal of BOD, AOX, and chlorate from softwood and hardwood bleached kraft effluents by one-stage and two-stage anaerobic treatment followed by aerobic posttreatment. Reproduced with permission from Tappi Press, Ref. 36

Table 8.3. Removal of COD, BOD, AOX, and chlorophenols by anaerobic-aerobic treatment of bleaching effluent[a]

Parameter	Influent	After anaerobic treatment	Aerobic outflow	Reduction (%)
COD_{Cr} (mg O_2/l)	3570	3010	1450	61
BOD_7 (mg O_2/l)	580	605	130	78
AOX (mg Cl/l)	175	122	56	68
Chlorophenols (μg/l)	313	186	73	77
2346-TeCP	6.1	2.2	1.8	
246-TCP	18.1	11.9	0.2	
24-DCP	9.6	10.3	2.2	
TeCG	48.3	18.8	7.7	
345-TCG	138.1	75.9	30.4	
456-TCG	22.4	13.4	4.9	
45-DCG	20.6	17.3	6.4	
TCS	13.9	8.1	5.1	

Reproduced with permission from Elsevier Science Ltd., Ref. 38.
[a] The sample was taken on day 50.

Table 8.4. Removal of chlorinated phenolic compounds in bleach-plant effluent by resting cells of polychlorophenol degraders *Rhodococcus chlorophenolicus* PCP-I and *Rhodococcus* sp. CP-2[a]

Compound	Initial conc[b] (μg/l)	% Removal in 2 days[c]	
		R. chlorophenolicus strain PCP-I	*Rhodococcus* sp. strain CP-2
Phenols			
PCP	1.4	100	100
2346-TeCP	5.8	100	100
246-DCP	27.3	33	16
24-DCP	13.7	19	0
Guaiacols			
TeCG	30.8	60	61
345-TCG	110.2	63	73
346/356-TCG	10.9	79	100
456-TCG	20.2	0	0
45-TCG	15.3	0	0
Syringols			
TCS	13.3	57	63

Reproduced with permission from Elsevier Science Ltd., Ref. 38.
[a] Cells were grown in DSM-65 medium and induced with PCP, washed, and resusupended in pulp-bleaching effluent to a cell concentration of 10^8 to 10^9/ml.
[b] The pulp-bleaching effluent (E + C stage).
[c] Removal of chlorinated phenolic compounds in sterile controls was less than 5%.

treatments to reduce chlorinated organic compounds in a fixed-bed reactor. AOX removal efficiency was primarily affected by increasing the AOX concentration and was only about 50% for 50 mg/l COD. BOD removal efficiencies were affected at higher AOX concentrations of 100 mg/l. AOX loading rate or hydraulic residence time were found to be more important limiting factors than the bleaching effluent AOX concentration. For AOX loading rates under 40 mg/l and an AOX concentration as high as 130 mg/l in the effluent feed, COD and BOD removal were about 80%.

Swedish researchers reported 35–40% COD and 42–45% AOX reductions in a 1.5-day anaerobic treatment.[41] They found that at least part of the higher molecular mass was dechlorinated. Since anaerobic microorganisms are believed to be unable to degrade the high molecular weight chlorolignin, the dechlorination of the higher molecular weight chlorolignin could be mainly due to physical and/or chemical actions in the aerobic process. The anaerobic dechlorination of chlorolignins is due to the combination of energy metabolism, growth, chemical hydrolysis, and probably adsorption and/or insolubilization.[41]

Eriksson and coworkers proposed a process based on UF and anaerobic and aerobic biological treatments.[42–44] The UF was used to separate the high molecular weight mass, which is relatively resistant to biological degradation. Anaerobic microorganisms were believed to be able to remove highly chlori-

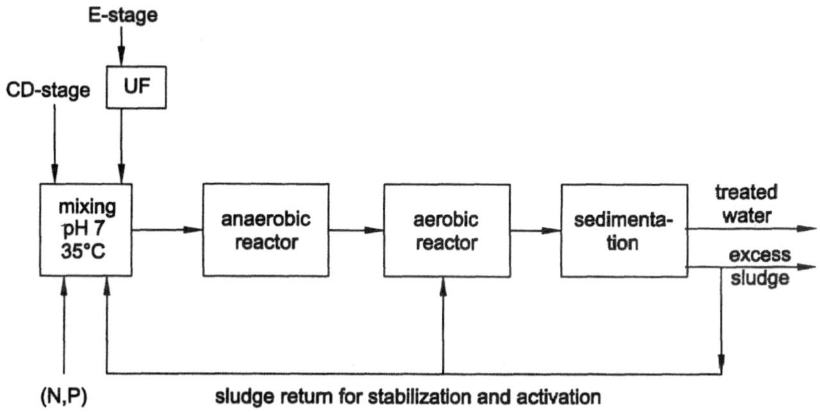

Fig. 8.3. Part of a system for purification of conventional waste bleach waters

Table 8.5. A comparison of purification results obtained with the new system and the conventional aerated lagoon technique

Purification of waste bleach waters		
Parameter	UF plus anaerobic/aerobic; predicted reductions(%)	Aerated lagoon estimated reductions(%)
BOD	95	40–55
COD	70–85	15–30
AOX	70–85	20–30
Color	50	0
Toxicity	100	Variable
Chlorinated phenols	>90	0–30
Chlorate	>99	Variable

Based on data from Ref. 44.

nated substances more efficiently than aerobic microorganisms. The remaining chlorine atoms were removed by aerobic microorganisms. The combined treatments typically removed 80% of the AOX, COD, and chlorinated phenolics, and completely removed chlorate. The principal for part of the purification system is given in Fig. 8.3 and the results obtained using this system are compared to those normally obtained in an aerated lagoon (Table 8.5).

Armentate et al.[45] investigated an integrated anaerobic and aerobic process for the biodegradation of chlorinated aromatic compounds. The sludge obtained from the anaerobic digestor of a commercial treatment plant was used to obtain an anaerobic consortium capable of partially dechlorinating 2,4,6-trichlorophenol. The clarified and sterilized effluent from the same anaerobic digestor was used as the medium for the anaerobic consortium. During the anaerobic process, 2,4,6-TCP was first dechlorinated to 2,4-

dichlorophenol and then to 4-chlorophenol. Stoichiometric amounts of 4-CP were recovered. Similar results were obtained when the anaerobic microorganisms were immobilized. After immobilization, the consortium was able to dechlorinate 150 μM of 2,4,6-TCP in 4 days. *Pseudomonas glatheri* and an indigenous culture, obtained from the same sludge used to produce the anaerobic enrichment culture, were shown to be able to degrade the 4-CP produced from the anaerobic dechlorination of 2,4,6-TCP.

Strehler and Welander[46] investigated the removal of AOX and COD from bleached kraft mill effluent in laboratory and pilot-scale aerobic-suspended carrier reactors and abiotic-thermoalkaline reactors. At pH 7.0, 37 °C, and HRTs of more than 3.5 h, a maximum COD removal of 55% was achieved in the suspended carrier process. The COD conversion rate at the minimum HRT was 2.6 kg COD/m^3/day. The suspended carrier treatment was operated successfully at pH 9.0 and 45 °C and at pH 7.0 and 50 °C with over 50% COD removal with an HRT of 4 h. The AOX removal achieved at pH 9.0 and 45 °C (50%) was significantly higher than the removal at pH 7.0 and 37 °C (39%), due to an increase in abiotic dechlorination at the higher pH and temperature levels. These researchers further studied a sequential thermoalkaline and biological treatment on a pilot scale. Thermoalkaline treatment at pH 10.0, 54 °C and HRT of 2 h followed by biological treatment at pH 8.0, 35 °C and a HRT of 4 h removed almost 80% of AOX and 50% of COD from the kraft mill effluents.

2. Fungal Treatment

The reason why bacteria show low efficiency in removing COD, AOX, and high molecular mass chlorolignins from bleaching effluent is that bacteria degrade the substrate by an intracellular enzyme system. The substrate must be able to pass the bacterial cell membrane in order to be degraded. In the light of this, the inefficiency of bacteria in degrading high molecular mass chlorolignins is understandable. Since most AOX and COD are due to the high molecular mass chlorolignins, low COD and AOX removal efficiencies are also to be expected.

An approach to degrading high molecular weight chlorolignins is to use white rot fungi. These fungi generate a ligninolytic activity to degrade lignin in a so-called secondary metabolism stage when one of several nutrients, nitrogen, phosphorus, or carbon is depleted. White rot fungi excrete an operating system including extracellular enzymes which can degrade high molecular mass chlorolignins effectively. So far, the widest applications of white rot fungi in wastewater purification have been concentrated on the decolorization/detoxification of bleaching or pulp mill effluents.[1]

The best-known white rot fungus is *Phanerochaete chrysosporium*. This fungus is known to secrete a family of enzymes which degrade both lignin and modified lignin. Both high- and low molecular mass chlorolignins generated during the pulp bleaching are significantly degraded.[47] The fungus reduces the COD by degrading the chlorolignins to CO_2 and chloride, decolorizes the bleaching effluent by destroying both the color bodies and chromophoric structure, and removes AOX by converting it to inorganic chloride.[48,49] Most of the low molecular weight chlorophenolics disappear after 1 day of fungal treatment.[50] This particular fungus is also known to be able to degrade refractory compounds such as TNT, PCB and lindane, DDT, chlorinated dioxin, and other difficult biodegradable compounds.[51,52]

Decomposition of lignin to CO_2 by the lignin-decomposing fungus *P. chrysosporium* requires a growth substrate such as cellulose or glucose. Growth on lignin as a sole carbon source is negligible. In addition to its requirement for a cosubstrate, the ligninolytic system (1) is produced even in the absence of lignin, (2) is expressed during secondary metabolism, (3) is triggered by carbon, sulfur, or nitrogen limitation, (4) is strongly repressed by glutamate and certain other amino acids, (5) is sensitive to the balance of trace metals supplied, (6) has a relatively narrow pH optimum (4–5), and (7) is markedly affected by oxygen concentration.[53–55] The fungal decolorization, like lignin metabolism, is a secondary metabolic event and is probably attributable to lignin metabolism since lignin and its degradation products are the main color source in the bleaching effluents. The optimum conditions favoring fungal growth are quite different from those favoring decolorization. The pH range for optimum growth is 4.3–4.8 and decolorization is greatly retarded below pH 4.0 or above 5.0 because of poor growth.[53] The optimum temperature for the growth of the fungus is 40 °C, whereas decolorization is not limited to the same narrow range of temperature, but takes place with little decrease in rate at temperatures as low as 25 °C.[53] The fungal decolorization requires oxygen and a cosubstrate, but, unlike fungal growth, the addition of a nitrogen source is not necessary. Eaton et al.[56] have outlined a process based on laboratory and bench-scale experiments for decolorization by *P. chrysosporium*. The fungi require a growth substrate for decolorization just as they do for lignin degradation. Investigations have demonstrated that the cellulose-rich primary sludge serves this purpose well. The fungus requires a fixed surface for pregrowth without agitation in a shallow medium for efficient color reduction. Fixed-film reactors of the rotating biological contactor (RBC) design, which are commercially available and are presently used in waste treatment, have proved to be effective. Color removal is greatly stimulated by an oxygen-enriched atmosphere. The RBC provides good aeration and can be enclosed for the addition of supplementary oxygen. A growth stage is necessary before decolorization begins. During this stage, nutrient nitrogen is depleted and the fungus becomes ligninolytic and able to decolorize. Based on the above considerations, a process, termed FPL/NCSU MyCoR,[57] has been developed. It has resulted from the cooperative research between the US Forest Products Laboratory and North Carolina State University. A fixed-film MyCoR reactor is charged with growth nutrients which can include primary sludge as the carbon source, and is inoculated with a suitable fungus. The sludge will provide some of the required mineral nutrients and trace elements as well as carbon. Nitrogen-rich secondary sludge can be also used to supply the nitrogen required for growth.

After the mycelium has grown over the reactor surface, it depletes the available nitrogen and becomes ligninolytic (pregrowth stage 2 to 4 days). The reactor is then ready for use. Operation for over 60 days has been achieved in bench reactors in a batch mode. The process converts 70% of the organic chlorides to inorganic chlorides in 48h while decolorizing the effluent and reducing both COD and BOD by about half.

Huynh et al.[50] used the MyCoR process for the treatment of chlorinated low molecular mass phenols of the E_1 effluent. It was found that most of the chlorinated phenols and low molecular mass components of the effluent were removed during the fungal treatment (Table 8.6). Pellinen et al.[58] have reported that the MyCoR process can be considerably improved in terms of COD removal by simply using less glucose as the carbon source for the fungi, *P. chrysosporium*. However, the dechlorination was reported to be faster at high glucose concentration. Yin et al.[49] studied the kinetics of decolorization of E_1 effluent with *P. chrysosporium* in an RBC under improved conditions. The

Table 8.6. Quantitation of chloroform-extractable aromatics obtained from bleach-plant effluent (E_1) before and after decolorization

Compound	RT[a]	Before decolorization (ppm)	After decolorization (ppm)
Benzoic acid	5.40	0.12–0.45[b]	None
Dichlorobenzoic acid[c]	5.48		None
2,4,6-Trichlorophenol	7.47	0.15–0.33	Trace-0.01
Vanillin	8.05	0.01–0.02	0.05–0.10
4,5-Dichloroguaiacol	9.18	0.08–0.25	Trace
Acetoguaiacone	9.22	Trace	None
6-Chlorovanillin	10.22	0.05–0.10	Trace
Trichloroguaiacol[c,d]	10.67	0.15–0.34	Trace
5-Chlorovanillin	11.05	Trace	None
4,5,6-Trichloroguaiacol	12.74	0.04–0.10	Trace
Tetrachloroguaiacol	14.06	0.08–0.15	Trace
Vanillyl alcohol	8.67	None	0.05–1.20
Veratraldehyde	9.11	None	0.08–1.30
Veratryl alcohol	9.37	None	10–20
4,5-Dichloroveratrole	9.59	None	0.10–0.30
4,5,6-Trichloroveratrole	11.66	None	Trace-0.02
3,4-Dimethoxyaetophenone	10.39	None	Trace-0.02
5-Chloroveratryl alcohol	11.84	None	Trace
6-Chloroveratry alcohol	12.20	None	0.50–0.80

Reproduced with permission from Tappi Press, Ref. 50.

[a] Retention time on gas chromatogram/capillary column-30 M with DB-5 as the liquid phase; oven temperature was held at 60 °C for 1 min, then programmed at a rate of 15 °C/min to 140 °C, and then 4 °C/min to 255 °C.

[b] Includes dichlorobenzoic acid.

[c] Isomer unknown.

[d] Quantitation of this compound was based on the GC response factor of 4,5,6-trichloroguaiacol.

kinetic model developed for 1 and 2 days retention times showed a characteristic pattern. The overall decolorization process can be divided into three stages, viz. a rapid color reduction in the first hour of contact between the effluent and the fungus followed by a zero-order reaction, and then a first-order reaction. The color removal rate on the second day of the 2-day batch treatment was less than that on the first day. The decolorization in a continuous flow reactor achieved approximately the same daily color removal rate, but the fungus had a larger working life than when in the batch reactor, thereby removing more color over the fungal life time. Pellinen et al.[59] studied dechlorination of high molecular mass chlorolignin in first extraction stage effluent with white rot fungus *P. chrysosporium* immobilized on RBC. The AOX decreased almost by 50% during the 1-day treatment. Correlation studies suggested that dechlorination, decolorization, and degradation of chlorolignin (as COD decrease) are metabolically connected, although these processes have different rates.

The combined treatment of E_1 effluent with white rot fungi and bacteria has been also reported. Yin et al. studied a sequential biological treatment using *P. chrysosporium* and bacteria to reduce AOX, color, and COD in conventional softwood kraft pulp bleaching effluent.[60] In six variations of the white rot fungus/bacterial systems studied, only the degree of fungal treatment was varied. In three of the six variations, ultrafiltration was also used to concentrate high molecular mass chlorolignins and to reduce effluent volume (and thus cost) prior to fungal treatment (Table 8.7). The best sequence, using ultrafiltration/white rot fungus/bacteria, removed 71% TOCl, 50% COD, and

Table 8.7. Sequential biological dechlorination schemes

Scheme number	Ultrafiltration (gal/ADT day)	Fungal treatment (gal/ADT day)	Bacterial treatment (gal/ADT day)
1		5900 C[a] 2400 E_1[b]	8300 fungal-treated
2		2160 C 2400 E_1	3740 C and 4560 fungal-treated
3		2400 E_1	5900 C and 2400 fungal-treated
4	5900 C 2400 E_1	2490 retentate	5810 permeate and 2490 fungal-treated
5	2400 E_1	504 retentate (2 days)	5900 C, 1896 permeate 469 fungal-treated
6	2400 E_1	504 retentate (3 days)	5900 C, 1898 permeate 436 fungal-treated

Reproduced with permission from Butterworth-Heinemann, Ref. 60.
[a] C Chlorination effluent.
[b] E_1 First alkaline extraction stage effluent.

Table 8.8. Treatability of sequential biological treatment

Scheme no.	Parameter		
	TOCl (kg/ADT)	COD (kg/ADT)	Color (kg/ADT)
Control[a]	6.26	54.19	239.80
1	1.19 (81%)[b]	44.00 (19%)	33.73 (86%)
2	1.49 (76%)	32.00 (41%)	81.69 (66%)
3	2.16 (65%)	29.77 (45%)	103.93 (57%)
4	1.99 (68%)	36.87 (32%)	62.02 (74%)
5	2.17 (65%)	29.79 (45%)	99.04 (59%)
6	1.84 (71%)	27.88 (49%)	83.50 (65%)

Reproduced with permission from Butterworth-Heinemann, Ref. 60.
[a] CE_1 load from conventional (CEDED) softwood kraft pulp bleach plant.
[b] Removal percentage.

65% color in the effluent (Table 8.8). Fungal treatment enhances the ability of bacteria to degrade and dechlorinate chlorinated organics in the effluent.

The degradation of model compounds – chlorophenols, and chloroguaiacols – in pure water solution by fungal treatment using an RBC has been studied by Guo et al.[61] It was found that at a concentration of 30 mg/l, 80–85% of chlorophenols and chloroguaiacols could be degraded after 3–4 h treatment.

In order to eliminate some of the problems associated with the RBC process, Prouty[62] proposed an areated reactor. The fungal life in the aerated reactor was longer and the color removal rate was significantly higher than with the RBC process in an air atmosphere. A preliminary economic evaluation of the RBC process indicated that the rate of decolorization and the life span of the fungus are the most critical factors.[63] Yin and coworkers[49] suggested that treatment of the E_1 effluent by ultrafiltration before RBC treatment would be economically attractive. Their study also suggested that combining ultrafiltration and the MyCoR system could maximize the efficiency of the MyCoR process and reduce the treatment cost, thereby making the process more economically feasible for industrial use.

Although the MyCoR process was efficient in removing color and AOX from bleaching effluents, it also had certain limitations. The biggest problem was the relatively short active life of the reactor. Therefore, several other bioreactors such as packed-bed and fixed-bed reactors were studied.[64–66] The use of a trickling filter-type bioreactor in which the fungus immobilized on porous carrier material was adopted in the MyCOPOR system.[65] For E_1 effluent with an initial color between 2600 and 3700 PCU, the mean rate of color reduction was 60% during a consecutive 12-h run. The mean AOX reduction value at a color reduction of 50 to 70% in 12 h was 45 to 55%. Cammorta and Santanna[66] developed a continuous packed-bed bioreactor in which *P. chrysosporium* was immobilized on polyurethane foam particles. The bioreactor operation at a hydraulic retention time of 5.8 days was able to promote 70% decolorization. In comparison with the MyCoR process, the fungal biomass could be maintained in this process for at least 66 days without any appreciable loss of activity.

To apply the MyCOPOR process on an industrial scale, relatively large reactors (diameter 70 and 100 mm; volume 4 to 16 l) were prepared and filled with polyurethane-foam cubes (1 cm^3) as carrier material. Long-term experiments were successfully carried out and it was decided to build a small pilot reactor at a large paper mill in Austria.[67] However, many aspects related to the operating conditions must be further investigated and improved. A disadvantage of these treatment processes is that *P. chrysosporium* required high concentrations of oxygen and energy sources such as glucose or cellulose as well as various basal nutrients, mineral solution, and Tween-80.[65] Recently, Kang and coworkers[68] developed a submerged biofilter system in which mycelia of *P. chrysosporium* were attached to media (net ring type) and used to dispose wastewater from a pulp mill. Maximum reduction of BOD, COD, and lignin concentrations were 94, 91, and 90%, respectively, in 12 h of hydraulic retention time.

Matsumoto and coworkers[69] demonstrated that RBC treatment of E_1 effluent was effective for the removal of organically bound chlorine as well as color. Removal of AOX was determined to be 62, 43, and 45%/day for the low molecular weight fraction of E_1 effluent, high molecular weight fraction of the same, and unfractioned E_1 effluent, respectively. After further optimization, 49% of the high molecular weight AOX was transformed to inorganic chloride in 1 day and 62% in 2 days. The chloride concentration increased simultaneously with decreasing AOX including dechlorination.

Fukoi and coworkers[70] determined the toxicity by the microtox bacterial assay of E_P (alkaline extraction with hydrogen peroxide) effluent and ultrafiltration fractionated E_P effluent before and after fungal treatment. The overall toxicity of unfractionated effluent was reduced; however, fungal degradation of higher molecular weight fractions led to an increase in toxicity because of the generation of lower molecular weight compounds after enzyme cleavage.

Another white rot fungus, *Coriolus versicolor*, has also shown good performance. This fungus removed 60% of the color of combined bleach kraft effluents within 6 days in the presence of sucrose. Decolorization of effluent was more efficient when the concentration of sucrose and inoculum was high. When the fungus was immobilized in calcium alginate gel, it removed 80% color from the same effluent in 3 days in the presence of sucrose.[71] The decolorization process affected not only the dissolved chromophores but also the suspended solids. The solids after centrifugation of the zero-time samples were dark brown, while the solids after a 4-day incubation were light brown. The beads with the immobilized mycelium remained light-colored throughout the experiments, with no indication of accumulation of the effluent chromophores. Recycled beads were found to remove color efficiently and repeatedly in the presence of air but not under anaerobic conditions. Biological reactors of the airlift type using calcium alginate beads to immobilize the fungus *C. versicolor* have been used to study the continuous decolorization of kraft mill effluents.[72] The effluent used contained only sucrose and no other nutrient source. An empirical kinetic model was proposed to describe the decolorization process caused by this fungus, but it shed no light on the chemical mechanism involved in the decolorization.

Bergbauer and Kraeplin[73] showed that *C. versicolor* efficiently degraded chlorolignins from bleaching effluents. More than 50% of the chlorolignins were degraded in a 9-day incubation period, resulting in a 39% reduction in AOX and 84% decrease in effluent color. In a 3-l laboratory fermenter, with 0.8% glucose and 12 mM ammonium sulfate, about 88% color reduction was achieved in 3 days. Simultaneously, the concentration of AOX dropped from 40 mg/l to 21.9 mg/l, a 45% reduction in 2 days.

Direct use of suspended mycelium of the fungus *C. versicolor* may not be feasible because of the problems of viscosity, oxygen transfer, and recycling of the fungus. The fungus was therefore grown in the form of pellets, thus eliminating the problems with biomass recycling and making it possible to use a larger amount.[74] Rate of decolorization with fungal pellets was almost ten times as high in batch culture as in continuous culture under similar conditions. The capacity for decolorization decreased markedly with increase in lignin loading.[74]

Bajpai et al.[75] reported 93% color removal and 35% COD reduction from first extraction stage effluent (7000 CU) with mycelial pellets of *C. versicolor* in 48 h in a batch reactor, whereas in a continuous reactor, the same level of color and COD reduction was obtained in 38 h. No loss in decolorization ability of mycelial pellets was obtained when the reactor was operated continuously for more than 30 days. Mehna et al.[76] also used *T. versicolor* for decolorization of effluents from a pulp mill using agriresidues. With an effluent of 18 500 color units, the color reduction of 88–92% with COD reduction of 69–72% was obtained. Royer et al.[77] described the use of pellets of *C. versicolor* to decolorize ultrafiltered kraft liquor under nonsterile conditions with a negligible loss of activity. The rate of decolorization was observed to be linearly related to the liquor concentration and was lower than that obtained in the MyCoR process. This could be due to the lower temperature used in this work and to the use of pellets with relatively large diameters, which could limit the microbial activity as compared to the free mycelium used in the MyCoR process. An effective decolorization of effluent having 400–500 color units/l can be obtained in presence of a simple carbon source such as glucose. In the repeated batch culture, the pellets exhibited a loss of activity dependent on the initial color concentration. Simple carbohydrates were found to be essential for effective decolorization with this fungus and a medium composed of inexpensive industrial by-products provided excellent growth and decolorization.[78]

Recently, Pallerla and Chambers[79] have shown that immobilization of *T. versicolor* in urethane prepolymers leads to significant reductions in color and chlorinated organic levels in the treatment of kraft bleach effluents. Color

Table 8.9. Performance of continuous bioreactor using polyurethane prepolymer-immobilized *T. versicolor*

Parameter	Value
Feed color, mean	2476 CU
Product color, mean	722 CU
Product color, SD	147 CU
Reduction percentage, mean	69%
Reduction percentage, SD	5%
Feed AOX (mg/l)	23
Product AOX (mg/l) mean	10.6
Product AOX (mg/l) SD	1.7
Reduction percentage, mean	53%
Reduction percentage, SD	5%

SD: Standard deviation.
Reproduced with permission from Tappi Press, Ref. 79.

reduction ranging from 72–80% and AOX reduction ranging from 52–59% is possible from a continuous bioreactor at a residence time of 24 h (Table 8.9). The highest color removal rate of 1920 CU/day was achieved at an initial color concentration of 2700 CU. The decolorization process was linearly dependent on the concentration of glucose cosubstrate up to a level of 0.8% by weight. The biocatalyst remained intact and stable after an extended 32-day operation.

Treatment of extraction stage effluent with ozone and the fungus *C. versicolor* has also been tried.[80] Both ozone treatment and biological treatment effectively destroyed effluent chromophores, but the fungal process resulted in greater degradation, as expressed by COD removal. Monoaromatic chlorophenolics and toxicity were removed partially by ozone and completely by *C. versicolor*. Molecular weight distributions showed roughly equal degradation of all sizes of molecules in both the treatments. The combination of a brief ozone treatment with a subsequent fungal treatment revealed a synergism between the two decolorization mechanisms on E_1 effluent. Effluent was pretreated with ozone (110–160 mg/l) or *C. versicolor* (24 h with 2–5 g/l wet weight fungal biomass). The pretreatment was followed by a 5-day incubation with *C. versicolor*. It was noted that partial color removal by ozone pretreatment allowed more effective removal by the fungus than that by fungal pretreatment. After 20 h, 46–53% decolorization was observed for ozone-pretreated effluents, compared to 29% for fungal treatment alone. The contribution of ozone seemed to be most important in the first 24 h following the pretreatment. Ozone pretreatment also produced a small improvement in the bioavailability of effluent organics to the fungus. A partial replacement of chlorine by ozone in the bleach plant or a brief ozone pretreatment of E_1-stage effluent should considerably reduce the low molecular mass toxic chlorophenolics. In addition, the use of ozone should also improve decolorization by subsequent fungal and possibly bacterial treatments.

A white rot fungus, *Tinctoporia borbonica*, has been reported to decolorize the kraft waste liquor to a light yellow color.[81] About 99% color reduction was achieved after 4 days of cultivation. Measurement of the culture filtrate by ultraviolet spectroscopy showed that the chlorine-oxylignin content also decreased with time and measurement of the culture filtrate plus mycelial extract after 14 days of cultivation showed the total removal of the chlorine-oxylignin content.

Another white rot fungus, *Schizophyllum commune*, has also been found to decolorize and degrade lignin in pulp and paper mill effluent.[82] The fungus was able to degrade lignin in the presence of an easily metabolizable carbon source. The addition of carbon and nitrogen not only improved the decolorizing efficiency of the fungus but also resulted in reduction of the BOD and COD of the effluent. Sucrose was found to be the best carbon source for the degradation of the lignin. A 2-day incubation period was sufficient for lignin degradation by this fungus. Under optimum conditions, this fungus reduced the color of the effluent by 90% and also reduced BOD and COD by 70 and 72% during a 2-day incubation. Esposito et al.[83] and Lee et al.[84] have reported that some white rot fungi show efficient decolorization of the E_1 effluent without any additional nutrients. Through a screening of several strains of ligninolytic fungi, the *Lentinus edodes* and fungus KS-62 were found to remove about 70–80% color in 5–7 days without any additional nutrients. Under similar conditions, *P. chrysosporium* and *T. versicolor* removed 13 and 29% color, respectively. To obtain a reasonable basis for evaluation of an industrial fungal treatment, Lee et al.[85] performed treatment of the E_1 effluent with the immobilized mycelium of the fungus KS-62. The fungus showed 70% color removal (initial color 6600 PCU) without any nutrients within 1 day of incubation with four times effluent replacement; however, the color removal started to decrease at the fifth replacement with the fresh E_1 effluent. The decolorization activity of the fungus was restored by one replacement of E_1 effluent containing 0.5% glucose, and the high decolorization was continuously observed for four replacements in the absence of glucose. Such decolorization activity was repeatedly obtained for 29 days of the total treatment period.

Duran et al.[86] reported that preradiation of the effluent, followed by fungal culture filtrate treatment resulted in efficient decolorization. Moreover, when an effluent preirradiated in the presence of ZnO was treated with *L. edodes*,[83] a marked enhancement of the decolorization at 48 h was obtained.[87] Duran et al.[87] proposed that the combined photobiological decolorization procedure appears to be an efficient decontamination method with potential for industrial effluent treatment. Kannan et al.[88] reported about 80% color removal and over 40% BOD and COD reduction with fungus *Aspergillus niger* in 2 days. Tono et al.[89] reported that *Aspergillus* sp. and *Penicillium* sp. achieved 90% decolorization in 1 week's treatment at 30 °C and at pH 6.8. Later, Milstein et al.[90] reported that these microorganisms removed appreciable levels of chlorophenols as well as chloroorganics from the bleach effluent. Gokcay and Taseli[91] have reported over 50% AOX and color removal from softwood bleach effluents in less than 2 days of contact with *Penicillium* sp. Bergbauer[92] reported AOX reduction by 68% and color reduction by 90% in 5 days with the coelomycetous fungus, *Stagonospora gigaspora*. Toxicity of the effluent was reduced significantly with this fungus. A few marine fungi have been also reported to decolorize the bleach plant effluents.[93] With *Trichoderma* sp. under optimal conditions, total color and COD decreased by almost 85 and 25%, respectively, after cultivation for 3 days.[94]

3. Enzymatic Treatment

A few enzymes have been also used to decolorize and detoxify effluents from bleach plants. Peroxidases and laccases are the most important of these. The use of microbial or enzyme-based treatment offers some distinct advantages over physical and chemical decolorization and AOX precipitation methods, in that only catalytic and not stoichiometric amounts of the reagents are needed, and the low organic concentrations and large volumes typical of bleaching effluents are therefore less of a problem. Also, both complete microbial systems and isolated enzymes such as peroxidases and laccases have been shown to reduce the acute toxicity by polymerizing, thereby rendering many of the low molecular mass nonchlorinated and polychlorinated phenolics less soluble.[95–97] In 1986, Field patented a method for the biological treatment of wastewaters containing non-degradable phenolic compounds and degradable nonphenolic compounds.[98] It consisted of an oxidative treatment to reduce or eliminate toxicity of the phenolic compounds followed by an anaerobic purification. This oxidative pretreatment could be performed with laccase enzymes and it was claimed to reduce chemical oxygen demand by 1000 fold. Milstein et al. described the removal of chlorophenols and chlorolignins from bleaching effluents by combined chemical and biological treatments.[99] The organic matter from spent bleaching effluents of the chlorination, extraction, or a mixture of both stages was precipitated as a water-insoluble complex with polyethyleneimide. The color, COD, and AOX were reduced by 92, 65 and 84%, respectively, for the chlorination effluent and by 76, 70, and 73% for the extraction effluent. No significant reduction in BOD of treated effluent was detected, but fish toxicity was greatly reduced. Enzyme treatment results in coprecipitation of the bulk mono- and dichlorophenols with the liquors of the chlorination and extraction bleaching stages. In 1988, a review on the use of enzymes for wastewater treatment in the pulp and paper industry examined the new possibilities of using enzymes like laccase, peroxidase, and ligninase for this effect.[100] Forss et al. examined the use of laccase for effluent treatment in 1989.[101] They aerated pulp bleaching wastewater in the presence of laccase for 1 h at pH 4.8 and subsequently flocculated with aluminum sulfate. High removal efficiencies were obtained for chlorinated phenols, guaiacols, vanilins,

Table 8.10. Comparison of the removal of chlorophenolics from a mixture in high molecular weight E_1 effluent by *T. versicolor* mycelium in 3 h and crude laccase in 30 min[a]

μM	Substrate	% Disappearance	
		Laccase	Mycelium
137	2-Chlorophenol	35.5 ± 1.4	61.5 ± 2.7
18	3-Chlorophenol	51.9 ± 4.3	2.0 ± 0.1
31	4-Chlorophenol	48.1 ± 2.9	41.6 ± 1.3
21	2,4-Dichlorophenol	45.0 ± 2.3	64.9 ± 1.8
26	2,4,6-Trichlorophenol	72.4 ± 5.2	83.8 ± 6.2
18	2,3,4,6-Tetrachlorophenol	40.7 ± 8.2	71.2 ± 2.7
24	Pentachlorophenol	34.0 ± 8.9	82.1 ± 3.2
52	4,5-Dichloroguaiacol	100	96.8 ± 6.1
32	4,5,6-Trichloroguaiacol	100	100
17	Tetrachloroguaiacol	46.5 ± 21.5	95.0 ± 7.6

Reproduced with permission from Elsevier Science Ltd., Ref. 105.
[a] Reactions were carried out in 20-ml high molecular weight E_1 effluent (dialyzed) containing 20 mM D-glucose, set to pH 4.6, shaken at 200 rpm, 25 °C in triplicate, using either 0.1 U/ml crude *T. versicolor* laccase or 0.5 g wet weight (~25 mg) washed *T. versicolor* mycelium.

and catechols. The use of laccase for wastewater treatment was patented by Call in 1991.[102] He claimed that wastewater from delignification and bleaching could be treated with laccases in the presence of nonaromatic oxidants and reductants and aromatic compounds. Almost complete polymerization of the lignins is obtained, which is 20–50% above the values attainable with the addition of laccase alone. About 70–90% lignin is developed into insoluble form, which is removed by flocculation and filtration. Lyr reported that laccase of *T. versicolor* partially dechlorinates PCP[103] and Hammel and Tordone reported that peroxidase from *P. chrysosporium* can partially dechlorinate PCP and 2,4,6-trichlorophenol.[104] Roy-Arcand and Archibald carried out a systematic study on direct dechlorination of chlorophenolic compounds in pulp and paper mill effluents by laccase from *T. versicolor.*[105] It was found that most of the laccase secreted by *T. versicolor* could partially dechlorinate a variety of chlorophenolics. The rate and extent of Cl^- release were found to be substantially affected by substrate and enzyme concentration and the presence of multiple laccase substrates. The dechlorination was found to be accompanied by extensive polymerization of the substrate. Table 8.10 shows the comparison of removal of chlorophenolics from a mixture in E_1 effluent by crude laccase and *T. versicolor* mycelium and Table 8.11 shows the effects of crude laccase on various chlorophenolics. These researchers also studied the effects of horseradish and *P. chrysosporium* peroxidases on the mixture of five chlorophenolics – pentachlorophenol, tetrachloroguaiacol, 4,5,6-trichloroguaiacol,

Table 8.11. Comparison of the effects of crude laccase (0.1 U/ml) on five chlorophenolics (initially at 160 μM) individually and mixed, with 30-min incubations

Individual substrates					Five-substrate mixture				
Substrate	Consumption		Cl^{-1} Release		Consumption		Cl^{-1} Release		t-test[a]
	(μM)	(%)[b]	(μM)	(%)[c]	(μM)	(%)[c]	(μM)	(%)[d]	(α = 0.05)
2,4,6-Trichlorophenol	83 ± 5.0	48.2	163 ± 22	34.0	75 ± 1.7	46.9	–	–	NS
4,5-Dichloroguaiacol	141 ± 0.4	88.1	237 ± 0.8	74.0	149 ± 6.0	93.1	–	–	NS
4,5,6-Trichloroguaiacol	124 ± 10.2	77.5	180 ± 10	37.5	160 ± 0	100	–	–	S
Tetrachloroguaiacol	107 ± 24.6	66.9	103 ± 3.8	16.1	90 ± 9.9	56.3	–	–	NS
Pentachlorophenol	28 ± 3.9	17.5	0	0[e]	70 ± 7.0	43.8	–	–	S
Total			683	25.1			540	19.9	

Reproduced with permission from Elsevier Science Ltd., Ref. 105.
[a] Consumption of substrate as single substrate and in mixed solution is compared by t-test applied on differences between two means. NS not significant.
[b] Percent of the initial chlorophenolic level removed (GC measurement).
[c] Percent of total organochlorine mineralized to Cl^-.
[d] Individual Cl^- releases unobtainable.
[e] No Cl^- evolved over 30 min. However, in a 17-h treatment there was 24% substrate disappearance and 6.6% Cl^- release.

Table 8.12. Effects of *P. chrysosporium*-secreted peroxidase and horseradish peroxidase and H_2O_2 on a solution of mixed chlorophenolics[a]

Substrate	Initial (μM)	*P. chrysosporium* peroxidase		Horseradish peroxidase	
		Final (μM)	% Degraded	Final (μM)	% Degraded
2,4,6-Trichlorophenol	107	43 ± 0.2	60	2.5 ± 0.2	98
4,5-Dichloroguaiacol	48	13 ± 1.4	73	<0.5	100
4,5,6-Trichloroguaiacol	18	6.2 ± 1.7	87	<0.5	100
Tetrachloroguaiacol	24	7.3 ± 0.7	70	0.9 ± 1.2	96
Pentachlorophenol	19	5.7 ± 1.4	70	17.7 ± 2.5	7
Total chloride released[b]		5.4 ± 4.1 μM		341.0 ± 2.2 μM	

Reproduced with permission from Elsevier Science Ltd., Ref. 105.
[a] Reaction mixture contained: <1.0% ethanol, 0.05 U/ml crude peroxidase in 50 mM Na tartrate pH 2.5, 0.4 mM H_2O_2; or 0.1 U/ml pure horseradish peroxidase in 10 mM Na phosphate pH 7.0, 1 mM H_2O_2. Mixtures were shaken (200 rpm, 25 °C) for 30 min, then Cl^- release measured and the solution extracted for gas chromatography as before.
[b] Percentages of organochlorine released as Cl^- by crude *P. chrysosporium* peroxidase and by purified horseradish peroxidase were 0.73 and 46%, respectively.

4,5-dichloroguaiacol, and 2,4,6-trichlorophenol.[105] Both peroxidase enzymes were found to degrade the majority of all substrates except PCP, but whereas the *P. chrysosporium* peroxidase was superior to both horseradish peroxidase and laccase in degrading PCP, it was inferior to horseradish peroxidase in degrading the other four phenolics (Table 8.12).

Paice and Jurasek[106] studied the ability of horseradish peroxidase to catalyze color removal from bleach-plant effluents. The color removal from effluents at neutral pH by low levels of hydrogen peroxide was enhanced by the addition of peroxidase. No precipitation occurred during the decolorization process. The catalysis with peroxidase (20 mg/l) was observed over a wide range of peroxide concentrations (0.1–800 mM) but the largest effect was between 1 and 100 mM. The pH optimum for catalysis was around 5.0. Compared with mycelial color removal by *C. versicolor*, the rate of color removal by peroxide plus peroxidase was initially faster (for the first 4 h) but the extent of color removal after 45 h was higher with the fungal treatment. Further addition of peroxidase to the enzyme-treated effluents produced no additional catalysis. Thus, the peroxide/peroxidase system did not fully represent the metabolic route used by the fungus. One working hypothesis has been proposed to explain the behavior of enzymes in the decolorization process.[106] Glucose is used by the cell to produce peroxidase. One of the extracellular enzymes often found in white rot fungi is peroxidase. It seems that this enzyme oxidizes the chromophores and so removes the color from bleaching wastewater. Ferrer et al. reported that immobilized lignin peroxidase decolorized kraft effluent.[107] Novel lignin peroxidases called Pulpases, produced by *P. chrysosporium* mutant strain SC 26 and described in two patents assigned to the Repligen Corporation, have been claimed to decolorize bleaching effluents.[108,109]

The covalent immobilization of laccase on activated carbon for phenolic effluent treatment was used in 1992 in an attempt to overcome the rapid inactivation that the enzyme usually suffers in such a process.[110] The carbon-immobilized laccase removed color from pulp mill bleach-plant effluent (extraction stage) at a rate of 115 color ISO units per enzyme unit/h and the removal rate increased with increasing effluent concentration. Although the stability of the laccase was improved by immobilization, its oxidation of the effluent was still very slow, making it inappropriate for color removal from pulping and bleaching effluents at industrial scale, and had to be performed on a recirculation mode.

Conclusions

Among the biological methods tried so far, the method using white rot fungi is quite effective in decolorizing, dechlorinating, and detoxifying bleach-plant effluents. One of the drawbacks associated with the fungal treatment has been the requirement for easily metabolizable cosubstrate like glucose for the growth and development of ligninolytic activity. To make the fungal treatment method economically feasible, there is a need to reduce the requirement of cosubstrate or identify a cheaper cosubstrate. Hence, efforts should be made to identify the strains that show good decolorization with less or no cosubstrate and can utilize industrial waste as a cosubstrate. Efforts should be also made to utilize the spent fungal biomass for preparing the culture medium required in the synthesis of active fungal biomass. If successful, the cost of treatment may be further reduced. As the lignin-degrading system of white rot fungus has a high oxygen requirement, use of oxygen instead of air as fluidizing media should be explored. Increasing the oxygen concentration in the culture atmosphere is expected to have a dual effect: it would lead to an increased titer of the lignin-degrading system and to increased stability of the existing system. A quantitative study of extracellular enzymes is also required in order to gain insight into the possible enzymatic mechanism involved in the degradation of lignin-derived compounds present in the effluents.

Use of white rot fungi can serve as a pretreatment method to bacterial treatment and to enhance the bacterial ability to remove organic chlorine and to degrade the relatively higher molecular weight chlorolignins. This process can be used as an alternative to internal process (modified cooking, oxygen bleaching, high level chlorine dioxide substitution, etc.) and conventional biological treatment.

Since the majority of AOX and color is in high molecular weight chlorolignins, research priority should concentrate on the fate of high molecular weight chlorolignins in biological treatment or in the natural environment. Since bacteria degrade significantly only those chloroorganics with molecular weights lower than 800–1000 Da, research is needed to decrease the chlorolignin molecular weight or to remove high molecular weight chlorolignins before biological treatment is applied, in order to enhance the biotreatability of bleaching effluents.

References

1. Bajpai P, Bajpai PK: Organochlorine compounds in bleach plant effluents – genesis and control. PIRA International U.K 1996; 127 pp.
2. Ander P, Eriksson KE, Kolar MC, Kringstad KP: Studies on the mutagenic properties of bleaching effluent. Sven Papperstidn 1977; 80: 454–459.
3. Annergren G, Kringstad KP, Lehtiren KJ: Environmental risks involved in discharging spent bleach liquors into receiving waters. Presented at the EuCePa Symposium Environmental Protection in the 90's. 1986, Helsinki, Finland May 9–22.
4. Anonymous; Estimating organic chlorine discharges. Environ. Sci. Technol. 1984; 18(8): 204–208.
5. Eriksson KE, Kolar MC, Kringstad KP: Studies on the mutagenic properties of bleaching effluent. Part 2. Sven Papperstidn. 1979; 82: 95–104.
6. Kringstad KP, Lindstrom K: Spent liquors from pulp bleaching. Environ. Sci. Technol. 1984; 18: 236–248.
7. Kringstad KP, Swanson SE: Bleaching and the environment. Proc. of 1988 International Pulp Bleaching Conference, Orlando, Florida June 5–9 1988; 5–9.
8. Gergov M, Priha M, Talka E, Valttila O, Kangas A, Kukkonen K: Chlorinated organic compounds in effluent treatment at kraft mills. Tappi J. 1988; 71(12): 175–184.
9. Lindstrom K, Mohamed M: Selective removal of chlorinated organics from kraft mill total effluents in aerated lagoons. Nord. Pulp Pap. Res. J. 1988; 3: 26–33.
10. Berry RM, Luthie CE, Voss RH, Wrist PE, Axegard P, Gellerstedt G, Lindblad PO, Popke I: Bleaching vol. 2 (Jameel H ed.) Tappi Press, Atlanta Georgia 1993; 759.

11. Eriksson KE, Kolar MC, Ljungquist PO, Kringstad KP: Studies on microbial and chemical conversions of chlorolignins. Environ. Sci. Technol. 1985; 19: 1219–1224.
12. Rennel J: TCF – an example of the growing importance of environmental perceptions in the choice of fibers. Nord. Pulp Pap. Res, J. 1995; 10(1): 24–32.
13. Hinsey NW: EPA's proposed cluster rule: the end of end-of-pipe. Tappi J. 1994; 77(9): 65–74.
14. Lewis J: Cluster bombs rock industry. World Pap. 1994; 219(7–8): 18–19.
15. Anonymous: Industry news: cluster rule signed. Tappi J. 1997; 80(12): 14.
16. Demel I, Oller HJ: State of the art of AOX avoidance in effluents from German paper mills. Wochenbl. Papierfbr. 1995; 123(21): 975–978.
17. Taguchi T: Environmental topics in 1993–1994 in Japan. 1995 International Environmental Conference, Atlanta, Georgia USA, Book 2 (7–10 May, 1995): 665–673.
18. Fulthorpe RR, Allen DG: A comparison of organochlorine removal from bleached kraft pulp and paper mill effluents by dehalogenating *Pseudomonas*, *Ancylobacter* and *Methylobacterium* strains. Appl Microbiol Biotechnol. 1995; 42: 782–789.
19. Melcer H, Steel P, Mckinley A, Cook CR: The removal of toxic contaminants from bleached kraft mill waste water with enhanced activated sludge treatment. Proc. Tappi International Environmental Conference 1995; 795–807.
20. Rogers IH, Davis JC, Kruzynski GM, Mahood HW, Servizi JA, Gordon JW: Fish toxicants in kraft effluents. Tappi J. 1975; 58(7): 136–140.
21. Leach JM, Mueller JC, Walden CC: Biological detoxication of pulp mill effluents. Process Biochem. 1978; 13(1): 18–26.
22. Voss R, Wearing JT, Wong A: Advances in the identification and analysis of organic pollutants in water. (Keith L.H. ed.) Ann Arbor Science, Ann Arbor 1981; Vol. 2: p.1059.
23. Valenzuela J, Bumann U, Cespedes R, Padilla L, Gonzalez B: Degradation of chlorophenols by *Alcaligenes eutrophus* TMP 134 (p JP4) in bleached kraft mill effluent. Appl Environ. Microbiol 1997; 63(1): 227–232.
24. Ataberk S, Gokcay CF: Removal of chlorinated organics from pulping effluents by activated sludge process. Fresenius Environ. Bull. 1997; 6: 147–153.
25. Mortha G, Mckay LR, Cadel F, Rouger J: AOX reduction in an activated sludge treatment of kraft bleaching effluent. 6th ISWPC. 1991; 1–2.
26. Bullock JM, Bicho PA, Saddler JN: The effect of high molecular weight organics in bleached kraft mill effluent on the biological removal of chlorinated phenolics. Proceedings of 1994 Environmental Conference. 1994; 371–378.
27. Deardorff TL, Willhelm RR, Nonni AJ, Renard JJ, Phillips RB: Formation of polychlorinated phenolic compounds during high chlorine dioxide substitution and bleaching. Tappi J 1994; 77(8): 163–173.
28. Jokela JK, Laine M, EK M, Salkinoja-Salonen M: Effect of biological treatment on halogenated organics in bleached kraft pulp mill effluents studied by molecular weight distribution analysis. Environ. Sci. Technol. 1993; 27(3): 547–552.
29. Eriksson KE, Kolar MC: Studies on microbial and chemical conversions of chlorolignins. Environ. Sci. Technol. 1985; 19(12): 1219–1224.
30. SSVL-85 Project 4 Final Report: Production of bleached pulp. Stockholm, Sweden 1985.
31. Graves JW, Joyce TW, Jameel H: Effect of chlorine dioxide substitution, oxygen delignification and biological treatment on bleach plant effluent. Tappi J. 1993; 76(7): 153 –159.
32. Altnbas U, Eroglu V: Treatment of bleaching effluent in sequential activated sludge and nitrification systems. Fresenius Environ. Bull. 1997; 6: 103–108.
33. Hakulinen R, Salkinoja-Salonen M: Treatment of pulp and paper industry waste waters in an anaerobic fluidized bed bioreactor. Process Biochem. 1982; 17(2): 18–22.
34. Parkar WJ, Farquhar J, Hall ER: Removal of chlorophenols and toxicity during high-rate anaerobic treatment of segregated kraft mill bleach plant effluents. Environ. Sci. Technol. 1993; 27: 1783–1789.
35. Lettinga G, Field JA, Sierra-Alvarez R, Vanlier JB, Rintala J: Future perspectives for the anaerobic treatment of forest industry waste waters. Water Sci. Technol. 1991; 24(314): 91–102.

36. Dorica J, Elliot A: Contribution of non-biological mechanisms to AOX reduction attained in anaerobic treatment of bleached kraft effluent. Proceedings of 1994 International Environmental Conference. Tappi Press, Atlanta Georgia USA 1994; 157–165.
37. Hakulinen R: The Enso-Fenox process for the treatment of kraft pulp bleaching effluent and other waste waters of the forest industry. Paperi Ja Puu-papper och Tra 1982; 5: 341–354.
38. Haggblom M, Salkinoja-Salonen M: Biodegradability of chlorinated organic compounds in pulp bleaching effluents. Water Sci. Technol. 1991; 24(314): 161–170.
39. Hall ER, Onysko KA, Parker WJ: Enhancement of bleached kraft organochlorine removal by coupling membrane filtration and anaerobic treatment. Environ. Technol. 1995; 16: 115–126.
40. Raizer-Neto E, Pichon M, Rouger J: Decreasing chlorinated organics in bleaching effluents in an anaerobic fixed bed bioreactor. Proceedings of the 4th Int. Biotech Conf. in Pulp and Paper industry, Raleigh, North Carolina 16–19 May 1989; 279–287.
41. Fitzsimons R, EK M, Eriksson KE: Anaerobic dechlorination/degradation of chlorinated organic compounds of different molecular masses in bleach plant effluents. Environ. Sci. Technol. 1990; 24(11): 1744–1748.
42. EK M, Eriksson KE: External treatment of bleach plant effluent. 4th Int. Symp. on Wood and Pulping Chemistry, Paris 1987.
43. EK M, Kolar MC: Reduction of AOX in bleach plant effluents by a combination of ultrafiltration and biological methods. Proceedings of the 4th Int. Biotech. Conf. on Pulp and Paper Industry, Raleigh, North Carolina, 16–19 May 1989; 271–278.
44. Eriksson KE: Biotechnology in the pulp and paper industry. Water Sci. Technol. 1990; 24: 79–101.
45. Armentate PM, Kafkewitz D, Lewandowski G, King CM: Integrated anaerobic aerobic process for the biodegradation of chlorinated aromatic compounds. Environ. Prog. 1992; 11(2): 113–118.
46. Strehler A, Welander T: A novel method for biological treatment of bleached kraft mill waste waters. Water Sci. Technol. 1994; 29(5/6): 295–300.
47. Eriksson KE, Kirk TK: Biopulping, biobleaching and treatment of kraft bleaching effluents with white rot fungi. In: Comprehensive biotechnology vol. 3 (Cooney CL, Humphrey AE eds.) Pergamon Press, Oxford, 1985; 271–294.
48. Sundman G, Kirk TK, Chang HM: Fungal decolorization of kraft bleach plant effluent; fate of the chromopheric material. Tappi J. 1981; 64(9): 145–148.
49. Yin CF, Joyce TW, Chang HM: Kinetics of bleach plant effluent decolorization by *Phanerochaete chrysosporium*. J. Biotechnol. 1989; 10: 67–76.
50. Huynh VB, Chang HM, Joyce TW, Kirk TK: Dechlorination of chloroorganics by a white rot fungus. Tappi J. 1985; 68(7): 98–102.
51. Chang HM, Vasudevan B, Joyce TW, Taneda H: Degradation of hazardous organics by the white rot fungus *Phanerochaete chrysosporium*. Presented at annual meeting of the American Chemical Society, New Orleans, LA. Aug. 30–Sept. 4, 1987.
52. Bumpus JA, Tien M, Wright D, Aust SD: Oxidation of persistent environmental pollutants by a white rot fungus. Science 1985; 228: 1434–1436.
53. Eaton DC, Chang HM, Kirk TK: Fungal decolorization of kraft bleach effluents. Tappi J 1980; 63(10): 103–106.
54. Kirk TK, Shimada M: Lignin biodegradation: the microorganisms involved and the physiology and biochemistry of degradation by white rot fungi. In: Higuchi T (ed.) Biosynthesis and biodegradation of wood components, New York, Academic Press 1985; 579–605.
55. Kirk TK, Farrell RL: Enzymatic combustion-the microbial degradation of lignin. Annu. Rev. Microbiol. 1987; 41: 465–505.
56. Eaton DC, Chang HM, Joyce TW, Jeffries TW, Kirk TK: Method obtains fungal reduction of the color of extraction-stage kraft bleach effluents. Tappi J. 1982; 65(6): 89–92.
57. Campbell AG, Gerrard ED, Joyce TW, Chang HM, Kirk TK: The MyCoR process for color removal from bleach plant effluents: bench-scale studies. Proceeding of the 1982 Tappi Research and Development Conference, Asheville, North Carolina, Tappi Press Atlanta 1982; 209–214.

58. Pellinen J, Yin CF, Joyce TW, Chang HM: Treatment of chlorine bleaching effluent using a white rot fungus. J. Biotechnol. 1988; 8: 67–76.
59. Pellinen J, Joyce TW, Chang HM: Dechlorination of high molecular weight chlorolignin by white rot fungus *Phanerochaete chrysosporium*. Tappi J. 1988; 71(9): 191–194.
60. Yin CF, Joyce TW, Chang HM: Dechlorination of conventional softwood bleaching effluent by sequential biological treatment. In: Biotechnology in pulp and paper manufacture (Kirk TK, Chang HM ed.) Butterworth-Heinemann, Stoneham, MA, 1990; 231–244.
61. Guo HY, Chang HM, Joyce TW, Glasser JH: Degradation of chlorinated phenols and guaiacols by the white rot fungus *Phanerochaete chrysosporium*. In: Biotechnology in pulp and paper manufacture (Kirk TK, Chang HM ed.) Butterworth-Heinemann, Stoneham, 1990; 223–230.
62. Prouty AL: Bench scale development and evaluation of a fungal bioreactor for color removal from bleach effluents. Appl. Microbiol. Biotechnol. 1990; 32: 490–493.
63. Joyce TW, Pellinen J: White rot fungi for the treatment of pulp and paper industry waste water. Tappi Environ. Conference, Seattle, April 9–11, 1990.
64. Lankinen VP, Inkeronen MM, Pellinen J, Hatakka Al: The onset of lignin modifying enzymes; decrease of AOX and color removal by white rot fungi grown on bleach plant effluents. Water Sci. Technol. 1991; 24(314): 189–198.
65. Messner K, Ertler G, Jaklin-Farcher S, Boskovsky P, Regensberger U, Blaha A: Treatment of bleach plant effluents by the MyCOPOR system. In: Biotechnology in pulp and paper manufacture (Kirk TK, Chang HM ed.) Butterworth-Heinemann, Stoneham, 1990; 245–253.
66. Commorta MC, Santanna Jr. GL: Decolorization of kraft bleach plant E_1 stage effluent in a fungal bioreactor. Environ. Technol. 1992; 13: 65–71.
67. Jaklin-Farcher S, Szaker E, Stifter U, Messner K: Scale up of the Mycopor reactor. 5th Int. Conf. Biotechnol. in Pulp and Paper Industry, Kyoto. 1992; 81–85.
68. Kang CH, On HK, Won CH: Studies on the treatment of paper mill wastewater by *Phanerochaete chrysosporium*. In: Biotechnology in pulp and paper industry (Srebotnik E, Messener K, ed.) Facultas-Universitatsverlag, Vienna, 1996; 263–266.
69. Matsumoto Y, Yin CF, Chang HM, Joyce TW, Kirk TK: Degradation of chlorinated lignin and chlorinated organics by a white rot fungus. Proc. of 3rd ISWPC, Vancouver, 1985; 45–53.
70. Fukui H, Presnell TL, Joyce TW, Chang HM: Dechlorination and detoxification of kraft E_P effluent by *Phanerochaete chrysosporium*. In: Biotechnology in pulp and paper manufacture (Kuwahara M, Shimada M ed.) Uni Publishers Tokyo 1992; 75–80.
71. Livernoche D, Jurasek L, Desrochers M, Dorica J, Veliky IA: Removal of color from kraft mill waste waters with cultures of white rot fungi and with immobilized mycelium of *Coriolus vesicolor*. Biotechnol. Bioeng. 1983; 25: 2055–2065.
72. Royer G, Livernoche D, Desrochers M, Jurasek L, Rouleau D, Mayer RC: Decolorization of kraft mill effluent: kinetics of a continuous process using immobilized *Coriolus versicolor*. Biotechnol. Lett. 1983; 5: 321–326.
73. Bergbauer M, Eggert C, Kraepelen G: Degradation of chlorinated lignin compounds in a bleach effluent by the white rot fungus *Trametes versicolor*. Appl. Microbiol. Biotechnol. 1991; 35: 105–109.
74. Royer G, Desrochers M, Jurasek L. Rouleau D, Mayer RC: Batch and continuous decolorization of bleached kraft effluents by a white rot fungus. J. Chem. Technol. Biotechnol. 1985; 35B: 14–22.
75. Bajpai P, Mehna A, Bajpai PK: Decolorization of kraft bleach effluent with white rot fungus *Trametes versicolor*. Process Biochem. 1993; 28: 377–384.
76. Mehna A, Bajpai P, Bajpai PK. Studies on decolorization of effluent from a small pulp mill utilizing agriresidues with *Trametes versicolor*. Enzyme Microb. Technol. 1995; 17(1): 18–22.
77. Royer G, Yerushalmi L, Rouleau D, Desrochers M: Continuous decolorization of bleached kraft effluents by *Coriolus versicolor* in the form of pellets. J. Ind. Microbiol. 1991; 7: 269–278.
78. Archibald F, Paice MG, Jurasek L: Decolorization of kraft bleaching effluent chromophores by *Coriolus versicolor*. Enzyme Microb. Technol. 1990; 12: 846–853.

79. Pallerla S, Chambers RP. New urethane prepolymer immobilized fungal bioreactor for decolorization and dechlorination of kraft bleach effluents. Tappi J. 1996; 79(5): 156–161.
80. Roy-Arcand L, Archibald FS: Comparison and combination of ozone and fungal treatments of a kraft bleaching effluents. Tappi J. 1991; 74(9): 211–218.
81. Fukuzumi T: Microbial decolorization and defoaming of pulping waste liquors. In: Lignin biodegradation (Kirk TK, Chang HM, Higuchi T ed.) Vol. 2, CRC Press, Boca Raton, Florida 1980; 161–177.
82. Belsare DK, Prasad DY: Decolorization of effluent from the bagasse based pulp mills by white rot fungus *Schizophyllum commune*. Appl. Microbiol. Biotechnol. 1988; 28: 301–304.
83. Esposito E, Canhos VP, Duran N: Screening of lignin degrading fungi for removal of color from kraft mill wastewater with no additional extra carbon source. Biotechnol. Lett. 1991; 13(8): 571–576.
84. Lee SH, Kondo R, Sakai K: Treatment of kraft bleaching effluents by lignin-degrading fungi III. Treatment by newly found fungus KS-62 without additional nutrients. Mokuzai Gakkaishi 1994; 40(6): 612–619.
85. Lee SH, Kondo R, Sakai K, Sonomoto K: Treatment of kraft bleaching effluents by lignin - degrading fungi V. Successive treatments with immobilized mycelium of the fungus KS-62. Mokuzai Gakkaishi 1995; 41(1): 63–68.
86. Duran N, Dezotti M, Rodriguez J: Biomass photochemistry XV: Photobleaching and biobleaching of kraft effluent. J. Photochem. Photobiol. 1991; 62: 269–279.
87. Duran N, Esposito E, Innocentini-Mei L, Canhos VP: A new alternative process for kraft E_1 effluent treatment, a combination of photochemical and biological methods. Biodegradation 1994; 5: 13–19.
88. Kannan K: Decolorization of pulp and paper mill effluent by *Aspergillus niger*. World J Microbiol. Biotechnol. 1990; 6(2): 114–116.
89. Tono T, Tani Y, Ono KJ: Microbial treatment of agricultural industrial waste. Part 1: Adsorptions of lignins and clarification of lignin containing liquor by mold. Ferment. Technol. 1968; 46: 569–574.
90. Milstein O, Haars A, Majcherczyk A, Trojanowski J, Tautz D, Zanker H, Huttermann A: Removal of chlorophenols and chlorolignins from bleaching effluents by combined chemical and biological treatment. Water Sci. Technol. 1988; 20(1): 161–170.
91. Gokcay CF, Taseli BK: Biological treatability of pulping effluents by using a *Penicillium* sp. Fresenius Environ Bull 1997; 6: 220–226.
92. Bergbauer M, Eggert C, Kalnowski G: Biotreatment of pulp mill bleachery effluent with the *Coelomycetous* fungus *Stagonospora gigaspora*. Biotechnol. Lett. 1992; 14(4): 317–322.
93. Raghukumar C, Chandramohan D, Michael Jr. FC, Ready CA: Degradation of lignin and decolorization of paper mill bleach plant effluent by marine fungi. Biotechnol. Lett. 1996; 18(1): 105–106.
94. Parsad DY, Joyce TW: Color removal from kraft bleach effluents by *Trichoderma* sp. Tappi J. 1991; 74(1): 165–169.
95. Bollag JM, Shottleworth KL, Anderson DH: Laccase-mediated detoxification of phenolic compounds. Appl. Environ. Microbiol. 1988; 54: 3086–3091.
96. Klibanev AM, Morris ED: Horseradish peroxidase for the removal of carcinogenic aromatic amines from water. Enzyme Microb. Technol. 1981; 3: 119–122.
97. Ruggiero P, Sarkar JM, Bollag JM: Detoxification of 2,4-dichlorophenol by a laccase immobilized on soil or clay. Soil Sci. 1989; 147: 361–370.
98. Field JA: Method for biological treatment of waste waters containing nondegradable phenolic compounds and degradable nonphenolic compounds. EP Patent 1986; EP 238148.
99. Milstein O, Trojanowski J, Majcherczyk A, Tautz D, Zanker H, Huettermann A: Removal of chlorophenols and chlorolignins from bleaching effluent by combined chemical and biological treament. Water Sci. Technol. 1988; 20(1): 161–170.
100. Hakulinen R: The use of enzymes in the waste water treatment of pulp and paper industry - a new possibility. Water Sci. Technol. 1988; 20(1): 251–262.

101. Forss K, Jokinen K, Savolainen M, Williamson H: Utilization of enzymes for effluent treatment in the pulp and paper industry. In: Proc 4th International Symposium on Wood and Pulping Chemistry Vol. 1, Paris, France 1987; 179–183.
102. Call HP: Laccases in delignification, bleaching and waste water treatment. Patent 1991; DE 4137761.
103. Lyr VH: Enzymic detoxification of chlorinated phenols. Phytopathology 1963; 47: 73–83.
104. Hammel KE, Tardone PJ: The oxidative 4-dechlorination of polychlorinated phenols is catalyzed by extracellular fungal lignin peroxidases. Biochemistry 1988; 27: 6563–6568.
105. Roy-Arcand L, Archibald FS: Direct dechlorination of chlorophenolic compounds by laccases from *Trametes versicolor*. Enzyme Microb. Technol. 1991; 13: 194–203.
106. Paice MG, Jurasek L: Peroxidase catalyzed color removal from bleach plant effluent. Biotechnol. Bioeng. 1984; 26: 477–480.
107. Ferrer I, Dezotti M, Duran N: Decolorization of kraft effluent by free and immobilized lignin peroxidase and horseradish peroxidase. Biotechnol. Lett. 1991; 13, 577–582.
108. Farrel RL: Use of rLDM 1–6 and othe ligninolytic enzymes. WO 87/00564, 1987.
109. Farrel RL: Industrial applications of lignin transforming enzymes. Philos Trans. R. Soc. Lond. A 1987; 321: 549–553.
110. Davis S, Burns RG: Covalent immobilization of laccase on activated carbon for phenolic effluent treatment. Appl. Microbiol. Biotechnol. 1992; 37: 474–479.

9 Purification of Process Water in Closed-Cycle Mills

Most of the pulp and paper mills are located near the major waterways and have access to a large, uninterrupted supply of water. After using the water for pulp and paper production, these mills discharge the used water into the waterways as waste. The paper industry is currently experiencing increasing regulatory pressure to reduce the volume and toxicity of its industrial wastewater. During the past few years, the concept of system closure has been gaining popularity in the forest products industry, mainly because of its potential to drastically reduce or even eliminate liquid discharges and the associated water-quality problems, to separate and recycle valuable resources, and to preserve energy that can be used to reduce cost of production and amortize capital costs. In addition to savings on chemicals and heat, it would also result in saving on capital and operating costs for effluent treatment, reduced use of freshwater, and the possibility of locating new mills independent of water supply or effluent recipient.

These prospects motivate the emergence of new and innovative zero-liquid discharge, or closed-cycle systems which, when implemented in a mill, will enable the recovery of clean process water from the effluent and recycle it back into the mill.

The effluent-free concept, first proposed by Rapson in 1967[1] and pioneered by Great Lakes Forest Products Ltd, Thunder Bay, Ontario, in 1978, involves the internal recycling of all the contaminated process streams, to eliminate all pulp mill effluents.[2] Any closed-cycle technology must use a process for the removal of problematic compounds from the system. Industrial waste streams are very often complex aqueous mixtures of inorganic and organic solutes and their purification requires a combination of technologies with different specificities.

The closed-cycle or zero-effluent mill is an attractive concept, since it responds to short-term environmental legislation as well as future legislation which could regulate additional effluent parameters and require tertiary effluent treatment systems, extensive testing of effluent contaminations, testing of effects on receiving waters, etc. Purification of industrial wastewaters for process water reuse represents a major step in water savings and in closing the chemical processes. In view of the increasingly higher demands on the quality of mill-effluent discharge, it appears advantageous to upgrade the secondary effluent by a suitable tertiary polishing stage and reuse the tertiary effluent in mill operations. This concept of effluent management might lead to complete closure of the effluent cycle in certain mills. The main issues associated with effluent reuse are effects of residual constituents on mill operations, product quality, and corrosion aspects.

In the interest of reducing the environmental impact of pulp mill effluents, many companies are examining ways to modify effluent composition and reduce effluent volume. In the extreme, reduced effluent volume means a closed effluent-free pulp mill operation. The best and lowest-cost closed-cycle configuration is specific to every mill. However, in all cases, it is necessary to determine the appropriate combination of in-mill and ex-mill measures to achieve zero discharge at minimum cost. The ultimate goal of a closed-cycle mill is to have minimal impact on the environment.

In-Mill and Ex-Mill Measures to Achieve a Closed Cycle

The only source of wastewater is the use of freshwater. This fact indicates that the only way of decreasing the amount of effluent is to replace freshwater by already used so-called whitewater, treated or not. The reuse of process water is, in most cases, possible without any treatment and this kind of recycling is a normal routine in all pulp and paper mills.

The base for selection of an adequate treatment system is good knowledge of the process, as well as of the raw materials and their influence on the water to be treated. In addition, the result of the treatment has to be defined clearly, taking into account how the different components react to different treatment methods and which components it is necessary to separate from the water to make it reusable.

The method for water treatment has to be chosen based on the quality demands of the treated water. For example, is it necessary to remove suspended solids, suspended solids and colloids, or all components? A coarse treatment classification can be made from the kind of treatment result achieved. These are the separation of (1) fiber, leaving pigments and other fines in the water (2) suspended solids, giving clear water (3) colloidal substances (4) dissolved substances (5) inorganic salts. In principle, the water-treatment methods available to achieve the established goal are: (1) mechanical separation, which can be amplified by using chemical coagulation and flocculation, (2) biological treatment, (3) membrane filtration, (4) evaporation, distillation, crystallization (freeze concentration).

Internal treatment of whitewater was originally introduced to retain otherwise lost fiber and filler, which had a positive effect on the procces economy. Later, the separation of the solids from the water led to high-quality clear filtrate which could be used in showers on the wire part of the paper machine. In the case of several paper machines in a mill, the water savings can be phenomenal. Over and above, energy is also saved, as the temperature is maintained. However, one has to be careful. Clear whitewater contains organic substances which act as carbon sources for bacterial growth. Elimination of primary suspended solids does not eliminate the possibility of secondary suspended solids. The bacterial growth (slime) can be very harmful, as it can cause spray nozzles to become plugged. The slime problem can be taken care of by using chemicals, biocides, dispersants, etc. It can also be controlled by increasing the temperature of the system to higher than 50 °C or maintaining a pH of 8.5–9, but both these methods are not possible in many cases.

Internal whitewater treatment has to be economically feasible. Earlier, the recovery of valuable solids justified a simple so-called save-all treatment; but now, the price of water has increased as stringent rules and limitations are applied on effluent quality and the demand for freshwater quality has increased. Water used for cooling and sealing purposes should be kept clean, but the situation is often the opposite. Whitewater is allowed to intrude into the system and the water is rejected as waste. In this case, it is easy to find the reason for pollution and take the necessary preventive measures. Clean cooling water can be reused. The retention-disturbing compounds should not be allowed to enter the paper machine system. In order to restrict the amount of disturbing components like anionic trash in the paper machine system in combination with low freshwater usage, it is necessary to isolate the pulping department from the paper mill. This can be achieved by high consistency pulp transfer in combination with indirect heat transfer. The filtrate from pulp dewatering is treated for efficient removal of dissolved and dispersed substances in the first place. The counter-current principle needs to be applied, which means that the clean (or treated) water is used on the paper machine and that excess, preferably clarified water, is pumped to the pulping department. The pulp is transferred to the paper machine at high consistency (15–35%) in order to reduce the transportation of disturbing agents. The most polluted water should be collected at the pulping department and, if so decided, treated biologically for reuse. While implementing this scheme, it is important to take care of the energy distribution through out the production line. Pulp at high consistency does not transfer heat, water does. The even distribution of energy has to be taken care of by heat exchangers, making it possible to maintain the correct temperature on the paper machine.

Separation of Suspended Solids (SS)

In cases when water has to be free from SS, both particle size and concentration have to be defined as, according to the definition, all particles above the size of 1–5 μm are SS. Separation of SS is dependent on the technique employed. Screening is used when only coarse material has to be separated, where as clarification is used when small particles are also to be separated. Sedimentation is a widely used clarification method. However, care has to be taken not to overload the system. Particles settle, according to Stoke's law, in the gravity field at a certain velocity depending on their size and density. In order to achieve good results, flocculation aids are used. Floatation is feasible if the solids to be separated have a lower density than water or when small amounts of oil-like droplets have to be eliminated from the water. Clarification by floatation is achieved by the rising force of air bubbles attached to the particles.

Water containing suspended solids can be filtered using a well-defined mesh or fine screen. The filtrate quality depends on the mat of fibers formed and usually water of different degrees of clarity is obtained from a drum filter or disk filter. Utilizing a bed of sand or a combination of different materials of different particle sizes, it is possible to effectively separate suspended solids from water. However, this kind of filtration needs backwashing because the capacity of this type of systems for solid accumulation is limited. Sometimes, the use of semipermeable membranes, especially the microfilters, is helpful in separating very fine suspended particles.

Removal of Colloidal and Dissolved Substances

One of the methods is chemical precipitation. By applying certain poly-valent metal ions, like Fe^{2+}, Fe^{3+}, Al^{3+}, Ca^{2+}, or Mg^{2+}, it is possible to destabilize a part of the colloids and achieve coagulation, which leads to adsorption on solids. Formation of metal-hydroxide enhances the separation and resulting solids are separated in conventional ways. The efficiency is dependent on the process behind the water to be treated. Lignin and fat-type colloids are partially separated but carbohydrates remain in solution. Efficiency, measured by COD or BOD, shows great variations. The application of metallic cations for coagulation is disturbed by complexing agents such as EDTA or DTPA used in the bleaching process.

Colloidal and dissolved substances can partially be removed from wastewater by membrane filtration, known as micro-, nano-, and ultrafiltration. The efficiency, measured as COD and TOC removal, depends on the membrane quality (its material and openness) and volume reduction factor, among other parameters. Following are the general observations about the performance of membrane filtration processes. (1) Typical flux for ultrafiltration (UF) is 5–10 m^3/m^2/day at a pressure of 1–5 bar, and for nanofiltration (NF), it is 1.5–2.5 m^3/m^2/day at 10–15 bar. (2) Reductions of dissolved COD by UF is 10–50% depending on the source of white-water or effluent; the corresponding TOC reductions are 15–50%. (3) Applying NF on the UF permeate, the reductions in COD and TOC are 65–95% and 70–90%, respectively, of the original values. (4) NF, also called loose reverse osmosis, removes inorganics, which is indicated by a reduction of specific conductance in the range of 40–85%. Thus, membrance filtration seems to be feasible as part of an internal water-treatment system. Ultrafiltration has been successfully implemented for the recovery of pigment and binder in connection with coating of board and paper.[3] Several types of whitewater and effluent have been studied on a pilot plant in connection with the Eureka project on Improved water reuse in the pulp and paper industry.[3]

It has been demonstrated that the biological treatment of whitewater improves the clarification considerably. The reason for this is that the organic compounds stabilizing the fines in suspension are consumed by the bacteria present in the bioreactor. The primary carbon sources are the carbohydrates and extractives, the group responsible for poor first pass retention, poor dewatering on the wire, and bad smell in the product. The introduction of internal biological treatment will improve the efficiency of conventional wastewater treatment systems. In addition, it will eliminate the possibility of secondary solids, as the carbon source is consumed in the bioreactor. However, the clear filtrate from the pulping has to be cooled before biological treatment if the temperature exceeds 30–35 °C, supplemented with phosphorous and nitrogen, and the pH has to be adjusted to neutral (7.0 ± 0.5). Several types of bioreactors and methods are available and the selection of optimum technology (aerobic or anaerobic) will depend on a case-by-case evaluation. In many mills, biological effluent treatment is already implemented, which can be easily utilized for this procedure.

In order to make the biologically treated water useful in the process replacing freshwater, suspended solids and possible color have to be removed. This is most efficiently achieved by UF. It has been experienced that the flux through the UF membrane increases considerably compared to water prior to biological treatment which enables higher capacity for a given filtration area at constant pressure drop and temperature. The bacterial biomass partially binds chlorides, which has been shown by the fact that the amount of chloride in dry matter was found to to be 1500–3000 ppm for biomass grown in a whitewater where the chloride concentration was 20–40 ppm.[4] However, in the conditions prevailing in biological treatment, heavy metals are normally precipitated as hydroxides, provided that the complexing additives like DTPA and EDTA are not used in excess. The biodegradability of these substances still remains questionable, which can be evaluated using the existing bioreactors.

With the combination of biological and efficient posttreatment (membrane or chemical precipitation), it is possible to replace freshwater in the process. However, there are few important requirements that need to be considered. For example: (1) Bioprocesses need nutrients in the form of nitrogen and phosphorus. These elements have to be added before the biological treatment in such a manner that the concentration in the treated water is lower than the incoming wastewater. (2) The biological process takes place at neutral pH. (3) The posttreatment is needed in order to separate secondary suspended solids (bacterial cells). With proper operation, the treated water will be free of components causing biological activity, retention and dewatering disturbances, and thus be of acceptable quality for recycling.

The above treatments do not remove inorganics from the water and the overall salt content, measured as specific conductivity, has to be kept at, or below, an acceptable limit by extracting part of the treated water as effluent. If waste heat of sufficient temperature is available, partial evaporation at low pressure can be feasible. The resulting salt solution has, of course, to be disposed of in an acceptable way. Alternately, the inorganic-rich stream may be cooled to crystallize out the salts.

Development of System Closure Process

1. Mechanical Pulp and Paper Mills

The yield of mechanical pulp, calculated on the dry wood basis, is normally in the range of 95–98%. This means that 20–50 kg/ton of wood is lost into the water phase, in the form of dissolved and dispersed substances. Simple carbohydrates and fats are dissolved and dispersed in the process water, leading to a certain concentration of potentially harmful substances. The extractives of wooden components containing fats (triglycerides), fatty acids (hydrolyzed fats), resin acids, sterols, lignans, and stilbenes, etc. amount to more than 10 kg/ton. Some of the substances dispersed in process water are detrimental. The term detrimental substances is commonly used in the paper industry to collectively describe all substances that may be present in a paper-making system and negatively influence the paper-making process. Fatty substances are considered to be among the most detrimental, as only these will form pitch deposits.[5]

The production of mechanical pulp often includes bleaching by hydrogen peroxide in alkaline conditions. The bleaching with about 20 kg H_2O_2/ton of pulp plus an equal amount of NaOH, in addition to an extra stabiliziing alkali in the form of sodium silicate, leads to further yield losses. Approximately 10–20 kg material/ton of pulp is dissolved in the bleaching process.

The majority of the dilute aqueous effluents generated by thermomechanical pulp (TMP) and chemi-thermomechanical pulp (CTMP) mills contain suspended solids, extractive compounds, lignin, cellulose and hemicellulose fragments, some low molecular weight compounds, metal ions, and residual chemicals. During the production and further treatment of mechanical pulp, wood components are released into the process waters of the paper mill. Since mechanical pulps are seldom washed, dissolved and colloidal substaces (DCS) are carried over to the paper machine, where they may interfere with the paper-making process. In most mills, efforts are being made to minimize water consumption due to stiffening environmental regulations. This is resulting in increased levels of DCS in process water systems.

Whitewater system closure for mechanical pulp and paper mills can be beneficial in saving fibers, filler, and fines, in minimizing freshwater and energy consumption, and in reducing effluent treatment costs. On the other hand, the closure of mechanical newsprint mills inevitably leads to increase in contaminant concentrations in whitewater systems. These contaminants may cause various operational problems in the paper machine as well as a decrease in product quality. Resin and fatty acids (RFA) are considered to be among the most detrimental, as their salts readily form deposits, causing pitch on the paper machine or on the product. Excessive fungal and bacterial growth produces slime that can cause local discoloration, weakening of the final product, and odor problems. Increased organic levels decrease the tensile strength.[6] These problems currently limit whitewater system closure. The development of satisfactory methods to purge troublesome contaminants from mill systems is criticial to closed-cycle development.

Several mills are currently attempting to achieve complete closure by treating the effluent streams by different techniques. So far, the only successful zero-effluent discharge mills are the two Canadian bleached chemithermomechanical pulp mills, viz. Millar Western at Meadow Lake, Saskatchewan,[7,8] and Louisiana-Pacific at Chetwynd, British Columbia.[9] They have been designed for zero-effluent operation. The mills utilize different schemes for effluent treatment and reuse. The basic design concept for the water recovery system at Meadow Lake is shown in Fig. 9.1. The combined mill effluent is treated by conventional gravity filtration followed by mechanical vapor recompression (MVR), evaporation, and steam-driven concentration. After evaporation of the effluent, the distillate from the evaporators comes back to the distillate equalization pond (DEP), where bioactivity destroys the small fraction of low molecular weight volatile organics left in the water storage reservoir, where it is stored until it reenters the mill for reuse in the process. A portion (about 60%) of the distillate goes directly back to the pulp mill as hot process water at 65 °C. The total hot distillate flow to the pulp mill is used as wash water on the back end of the mill with full counter-current flow through to the front end. The mill can run for 20 days from the clean water inventory in the water storage reservoir. This process has been in operation since February 1992. A net loss of water exists primarily because of evaporation from the ponds, plus a cooling tower and a heat recovery stack. Makeup water is brought in at an average rate of approximately 2 m^3/ton pulp. Since there are no discharges to receiving waters, there are no concerns with meeting the government requirements.

MVR technology was selected for the following reasons: (1) it was more cost-effective; (2) it provided greater flexibility by enabling either parallel or series operation; (3) it did not require high shell side temperatures and thus was judged to have a lower tendency for scaling; (4) it did not produce a large amount of low-pressure steam as an end product that would have to be condensed or vented. The concentrate is incinerated in the recovery bioler, where the inorganic material comes out in the form of smelt containing 85–90% sodium carbonate. After converting the smelt into the form of ingots, it is disposed of in landfilling, but can be recausticized to produce sodium hydroxide or can be electrohydrolyzed to produce caustic and sulfuric acid.

The Chetwynd mill has played a pioneering role in closed-mill technology based on freeze crystallization, first coming online in June 1991.[9] Although it worked, it was not deemed sufficiently reliable due to some problems hampering the crystallization process, especially the formation of ice layers in the

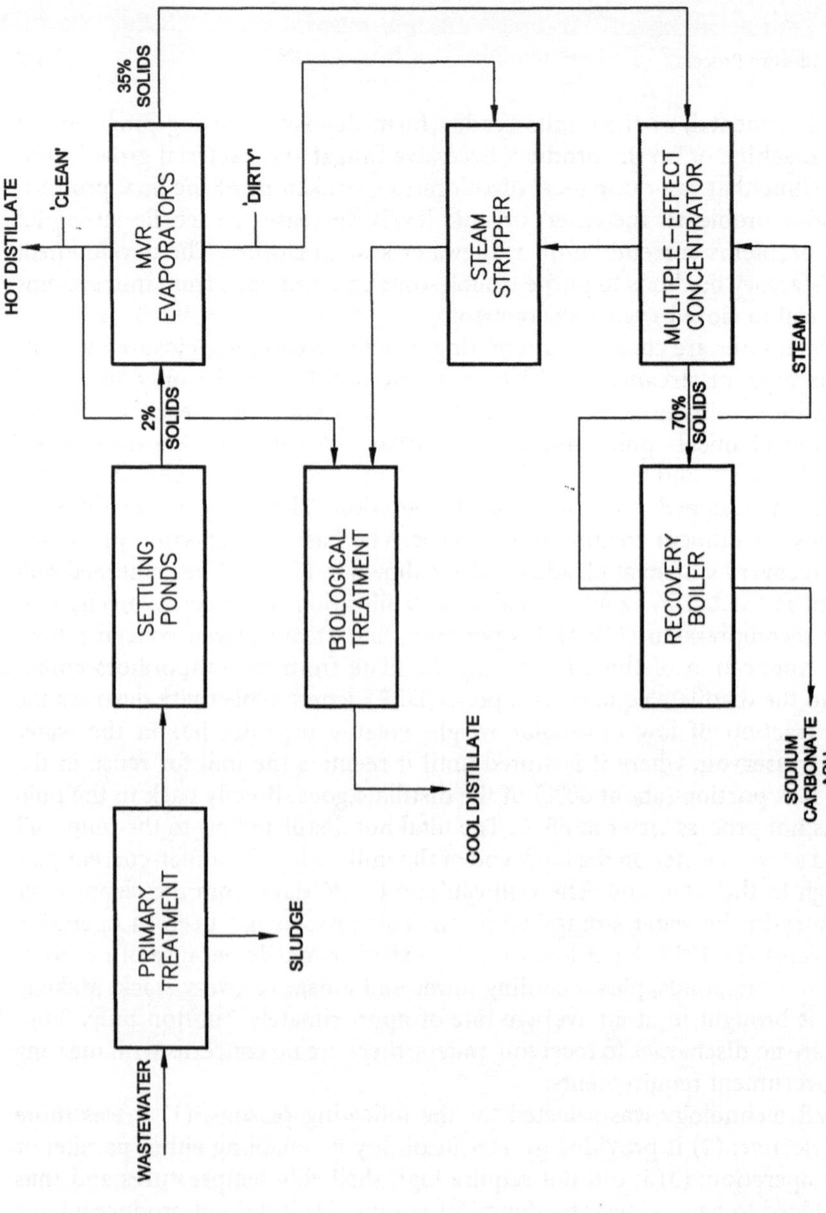

Fig. 9.1. The zero liquid-effluent process concept at Meadow Lake. Reproduced with permission from Tappi Press, Ref. 8

crystallizer tubes reducing the heat transfer to zero and plugging the tubes. Therefore, in March 1993, the system was shut down and the crystallizer was converted to a mechanical vapor recompression evaporator similar to the Millar Western Pulp Mill at Meadow Lake. The revised system came online in July 1993. The evaporators produce clear, hot product water for mill use. The condensed steam is sent to the distillate equalization pond for aeration and bacterial treatment. The pond has a capacity of 4 million gal, with a retention time of 11 days. Water from the clean water pond is further clarified before returning to the mill. Natural evaporation from the distillate equalization pond is the primary source of the mill's water loss. Other losses occur at the Venturi scrubber on the hog fuel boiler and with sludge removal. Makeup water from the wells is about 1 m^3/ton of pulp.

An uncoated groundwood speciality paper mills, Hennepin Paper Co. Little Falls, MN, undertook a scheme in July 1990 to eliminate the discharge of industrial wastewater to the Mississippi River.[10] The wastewater recycling system consisting of pumps, surge tank and filtration system is shown in Fig 9.2. The process wastewater is treated in the activated sludge treatment plant. To achieve the stated goal of absolute zero discharge, a wastewater polishing system treats excess wastewater. The polishing system consists of a sand filter and an activated-carbon filter that draws water from the surge tank. The sand filter removes solids from the continuous slip-stream and then returns the filtered water to the freshwater tank. The activated carbon filter removes the color and the dissolved organics in the wastewater. Polished water undergoes either rerouting to the closed-loop cooling water system to replace water makeup or discharge to the river. Together, these filters remove the suspended solids and color in the wastewater. Eventually, this might replace several river water or city water usage points within the mill. There were reductions in BOD and TSS discharges of 51% (from 269 371 to 131 215 lb) and 79% (from 468 394 to 981 236 lb), respectively. By pumping increased amounts of treated wastewater into the freshwater tank using the closed-loop system, the mill's process and cooling water consisted of nearly 100% recycled wastewater. This has also created some unexpected operational problems. The higher levels of SS and biological contaminants contained in this wastewater increased the wear on pumps and plugged pipes and showers. In addition, the conductivity of the process wastewater increased during recycling. This created a more corrosive environment for much of the piping and equipment. Installing a closed-loop system to provide cooling water for the heat exchangers ultimately solved the biofouling problems. This system uses river water for makeup. In addition, blowdown from the closed-loop cooling system to prevent solids buildup is now used for headbox and breast roll showers. By installing the closed-loop cooling system, the volume of water going through the wastewater treatment plant decreased and treatment efficiency rose. The quality of wastewater now generally does not cause severe biofouling problems. There is no cost justification for disinfection.

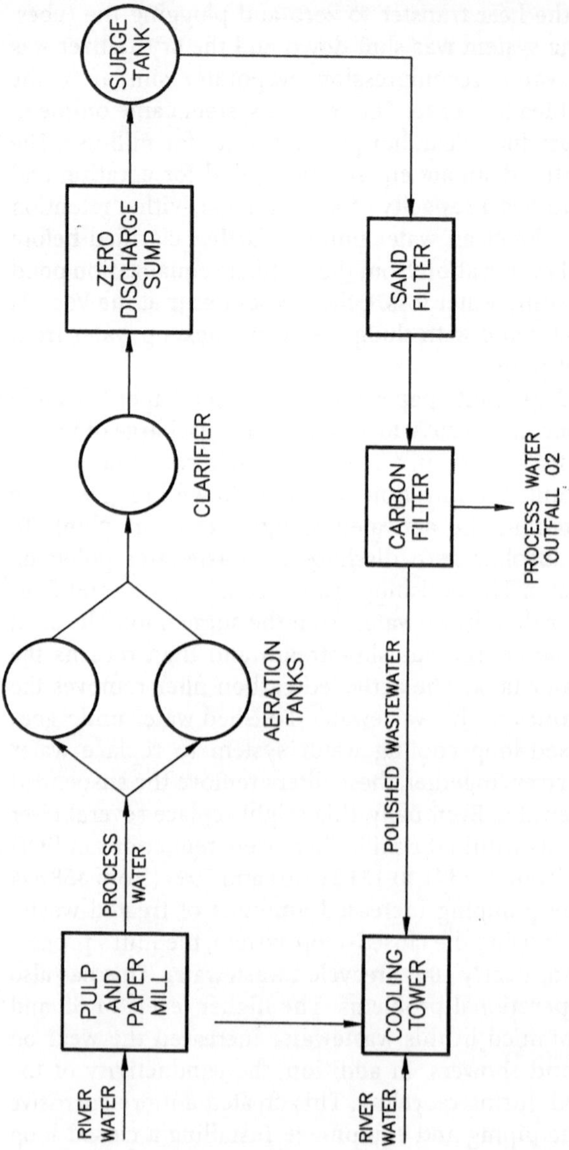

Fig. 9.2. Closed-loop wastewater recycling system. Reproduced with permission from Tappi Press, Ref. 10

Extensive recirculation has been incorporated, with water passing through the hierarchy of cold to warm and clean to dirty, in the Australian Newsprint Mills Ltd. Albury, on the New South Wales-Victoria border, which used 75% TMP and 25% recycled fiber.[11] As the mill is located near the head waters of the Murray River, Australia's most important waterway, the environmental impact required close consideration. The reduction in water consumption to a level of 24 m^3/ton has been brought about by increased use of process whitewater, closing up of cooling water circuits with temperature control, and reuse of treated wastewater in a number of applications including forest irrigation. Treated wastewater is used to wet down logs in the woodyard, for chemical makedown in the wastewater treatment plant, showers on sludge belt presses, wastewater treatment hose stations, and lawn and garden irrigation. Approximately 25% of the mill freshwater requirement is for cooling purposes and this normally bypasses the wastewater treatment plant as it is not contaminated. Several of these circuits are now on temperature control, which reduces freshwater makeup. The overall result is a reduction in freshwater use. The raw water coming into the mill is supplemented with reused treated wastewater so that freshwater is 15% treated wastewater. Reduction in water consumption has also reduced the loading on the wastewater treatment plant. However, increased recycling of whitewater and reuse of treated wastewater results in increased total dissolved solids, which can limit its suitability for irrigation, and ultimately can impose a ceiling on the ability to conserve further water, unless desalination is economically feasible. The mill has not yet reached that ceiling.

Recently, Lagace et al.[12] developed a physicochemical process for treating the effluent of a TPM newsprint mill, so that the mill could operate without any effluent discharge. It was found that an optimum combination of magnesium oxide, lime, and potassium phosphate was the most effective in removing the contaminants from the effluents. This method proved to be effective in removing extractives almost completely, and achieved low levels of color and turbidity. This physicochemical process shows promise as part of a closed-cycle treatment process for TMP newsprint mills, if a suitable and cost-effective process for sludge disposal can be developed. The proposed zero-effluent scheme is shown in Fig. 9.3. In this, a biological treatment is required to remove the low molecular weight organic contaminants that are not removed by the physicochemical treatment. The short-chain organic material is readily metabolized in a relatively low-cost trickling filter. The trickling filter would also act as a cooling tower to cool down the effluent as heat builds up with zero discharge implementation. pH neutralization would be required prior to biological treatment. An ion exchange unit might become neccesary for sodium removal after a high degree of recycling has been achieved because sodium is not removed in the physicochemical process and could build up to levels that are harmful to paper-making operations or product quality. It would be used to treat a small portion of the mill effluent to bring down the sodium to an acceptable level.

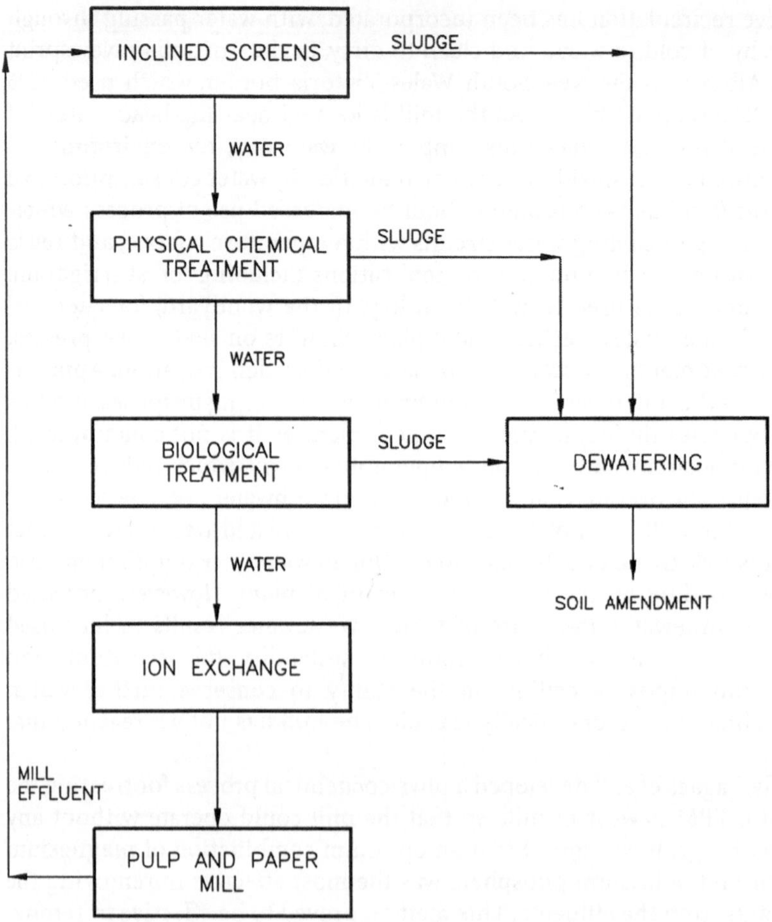

Fig. 9.3. Proposed treatment scheme. Reproduced with permission from Pulp and Paper, Canada, Ref. 12

The potential feasibility of recycling biologically treated effluent from a newsprint mill back to the mill through the freshwater purification plant was evaluated by Dorica et al.[13] at a mill site using full-scale primary treatment, followed by pilot plant-activated sludge treatment. Coagulant-aided primary treatment plus biological treatment of newsprint mill effluent removed 95–99% of both BOD and resinous fatty acids and 91% of COD. The final treated effluent samples were nontoxic to trout (96 h $LC_{50} > 100\%$). However, the final treated effluent exceeded the intake surface water in terms of COD, color, sulfate, aluminum, calcium, chloride, and silica. The final treated effluent was proposed to be recycled to the newsprint mill. It was assumed that half of the freshwater feed to the water purification plant would be replaced by biologically treated effluent. Upon dilution of final treated effluent with surface water, the residual levels of most metals and ionic species decreased in accordance with a straight-line dilution response. However, a synergistic removal (9–30%) was observed with COD, K, Al, Fe, and Mn, and was probably caused by interactions between the dissolved components of the effluent and surface water.

An integrated in-mill and out-mill analysis of water and effluent management options used by Bowater Mersey Paper Co. Ltd., Liverpool, NS, Canada (a TMP-based newsprint mill), have been described by Wilson and Purdue,[14] to develop an efficient and economical approach to comply with effluent regulations. It involves flow segregation – one stream for contaminated process effluent, and other for noncontact cooling waters and vacuum seal waters. The noncontact cooling waters from the steam plant and paper mill were collected and directed to the vacuum pump seal water chest that provides seal water to all of the vacuum pumps in the mill. This resulted in the reduction of mill water use by 9 m^3/ton. The flow segregation, reuse, and flow reduction measures were implemented as part of primary and secondary effluent treatment. The flow of combined process effluent was reduced from 75 m^3 to 23 m^3/ton in the process sewer and 30 m^3 in the clean sewer. This had a dramatic effect on the design and cost of external effluent treatment.

Based on a theoretical study, it has been shown that it is possible to build an effluent-free newsprint mill.[15] There is no difference in capital costs and only a minor difference in operating costs between a noneffluent and a conventional mill. The treatment facilities include biological treatment, flocking with chemicals, and filtering through a sand bed filter and ultrafiltration for other process waters. A water consumption of 7–10 m^3/ton of paper is sufficient for a newsprint mill. However, the mechanical pulp must be washed, otherwise dissolved organic material will build up in the whitewater, causing pitch and slime problems and disrupting the first pass retention. The water circulation can also be closed at an existing mill, but this is unlikely to be profitable if the mill already has an external effluent treatment system in operation.

Lindholm and Jantunen[16] have proposed a solution for effluent-free operation of a high-yield pulp (mechanical) mill which consists of internal or external biological treatment of whitewater filtrate or effluent combined with absolute removal of suspended solids from the water. Water treated in this way can replace freshwater through out the process. However, increased process water recycling will cause an accumulation of inorganic salts in the system. The potential danger of the increased salt concentration can be managed, for instance, by treating a side flow of water by evaporation or nanofiltration.

Potential of ultrafiltration (UF) and/or biological treatment to remove contaminants like TDS, COD, TOC, and resin and fatty acids from simulated whitewaters of TMP and CTMP mills have also been studied.[17] Biological treatment in a 10-l sequencing batch reactor (SBR) operated at 20 or 30 °C was found to be effective for the removal of RFA (>90%), total and soluble COD (>70%). However, only 35% of the TDS content was reduced in the bioreactor.

Furthermore, ultrafiltration of biologically treated process water resulted in additional TDS, soluble COD, and TOC removal.

In a pilot-scale effluent detoxification study at an integrated newsprint mill, following the anaerobic treatment system, an aeration basin is included to further metabolize dissolved organics which are not effectively removed by anaerobic treatment alone.[18] The biologically treated effluent is subsequently treated using UF and reverse osmosis (RO) membrane technologies. The UF membrane system removes medium to high molecular weight organics from the effluent, and fine suspended solids carried from the biological treatment. The UF concentrate can be returned to the anaerobic treatment system for further biodegradation of organics. Some of the permeate from the UF system can be used directly back in the mill without RO treatment. However, to have a completely closed system, an evaporation/drying system is required. This produces a clean distillate for direct reuse in the mill as hot process water.

Based on the previous study,[17] a membrane biological reactor (MBR) treatment of recirculated newsprint whitewater was studied by Tardif and Hall[19] using a lab-scale MBR consisting of a 10-l aerobic reactor and a cross-flow tubular UF unit (Permaflow-1 of 75 kDa molecular weight cut off) as shown in Fig. 9.4. MBR is a novel process which integrates biological oxidation with ultrafiltration for treating wastewaters, and has been used to describe different systems in which both biological treatment and membrane filtrations are used. MBR consists of an aerobic suspended growth biological reactor in which a high concentration of biomass is maintained by separating the solids from the liquid with an ultrafiltration membrane. The system is similar to an activated sludge system except that the settling stage is replaced by an ultrafiltration stage. It was operated at temperatures ranging from 40–55 °C and HRT of either 2.8, 1.1, or 0.7 day, applying 11.5 psi transmembrane pressure. The MBR was found to be stable under the varied operating conditions and was particularly effective at removing over 99% of RFA (on average), 81% total COD, 76% soluble COD, and 74% dissolved organic carbon. The reduction of the whitewater cationic demand was good (57%) while the TDS removal was only fair (25%). However, 67% of the volatile fraction of the influent TDS was removed. The high removal efficiencies of RFA may be due to the fact that all suspended solids were retained by the UF membrane. RFA have a tendency to be adsorbed by the biosolids, which cannot pass through the UF membrane. This and other inherent advantages of the MBR suggest much potential for this process to treat recirculated whitewater internally in pulp and paper mills.

The interactions of dissolved and colloidal organic substances with a high-charge density, low molecular weight poly-diallyldimethylammonium chloride (DADMAC), in suspensions of unbleached and peroxide-bleached TMP have also been investigated with a view to improve the whitewater quality for recycling.[20] The optimum polymer dose for destabilization of the colloidal substances was nearly twice for bleached TMP compared to the dose required for suspensions of unbleached pulp. Poly-DADMAC formed aggregates according to the patch flocculation or charge neutralization mechanism with colloidal

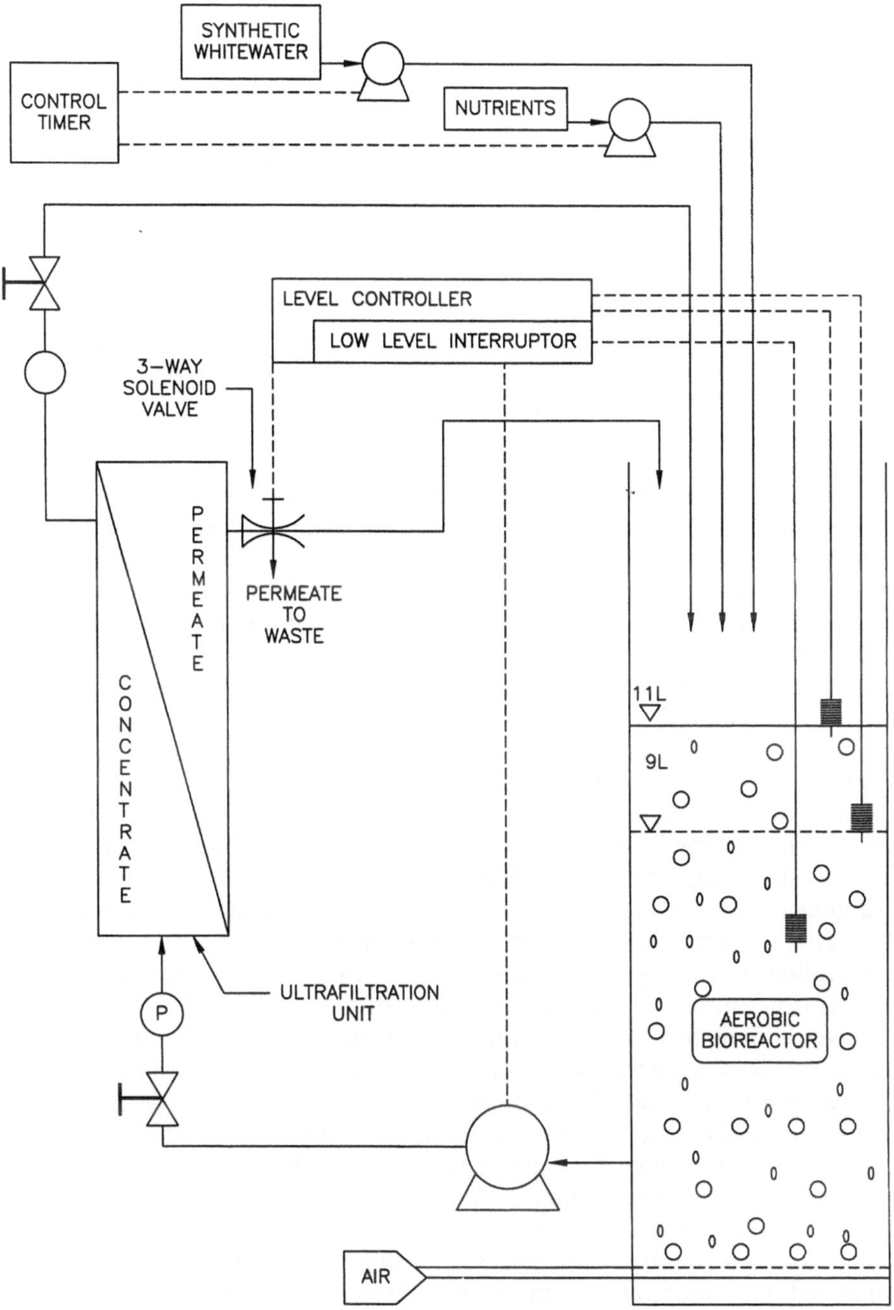

Fig. 9.4. Aerobic membrane biological reactor. Reproduced with permission from Tappi Press, Ref. 19

lipophylic substances and dissolved anionic hemicelluloses. The unstable aggregates formed could be removed by centrifugation. The colloidal substances were destabilized when the electrophoretic mobility of the suspension was close to zero.

The use of clathrate hydrates for the concentration of effluents from four different high yield mechanical pulp mills, viz. bleached TMP, unbleached TMP, combined bleached CTMP/TMP, and bleached CTMP, have been studied on a laboratory scale to assess the feasibility of water recovery and to generate the preliminary data.[21] Clathrate hydrate concentration is a crystallization process similar to freeze concentration. In clathrate concentration, the ice-formation step is replaced by clathrate hydrate crystal formation. Clathrate hydrates are ice-like crystalline nonstoichiometric compounds.[22] They are formed from a mixture of water and low molecular weight substances such as carbon dioxide, methane, ethane, and propane. Hydrates can form at temperatures several degrees above the normal freezing of water, thereby decreasing the energy requirements compared with freeze concentration. These temperatures, however, are not high enough to cause corrosion, scaling, or loss of volatile substances, as is the case with evaporation. Based on environmental, economical, and operational considerations, a C_3H_8 or a C_3H_8-CO_2 mixture is a suitable clathrate hydrate former to use in the process. The conditions in the TMP and CTMP effluents did not differ significantly from those in water but in the bleached CTMP effluent greater pressures were required to form hydrate. In spite of the complexity of the effluents, the characteristics of hydrate nucleation are similar to those in water. Clathrate process could operate at 8–10 °C above the temperature of the freeze concentration process.

Technoeconomic assessment of different closed-cycle technology alternatives for an existing TMP-newsprint mill has been carried out by Gerbasi et al.[23] The following three technologies were assessed by them: biological-membrane treatment, freeze crystallization, and evaporation concentration. According to them, the freeze crystallization process is more expensive than evaporation, while the biological-membrane treatment scheme, assuming the RO concentrate is discharged to a receiving water body, is comparable. However, adding the cost to evaporate and dry the RO concentrate stream will render this option significantly more costly than evaporation. Consequently, both the freeze crystallization and biological membrane options do not indicate any cost advantages over evaporation. Given the questionable technical feasibility of the biological-membrane system and the operating difficulties encountered with freeze crystallization, the evaporation based closed-cycle process is the preferred closed-cycle option. However, an evaporation-based closed-cycle process is by no means an economically attractive alternative compared to biological treatment and discharge for an existing TMP-newsprint mill. Promising closed-cycle technology alternatives will most likely involve treatment schemes which have the following general characteristics: they should be relatively insensitive to effluent flow rates, and they should treat effluents to remove only the level of contaminants necessary for the pulp and papermaking process.

2. Kraft Mills

Environmental regulations are forcing pulp producers to eliminate or significantly reduce most types of chlorinated organic discharges. In the United States, the proposed AOX limit for existing non-TCF bleached kraft mills being 0.156 kg/metric ton pulp.[24,25] The chemical oxygen demand (COD) limit is proposed to be 25.4 kg/metric ton pulp. Bleach plant closure, through filtrate recycle, to the recovery cycle, is becoming attractive in light of these new regulations. Complete closure, however, is difficult with chlorine-based sequences, because the resulting bleach liquor chloride level is a threat to the recovery boiler. Mills with oxygen delignification and a low chlorine dioxide charge may be able to close the bleach plant if chloride levels in the recovery cycle are monitored carefully. Fiber line and bleach plant process changes that reduce elemental chlorine demand, such as extended cooking, oxygen delignification and high chlorine dioxide substitution, have been used for several years; but lower AOX limits, changing markets surrounding the use of chlorine chemicals, and the drive for bleach plant closure to reduce COD emissions have focused interest on TCF bleaching. The TCF bleach plant uses no chlorine-based bleaching chemicals, eliminating concerns about dioxins and furans and the more general measurement of chlorinated organic compounds, AOX. An added benefit of TCF processes is the high potential for complete reuse of bleach plant filtrates in the recovery cycle.

The North American industry is destined to focus on environmental compliance in the near future and mill closure further into the future. In contrast, the Scandinavian response to the European demand for green products has been to rapidly convert a portion of output to totally chlorine-free (TCF) pulp production. In some cases, the success of this approach, coupled with environmental marketing, is leading companies such as Sodra Cell towards becoming exclusive TCF pulp producers.

The Scandinavian approach focuses more on in-plant technologies, such as oxygen delignification, and less on out-plant treatment facilities. In this context, mill closure is progressively being implemented in both TCF and elementally chlorine-free (ECF) bleach plants on both sides of the Atlantic. There are two basic approaches to mill closure that relate to the building of the bleaching effluent, viz. integrated recovery and separate recovery (Fig. 9.5).[26] Superimposed on these approaches are the bleaching options that can be applied. Both ECF and TCF routes are being pursued in North America and elsewhere. In these two bleaching routes to closed cycle, ECF and TCF, the risks are taken in different ways. For the ECF route, the risk is in the process. Extensive work has shown that on a small scale, the ECF route is feasible and

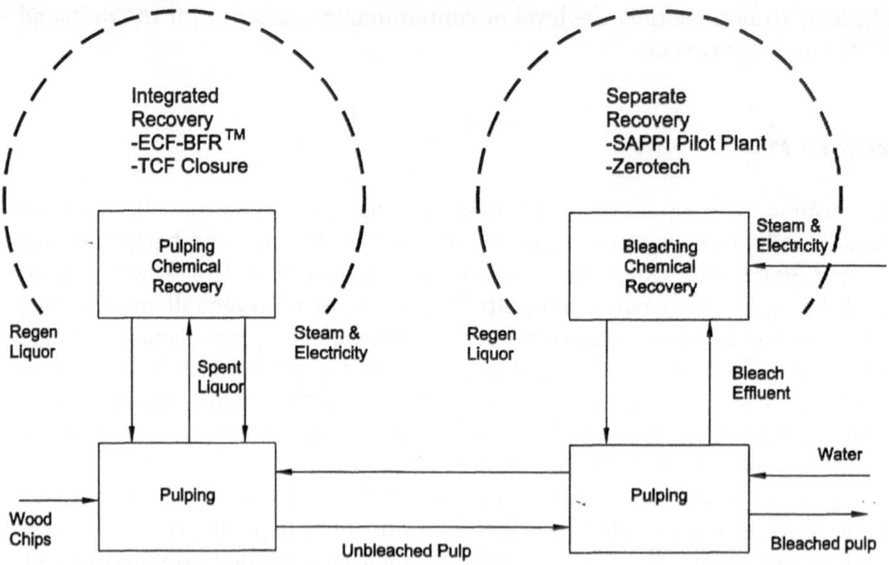

Fig. 9.5. Closed-cycle approaches. Reproduced with permission from Appita, Ref. 26

the risks minimized. However, it is not until full mill-scale operation that this becomes a viable reality and the risks of scaling, corrosion, and equipment suitability are solved. The risk of converting a TCF pulp mill to closed cycle is predominantly in product quality. The attainment of full brightness for TCF pulp, particularly softwood, has been difficult and the added complexity of closed-cycle operation with the counter-current filtrate streams needs careful evaluation. Also, there are major questions on chemical consumption.

The first step taken by the National Swedish Environmental Protection board towards a closed kraft mill was oxygen bleaching along with other internal measures, screen room closure, improved washing, condensate stripping, and increased chlorine dioxide substitution.[27] Swedish pulp mills are currently recycling the E-stage liquors to the recovery cycle. With O_2 prebleaching and high chlorine dioxide substitution in the chlorination stage, E-stage effluent with and without membrane treatment has been burned in the recovery furnace.[27] Chlorine was expelled from the system by liquor losses and flue gases to such an extent that accumulation never occurred. No corrosion effect was observed. The only problems experienced were the deposit and plugging in the recovery boiler and dilution to black liquor.

Anderson and Lindberg[28] and Lund[29] reported a nonpolluting bleach plant by counter-current washing and decolorizing and reusing the extaction-stage effluent as wash water in the chlorination stage. The entire effluent flow was treated by ion exchange resin and condensate was stripped. The system reduced 90% of the effluent COD and color, 60% of the BOD, and significantly removed chlorinated phenols and guaiacols. The treatment system required half of the energy cost needed for an aerated lagoon treatment. A full-scale installation for a 250 ton/day kraft bleach plant has been in operation since 1973.

Dorica et al.[30] proposed a complete effluent recycling system based on ultrafiltration and reverse osmosis membrane techniques. The concept was based on E_1 effluent color removal by ultrafiltration, chloride removal from ultrafiltration filtrate, and C-stage effluent by reverse osmosis and separation of chloride from the ultrafiltration/reverse osmosis concentrates by diafiltration or reverse osmosis. The chloride-lean E_1 filtrate and C-stage effluent were recycled. The chloride-rich diafiltration filtrate had a potential to be used for generation of sodium chlorate and chlorine dioxide, while the organic material in the concentrate could be incinerated. The cost of ultrafiltration and reverse osmosis was higher than that of the Rapson and Reeve closed-cycle bleach plant, which is described next.

A closed-cycle concept was tested in mill scale at the Canadian Pacific Forest Products mill (previously Great Lakes Paper Co.) in Thunder Bay, Ontario, Canada, from 1977 to 1985 (Fig. 9.6). The process was proposed by Rapson and Reeve.[1,31–36] The key elements of the system include counter-current washing in both the bleaching and pulping stages, closed screen room and woodyard, high chlorine dioxide substitution (70%), condensate stripping, and salt precipitation and separation. The sodium chloride obtained from the closed system was used to generate the chlorine and chlorine dioxide required for pulp bleaching. The cost of the closed cycle for a 1000 ton per day DC E_1D E_2D bleach plant was US$ 13.8 million based on 1982 prices. Although the effluent-free objective was not achieved, this pioneering work resulted in the evaluation of many new concepts and served to illustrate the difficulties that must be overcome in achieving bleach plant closure. Based on EFM (effluent-free mill) demonstration experiences, Reeve reported several developments that would be needed to eliminate bleached kraft mill effluent.[37] Five of the developments relate directly to the successful recovery of bleach plant filtrates. First, chloride concentrations in the recovery process liquor cycles must be kept low, as these cause a number of problems including accelerated corrosion of evaporators and boiler tubes and recovery boiler operating difficulties. The experience at Thunder Bay demonstrated that the chloride loading to the recovery cycle must be significantly reduced, if successful filtrate recovery is to be achieved. Second, salt recovery must be accomplished with minimal evaporations. Third, the net flow of bleach filtrate to the recovery cycle must be low to minimize additional evaporation needs. This is important in making the recovery of filtrates economically viable. Fourth, bleach chemical consumption, due to higher levels of dissolved organic matter, must be minimized. Finally, minor wood components such as potassium, calcium, and pitch must be effectively removed from the system. Deposition of pitch and calcium on process equipment and in pipelines was a problem encountered during the EFM demonstration. Potassium adversely affects recovery boiler operation if allowed to accumulate

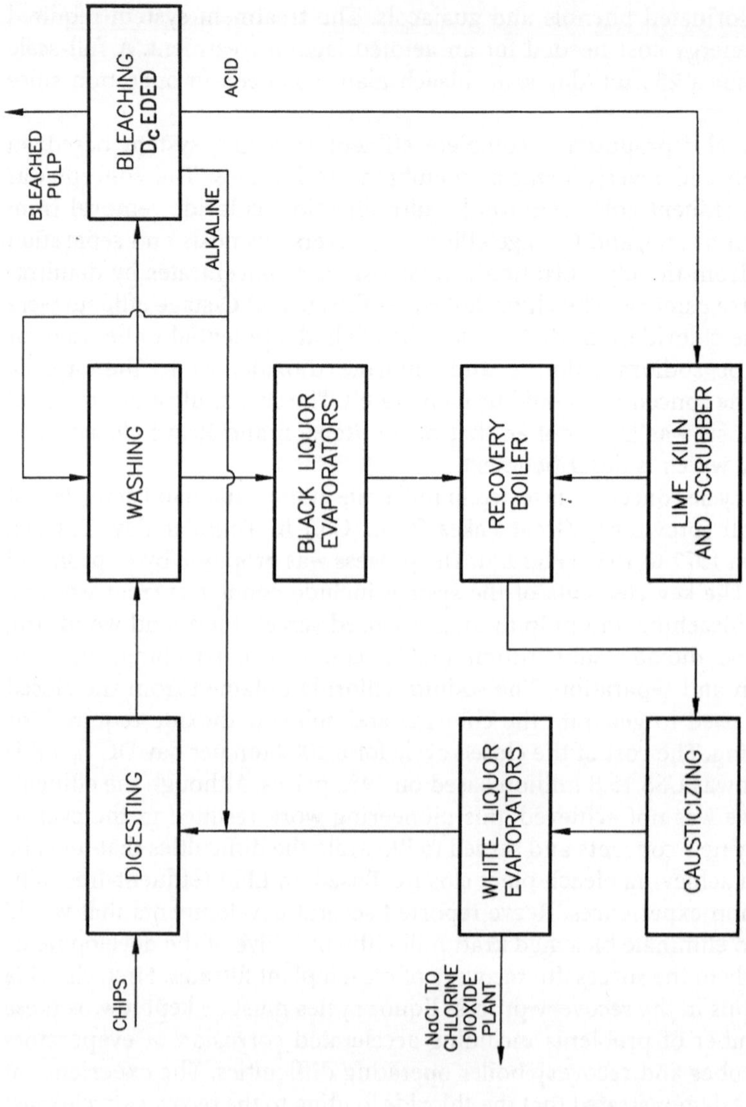

Fig. 9.6. Closed-cycle operation at Thunder Bay. Reproduced with permission from Tappi Press, Ref. 8

in the system. Development of suitable methods to remove these impurities from the process would be required to minimize their impacts on mill operations.

The information on mill-scale activities in bleached kraft mills which are oriented towards complete process closure was reviewed.[25] Seventeen mills which use innovative processes and which have already reduced effluent discharges to well below those considered normal in modern mills were tested. Three are in the US, but none are in Canada. Most use TCF bleaching and the other ECF. Effluent discharges mentioned were generally below 20 m^3/t containing under 20 kg/t COD. AOX discharge in such mills are generally so low as to be of little environmental significance or interest. This review did not include the several Brazilian mills using oxygen/ozone TCF processes which are well known to have low effluent discharges but have not announced programs aiming at zero discharge. Champion International Corporation (Canton, NC) has invested heavily in, and is committed to, ECF technology. It has nine bleached kraft mills in the Americas, seven of which either use or are committed to using oxygen delignification and 100% chlorine dioxide substitution bleaching. Champion has developed a Bleach Filtrate Recycle (BFR) process specifically to enable ECF mills to return bleach filtrates to the recovery cycle, which is the most significant challenge in enabling closed-cycle operation.[38] They consider the process to be technically feasible and are commercially demonstrating it with an initial installation in their Canton, NC mill. The BFR process encompasses a metals-removal process to enable metals removal from the first D stage (D_1) filtrate, reuse of this filtrate for bleach plant washing, fiberline filtrate flow changes, water conservation measures, and a chloride recovery process for removing chloride and potassium from the recovery boiler precipitator dust (Fig. 9.7). Initial installation of BFR is on the softwood fiberline at Canton. Early estimates of the cost of BFR implementation were US$ 15–30 million for Canton and are estimated at US$ 30 million for a 1000-tpd system. Installation of the required instrument has been done and the demonstration of the full system is expected. The demonstration will consist of two phases. During phase I, the D_1 and E_{OP} stage filtrates will be recovered and the D_2 stage filtrate will be sewered. Phase II will involve closure of all bleach stage filtrates with the D_2 stage filtrate recycled back to the E_{OP} stage washer. The BFR process avoids chloride buildup through a combination of chloride removal process, oxygen delignification, D_{100}, and E_{OP} processes. The problems with scale and loss of bleaching effectiveness due to metal buildup are not expected, due to the metal removal process.

A new bleach plant designed for TCF bleaching and extensive closure of the process water system was started up at the end of May 1995 in the SCA Graphic Sundsvall kraft mill at Ostrand in Central Sweden.[26] The design capacity is 1250 t/day. The mill produces bleached hardwood pulp, mainly integrated with fine paper production in the adjacent Wifsta Paper mill, and softwood kraft pulp for both fine paper at Wifsta and wood-containing printing papers (LWC

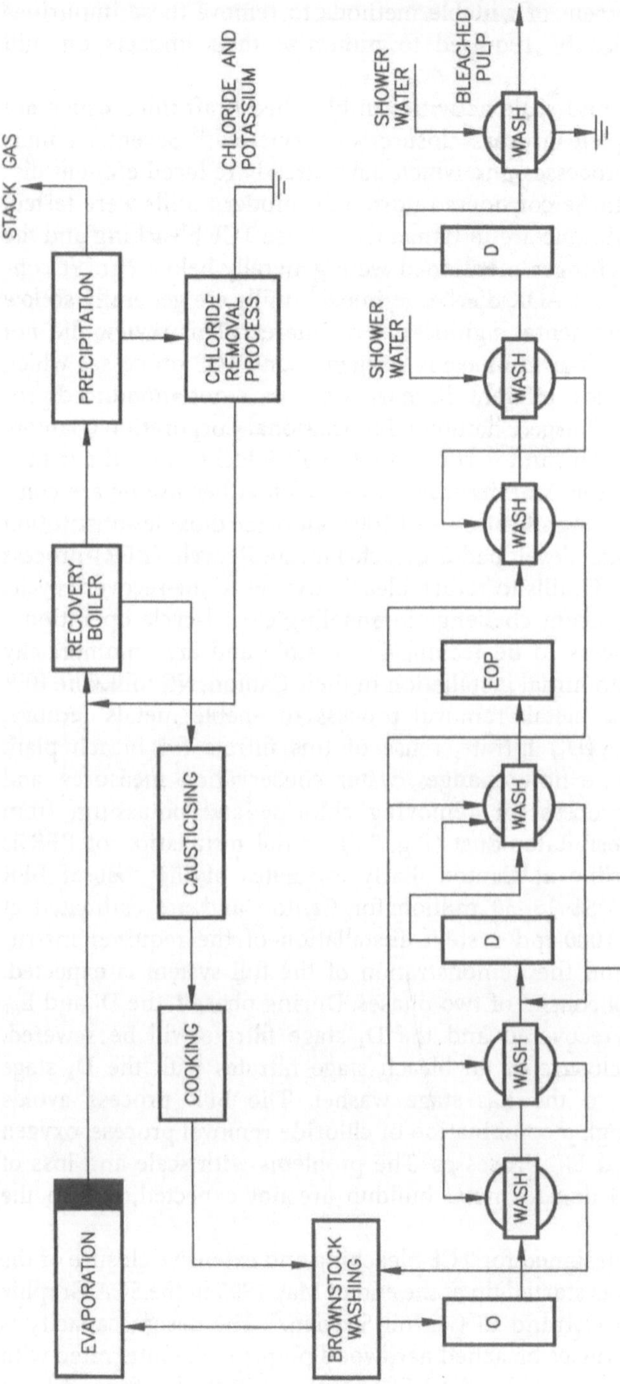

Fig. 9.7. Overview of the Champion BFR process. Reproduced with permission from Tappi Press, Ref. 38

and SC papers) and tissue within the business group. The bleach plant has been designed for extreme closure of the water loops and this involves better washing between the stages and a large buffer capacity of the liquor tanks, counter-current washing in different ways, preferably a somewhat modified jump-stage variant, and two spill systems for spent bleaching liquors of different degrees of cleanliness. The plant has been run more or less continuously at a very high degree of closure since August 1995. Serious scaling has been experienced. However, pulp properties have not been significantly influenced by the closure although certain targets in the bleach plant have had to be changed.

Union Camp Corporation (Franklin, VA) has taken an active interest in effluent minimization from its operations, as it has a number of large bleaching facilities on small rivers.[26] In replacing two aging bleach plants, Union Camp selected an OZ(EO)D sequence over an ODED sequence as it met the needs for effluent recycle. With this sequence, Union Camp pioneered and installed a high-consistency ozone-bleaching line designated C-free. Startup was in September 1992 of the 900 ADt/d softwood bleach line. Both the O and Z stages are at high consistency and an 84.5 GE brightness pulp is produced. Filtrates from O, Z, and EO stages are sent back to the recovery via countercurrent washing with a bleed of part (2.1 m^3/t) of the Z stage to prevent scaling. The final D stage (7.3 m^3/t) is open to sewer. The mill can run up to 88 GE brightness. Union Camp has conducted many trials while the acid (Z) purge is reduced or eliminated using three strategies: (1) modification of the chemistry, (2) use of additional processes, and (3) manipulation of the internal filtrate flow configuration. Union Camp technology has been licensed for installations at SCA (Östrand), Sappi (Ngodwana), and Consolidated Papers (Wisconsin Rapids, WI).[26]

Weyerhaeuser Paper Company (Tacoma, WA), which is the world's largest producer of market pulp, will soon complete modernization projects at three bleached pulp and paper facilities: Plymouth NC; Long View, WA; and Kamloops, BC. Most efforts have been directed at upgrading the pulping and bleaching systems to improve environmental performance and to reduce costs. Weyerhaeuser is supporting the demonstration of technology that may be useful in closed-cycle implementation.[26] This includes pilot operation of the MTCI pulsed bed black liquor gasifier at their New Bern, NC, kraft mill. Weyerhaeuser's minimum-impact mill model is one that has accepted full accountability for the impact of its operations on the environment, is committed to continuous reduction of air, water, and solid waste discharges; achieves discharge levels that are recognized as being minimum impact using sound science; and maintains product performance and purity. In order to monitor and benchmark performance towards a minimum-impact mill, over 30 parameters relating to water, air, and solid waste discharges are used.

Louisiana-Pacific corporation (Samoa, California) is unique in North America in its Swedish-style embrace of in-plant TCF and closed-loop bleaching technologies instead of external treatment as the means to satisfy environmental needs.[2] The LP mill was exempted from the clean water act requirement that bleached kraft mills install external effluent treatment, and proceeded with the in-plant modifications to reduce effluent parameters. The mill has an extensive liquor spill collection and recovery system. Since January 1994, Samoa has produced only TCF pulp and recently demolished their chlorine dioxide generators. Mill production is 650 ADt/day. Samoa uses a continuous digestor, built in 1965 with a 700 ADt/day capacity with use of soluble AQ and a retrofit to Lo-Solids to consistently achieve a softwood kappa of 16. The stock is washed in three vacuum filters, followed by closed screening, installed in 1987, and medium consistency single-stage oxygen delignification to kappa 8–10. Oxygen delignification is followed by a CB fillter and vaccum washer. The bleach plant uses towers originally designed for a CEDED sequence and runs using a Q(EOP)PPP sequence. The mill has a new low-odor boiler, installed in 1990, which meets all mill steam and electric power demands and exports both steam and electricity to neighboring domestic and manufacturing users. Steam stripping of foul condensates is used and on startup of this system in 1994, the mill COD was halved. The mill has an advanced NCG collection system with cofiring of NCGs in the lime kiln and a dedicated incinerator with the ability to transfer load without venting, should either combustion source fail. The only bleach plant effluent discharged is from the chelation stage, and this was determined to be 6.76 m^3/t. This is the lowest bleach effluent flow for a continuously operating bleach plant known. Figure 9.8 shows filtrate recycle in the LP bleach plant. LP is currently investigating and implementing means to close the bleach plant and expects to be operating a effluent-free process. The main challenges are to prevent harmful mineral buildups within the process; develop methods to prevent spills under all operating conditions, hydraulically balance recycled filtrates and make optimum use of chemical residuals and recycle the metal-containing stream to a combination of pulping liquor preparation, directly to evaporation or to the fiber line pulp washers. Mill trials of partial Q stage filtrate recycle at LP have shown that for peroxide-based TCF bleaching, low brown stock metal content is required to enable higher levels of filtrate recycle. LP successfully carried out a Lo-solids cooking trial in October 1995 which reduced brown stock manganese by 21% and the calcium by 23%. They believe that continuous operation of Lo-solids cooking will ensure optimal Q stage filtrate recycle.

Husum mill of MoDo paper AB, Sweden began trials with closed prebleaching for hardwood in 1993, and now regularly produces hardwood pulp in a closed-loop fashion, marketed as MoDo balans.[25] Bleach chemicals used include oxygen, ozone, and peroxide, and currently the sequence includes 3 to 5 kg ClO_2/ADt to decrease kappa number variations. Recycle of the ZP filtrates to brownstock results in increased chemical consumption during bleaching. Direct counter-current filtrate recirculation is used in the bleach plant, with fresh caustic used in the Eo and P stages. At this time, the first two stages of bleaching are closed, with filtrate sewered from the last two bleaching stages, for regular hardwood production. For approximately 25% of the time, the line

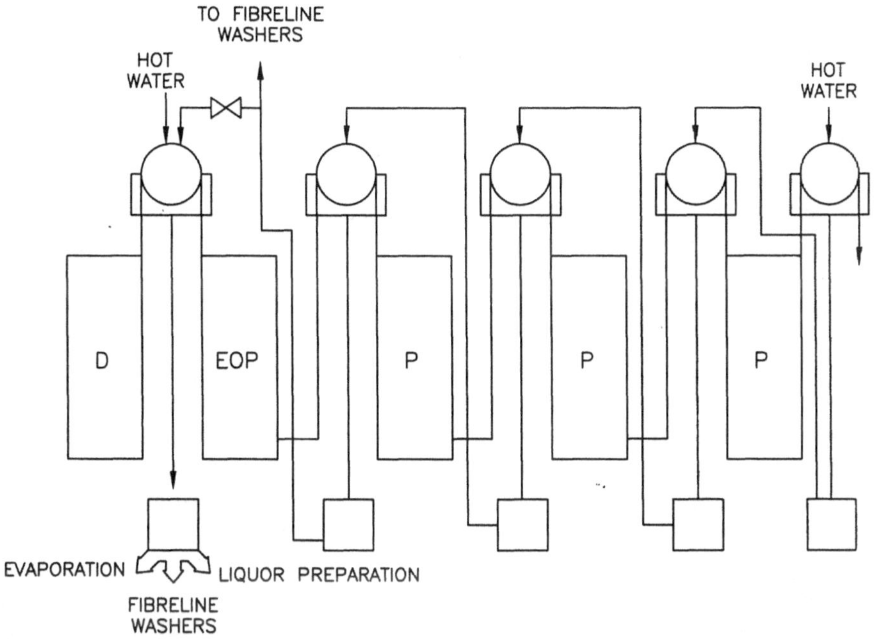

Fig. 9.8. Louisiana Pacific Bleach Plant Filtrate Recycle. Reproduced with permission from Appita, Ref. 26

is run counter-currently, with no filtrate sent to sewer. The hardwood line averages 5 m^3/ADt effluent, on a monthly basis. With closed-cycle operation there are scaling problems, which have resulted in reduced productivity. The scaling problems are addressed through intensive mechanical cleaning.

Over the past few years, Sodra Cell has made a concerted effort to communicate its goal of becoming 100% TCF at all three of its bleached kraft mills.[26] Initial focus for closeup is the Morrum mill, although all mills are moving towards closure in a step-wise fashion. Sodra has an 8-day aerated lagoon for effluent treatment at its Monsteras mill and primary clarifier at its Morrum and Varo mills. Monsteras has a capacity of 335000 ADt/year and produces ozone-based TCF softwood and hardwood pulp in campaigns on one line. Monsteras started its ozone bleach plant in September 1992. Sodra is expanding production capacity at Monsteras by 60% for a daily production capacity of 1700 t. The modernized system will include black liquor impregnation and ITC in the continuous digestor, ozone bleaching and use of a wash process in the bleach plant. Based on information from Kvaerner, the modernization will allow the bleach plant to operate effluent-free in the near future. Sodra is rebuilding its Morrum mill for TCF and expanding production by 44000

ADt/year to 415 000 ADt/year. The project includes a new bark press and shredder, replacement of the existing digesters with new RDH batch digesters, a complete rebuild of both bleach lines, increased evaporation and recovery boiler capacity, a cooling tower for excess heat, and a new effluent treatment plant and system for collecting and stripping volatile malodorous gases. They are also installing several large buffer tanks to facilitate closeup, and evaporator capacity has been increased by 20%. Sodra predicts that the effluent COD will drop down to 15 kg/ADt and that bleach plant effluent will be reduced to $8\,m^3$/ADt.

Stora Billerud is working generally in the area of water reduction and closing the water cycle in all of their mills.[26] Eka Nobel is running trials with a process in which bleach effluent at an intermediate concentration (concentrate from the first of two evaporation stages) is flocculated and the sludge produced is returned to the recovery boiler. The filtrate is then electrolyzed to produce a brine for reuse in the chlorate plant. Initial mechanical difficulties with the evaporation unit included plugging of the inlet distributor and these have rectified with a modern design. Evaporation of alkaline, acidic, and a mix of both acid and alkaline have been trialed, along with a range of different membrane polymer profiles and thickness profiles.

The Gruvon mill bleaches part of their production with a DEDED sequence and has reportedly reduced its bleach effluent flow to around $12\,m^3$/ADt. Gruvon is also using carbon dioxide addition in the brownstock line to improve washing of both COD and soap. The Skoghall mill has a pilot-plant pressurized Chemrec gasifier located on site and also has pilot-plant facilities for a cooperative research program. The program focus is on the ECF closed-cycle and partners in the research include Jaakko-Poyry, Eka Nobel, Metsabotnia, and Kymmene. The approach used consists of evaporation of bleach plant effluents to optimum concentration, purification of the condensates so that they can be used as process water, incineration or thermal oxidation of the remaining solids in an acceptable way, and the final treatment of the inorganic slag obtained by incineration.

Another mill in Sweden-Vallvik is installing a pressurized peroxide stage for 800 ADt/day softwood to make full brightness TCF. The mill has shown some interest in water systems closure and has been performing trials using peracetic acid in bleaching.[26]

Finnish industry has adopted efficient effluent treatment technologies and is also competing with Sweden's in-plant and TCF technologies, predominantly for market reasons. Metsa-Rauma (MR) is constructing a new 500 000 Adt/year TCF softwood pulp mill in Southern Finland. MR expects the TCF sequence to offer advantages both in market place and in reducing bleach plant flows.[3] The mill will incorporate Sunds Super-batch digesters and an Ahlstrom oxygen delignification system. Bleaching will use oxygen, peroxide, and medium consistency ozone bleaching technology. The mill expects to have a total effluent of 10 to $15\,m^3$/ADt at startup with provision to decrease to 5. It then aims to reduce the remaining effluents using a stepwise approach. Alkaline filtrates will be recirculated first, with the goal of closing the acidic filtrate cycle within a couple of years. The new MR mill may be seen as the first of a new generation of mills built for TCF and water systems closure. The equivalent new generation ECF mills could be seen as Alberta-Pacific in Alberta and Bahia-Sul in Brazil. Metsabotnia mill has been steadily increasing TCF production at both Kemi and Kaskinen mills since first producing TCF in 1991 with enzymes, oxygen, and peroxide at 82 ISO brightness for softwood and 85 ISO for hardwood. Kaskinen's objective is to develop a completely closed cycle.[26] Total mill effluent flow is around 50 m^3/ADt and over the past 10 years, the COD, BOD, and AOX loads have been reduced by 50–90%. Partial return of alkaline filtrates was achieved in 1992. There are ongoing trials towards achieving closed cycle both for Kaskinen and to support the development of new Rauma mill. Kymmene, Wisforest capacity is 580 000 ADt/year bleached hardwood and softwood pulp of which TCF capacity is 280 000 ADt/year. It has secondary treatment and was the first mill in Finland to implement both oxygen delignification and ozone. It has been reducing its bleach filtrate effluent streams over the past several years and current process water use is about 25 m^3/ADt.[26] The mill would like to cut this to half. The ECF / TCF fiber line uses six conventional and two superbatch digesters followed by oxygen delignification. The ECF bleach sequence used is O(ZD)(O/EO)(ZD)(EPD) and the TCF sequence is O(ZQ)(OP)(ZP). Process water use averages 25 m^3/ADt. While on TCF production, the mill uses 15 m^3/ADt and is trying to attain a level of 10 m^3/ADt. The mill is also working towards closure of air emissions with all strong odors and 85% of weak odors collected. The target for sulfur dioxide emissions was 0.15 kg/ADt and the total of all gaseous sulfur compound emissions was less than 0.3 kg/ADt by the end of 1995.

Mills in South Africa and Australia have been leaders in low water use due to their arid climates. There is active closed-cycle research in New Zealand and Australia. The new Chilean and Brazilian mills environmental performance is world class. Specifically, environmental performance at the Brazilian mills Riocell and Bahia-Sul are among the world's best.[26]

In South Africa, Sappi's pulp paper and board mill, Ngodwana, Transvaal, has been at the forefront of efforts to reduce water use for the past 10 years.[3,27] It is located in an arid area and stores water for up to a year for use. Effluent cannot be discharged back into the river due to sensitive agricultural uses downstream. The mill irrigates all of its effluent on pastures. The mill manufactures 1500 t/day of final product. This comprises bleached pulp, unbleached pulp, newsprint, and liner board. Kraft pulping capacity is 1220 ADt/day and bleaching capacity is 550 ADt/day. The mill water use is 33 million l/day and effluent flow to irrigation is 24 million l/day (16 m^3/t prod.). Concern for the long-term viability of irrigating chloride-containing effluents initiated an extensive research program into alternatives. These include investigation into Rapson-Reeve concept and a number of other closed-cycle processes. Sappi developed a bleach effluent recovery process suitable for chlorine- and chlorine dioxide-containing bleach sequences, based on separate effluent concentration and incineration using a magnesium-based recovery process. This was successfully piloted and installation was initiated at Ngodwana. The project was halted with the emergence of commercial ozone and TCF bleaching, and Sappi is now proceeding with ozone installation. Sappi has licensed the Union Camp C-Free high-consistency ozone process and is installing an ozone-based ECF/TCF bleach plant. Effluent from the plant will continue to be used for irrigation, as maintenance of the kikuyio grass vegetation is important in immobilizing existing chloride-containing soils.

Mannisto et al. have examined the technical and economic consequences of converting to low-effluent pulp production for three hypothetical bleached kraft mills: an old mill, a modern mill, and a greenfield mill.[39] Although liquid-effluent-free production of bleached kraft mill was found to be technically possible, it was expensive. It was found that total cost of converting a mill to low-effluent production would increase the cost of the pulp manufacture by US$ 50–150/air-dried ton. In the case of a greenfield mill designed for liquid-effluent-free operation, the cost of producing a ton of air-dried pulp is expected to be US$ 20–60 higher than for a conventional greenfield mill. Development of new technologies may reduce these cost differences. It was suggested that if the liquid-effluent-free bleached kraft mill is to become a reality, the industry will need to focus its efforts on several key areas, which include (1) improving the quality of nonchlorine bleach sequences, (2) developing methods for removing and managing nonprocess elements (chloride, metals, nutrients), (3) minimizing increase in solid waste generation and air emissions, (4) managing the mill chemical balance, (5) managing upset conditions, and (6) developing design standards.

Brook et al. used process simulation and mill data from Louisiana-Pacific's Samoa, CA, market pulp mill, to evaluate the conversion of a conventional oxygen delignified softwood bleach plant to totally chlorine-free bleaching based on hydrogen peroxide.[40] During a recent mill TCF closeup run, substantial reductions were made in water use, steam consumption, and sewer loading. Thirty-one% of the bleach plant organic dissolved solid material, normally lost to the sewer during conventional chlorine bleaching, was recycled to the recovery boiler. An economic evaluation comparing the TCF process, conventional chlorine bleaching, and 100% ClO_2 substitution was done. TCF bleaching costs were found to be about US$ 26–29/AD metric ton pulp more than ECF bleaching.

Studies by Prasad et al. have shown that high-kappa kraft pulping coupled with multiple oxygen stages and TCF bleaching can be used to close a pulp mill without overloading the recovery system and with a production increase under certain circumstances.[41] Pulp produced at a kraft kappa number of 40–50 followed by multiple oxygen stages and TCF bleaching when compared to a conventional kraft pulp (kappa no. 30) followed by one oxygen stage and ECF bleaching showed a similar or lower cost and a lower boiler throughput at constant production. Under constraints of limited digestor capacities or limited boiler thermal capacity, production can be increased with a multiple oxygen/TCF closed system relative to a conventional kraft-oxygen system.

A process has been designed to remove chloride and potassium from electrostatic precipitator dust.[42] In a crystallizer/evaporator, sodium sulfate crystals are formed to be separated and returned to the kraft liquor system. The filtrate, which is concentrated in chloride and potassium, is in part purged to control the chloride and potassium concentration in the system.

Covey and Nguyen have shown what DARS process can be integrated with a chlorine-based bleach plant for waste treatment and recovery of active chemicals that can be reused in the pulping or bleaching process.[43] It is possible to treat the effluent from an ECF bleach sequence with the DARS fluidized-bed reactor by providing means to extract and discharge chloride from the chemical cycle. At the typical operating temperature of the DARS process, sodium chloride is volatile and accumulates mainly in the dust of the reactor. The chloride can be leached and separated from the other careful chemicals by evaporative crystallization. This approach can maintain the chloride in the liquor circuit at the tolerable level.

Techniques of nanofiltration and electrodialysis have been shown to be technoeconomically feasible to purify the E_1 effluent and to recycle it as process water.[44] The Selenga Pulp & Paper Company in Russia is the world's first zero-discharge unbleached kraft mill.[45] Zero discharge was accomplished by employing activated sludge, chemical precipitation, and reaeration waste treatment methods.

Saunamaki compared the biodegradability of wastewaters from TCF bleaching with those of wastewaters from chlorine-gas bleaching (CG) and ECF bleaching.[46] Activated-sludge treatment of TCF wastewaters reduced COD by 55–65% compared with 34–45% for wastewater containing chlorine compounds. Chemical flocculation of biologically treated TCF wastewaters with aluminum sulfate resulted in 85% total removal of COD. Residual peroxide in TCF wastewaters made them highly toxic, although biological treatment removed this toxicity. Treated TCF wastewaters had higher levels of heavy metals and nonprocess elements than CG and ECF wastewaters.

The Russians have developed tertiary treatment systems to minimize the impact of effluent on receiving waters.[47] Tertiary treatment at the Baikalsk Pulp and Paper Co. and Svetogorsk Pulp and Paper Co. are examples of effective tertiary treatment technology unique to the pulp and paper industry. The Baikalsk three-stage, biological, chemical, and sand filtration treatment is highly effective in reducing BOD, suspended solids, and color. Svetogorsk's sand filtration/flotation tertiary technology is effective in removing suspended solids and BOD.

A variety of process condensates are generated within the digestor and liquor evaporation areas of a kraft process. These condensates contain various concentrations of reduced sulfur compounds, terpenes, and related wood derivatives, black liquor carryover, and semivolatiles such as methanol, acetone, and methyl-ethyl ketone. Reuse of condensate is an important component of water-usage and heat-conservation programs in a kraft mill operation. Increasing the quantity of condensates reuse, and for some reuse applications, improving the quality of condensates via treatment, offers the potential for: (1) further reduction in water usage rates and (2) reductions in atmospheric emissions of volatile organic compounds such as methanol from a unit process in which condensates are reused. It is likely that present limitations on the quantity of condensates used in a mill are, in part, related to the quality of the remaining available condensates. Thus, further reuse of condensates may require some pretreatment.

Biological treatment of pulp mill condensates in a stand-alone treatment process offers a means to recover large quantities of high-quality condensates. Recently, references related to aerobic and anaerobic treatment of kraft condensates have been reviewed.[48] Laboratory-scale research suggested that most of the components of kraft condensates were readily treatable under aerobic conditions. Methanol concentrations of up to 5.5 g/l can be tolerated under steady-state operation. Anaerobic treatment may be feasible for relatively high-strength condensates. No serious treatability difficulties were identified in any of the anaerobic studies carried out under steady-state conditions. The best anaerobic treatment performance was described for a fixed-bed reactor coupled to a membrane filtration process for effluent polishing. With this process, ultra-concentrated condensate was treated under thermophilic conditions at high organic loading rates. One full-scale anaerobic facility has been described for methanol-containing wastewater, operating under transient industrial discharge conditions. This hybrid anaerobic treatment process achieved COD removal efficiencies of 95% at moderate organic loading rates of 2–4 kg COD/m^3/day and HRTs of 1.4 to 4 days. The effluent suspended solids levels from the hybrid process usually exceeded the 20 mg/l criterion.

Barton et al. selected two aerobic biological treatment processes:[48] pure oxygen-activated sludge and an aerobic membrane reactor technology. The aerobic membrane system includes a suspended growth aeration basin with high MLSS concentrations (>20 g/l) followed by an ultrafiltration membrane system for separation of biomass from the treated effluent. The study indicated that residual BOD of 25 mg/l was achieved by activated sludge treatment in a reaction period equal to or less than 24 h. The kinetics for BOD reduction indicate that an MLSS of 3 g/l would require a reaction time of about 16 h to achieve a BOD of 25 mg/l under the conditions of study. A greater level of activated sludge would be expected to reduce the treatment time. High degrees of methanol reduction were provided by the activated sludge treatment. Both reactors achieved residual methanol levels of less than 0.3 mg/l following the 24-h biooxidation period, resulting in a greater than 99% reduction of methanol. The result of this study suggested that (1) biotreatment of kraft mill condensates to an acceptable quality for reuse is feasible and (2) the capital and operating costs for the treatment of condensates are comparable to the costs for collecting and incinerating vent gases from vacuum drum brown stock washer systems.

3. Sulfite Mills

Not much information is available on closed-cycle activities in sulfite mills. In Germany, five of the six sulfite pulp mills are processing 100% of their capacity totally chlorine-free and only one mill is bleaching pulp for specialties with the aid of the ClO_2. PWA Waldhof Manheim Pulp and Paper mill has been able to achieve closed cycle by using a single-stage bleaching process – OPMgO (hydrogen peroxide reinforced oxygen delignification with magnesium oxide).[49] The main features of this process are: (1) dissolved organic compounds are returned to the recovery cycle, (2) magnesium oxide is recycled into the process and partly replaces the demand for makeup magnesium oxide, and (3) dissolved lignin and other organics are used to produce combustion energy. With this novel system, the mill was able to decrease significantly the effluent load from the bleaching plant in terms of COD.

MoDo paper mill in Sweden is also operating closed-loop TCF bleaching at their Domsjo sulfite mill.[26]

Lenzing AG in Austria has produced sulfite viscose pulp using ozone and peroxide since 1992. Most of the effluent from the bleach plant is now completely recycled and the mill has run several trials in the fully closed-cycle mode. At the Tappi 1993 Pulping Conference, the mill reported that it is well on its way to complete closure.[8]

4. Paper Mills Utilizing Secondary Fibers

The current climate of increasingly stringent effluent regulations is causing many recycled fiber paper board mills to consider closure of their process water circuit as a cost-effective investment to meet effluent regulations.[50,51] The concept of a zero-discharge mill is attractive both from the standpoint of biological treatment plant savings, as well as the complete elimination of environmental impact due to the mill effluent. Paperboard and roofing felt mills using recycled fiber as furnish are good candidates for zero-discharge operation because they typically do not have extensive treatment facilities already in place and their product quality standards are less demanding than for many other paper products.

Most Canadian recycled paper board mills have already implemented extensive water and energy conservation techniques such as segregation of clean water from whitewater, alkaline paper board making, which enable the reuse whitewater for vacuum seals and good heat exchange networks. These measures have helped to reduce the effluent discharge to about 6 m^3 of effluent/ADt of board. A large fraction of whitewater is recycled and reused. Freshwater is needed for preparation and dilution of chemicals, selected high pressure showers, and gland seals. Since paper board mills often change from white top liner to fully brown board, two whitewater systems are used. When the mill reverts from the fully brownstock to white top liner stock, a substantial amount of brown whitewater has to be purged and freshwater added for starting the white liner.

There are several references in the literature to closed whitewater systems in recycled paper board mills.[52–55] Green Bay Packaging Inc. implemented a closed cycle system in the 1970s using reverse osmosis as the method for purging inorganic matter.[52] However, the mill is now operating in a closed-cycle mode without reverse osmosis or any other expensive technology.[53] The mill has achieved systems closure primarily by extensively replacing freshwater with process water and a proper selection of chemical additives. The Haltown Paperboard Company mill has used a similar approach for mill systems closure.[55] At the present time, of the 30 paper board mills in the US, about 15 mills operate with relatively low effluent discharge[55] (mean = 18 m^3/ADt and the minimum = 0.4 m^3/ADt). However, details on the technologies selected, choice of water reuse locations, effluent discharge conditions, and effects of system closure on production are not available. Little evidence exists for why a particular method of water reuse was successful or unsuccessful. In cases where high degrees of system closure were achieved, corrosion and bacterial growth in the paper machine area were observed.[54] Several mills are currently attempting to achieve complete closure by treating the effluent as the freshwater source.[56]

Reuse of whitewater for dilution of wet-end chemicals and cleaning felts was investigated in the laboratory, and successful strategies were implemented at a recycled paper board mill.[57] Whitewater was successfully used for second-stage dilution of starch and bentonite clay before addition to the headbox stock. Screened whitewater (400 mesh) was found to be suitable for cleaning felts at high pressure. The implementation of screened whitewater reuse in a board mill enabled the mill to reduce its daily minimum effluent discharge from 5.5 m^3/ADt of board to about 2.3 m^3/ADt of board without affecting the production. The mill's average effluent discharge was reduced by about 40% from 8.5 m^3/ADt to 5.1 m^3/ADt.

St. Laurent Paperboard Inc., a corrugated medium mill, situated at Matane, QC, Canada, has successfully undertaken and attained zero process effluent discharge.[58] The process effluent flow was steadily decreased from 6000 m^3/day to zero in October 1995. The mill presently meets all present and future environmental regulations. Chemical dosages and injection points are being optimized. Although the chemical dosages have increased, the incremental cost is just a fraction of what it could cost to operate a secondary treatment plant. Also, zero effluent has resulted in an OCC yield increased from 85 to 92% related to the higher overall retention of fines/clay into the paper. The reuse process water also resulted in energy saving of about 5%.

Stuart and Lagace reviewed typical process modifications that are required to implement zero discharge operation at paper board mills.[59] The process equipment items used to achieve this are not sophisticated, and generally include primary treatment, adequate water storage, stainless steel metallurgy, and often a segregated cooling water system. Chemical programs must achieve consistently high first-pass retention of fines, clays, size, and colloidal contaminants such as pitch. Recycled water must replace freshwater for most applications, including gland water and paper machine shower water.

Increased whitewater recycling results in a buildup of organic and inorganic contaminants in the whitewater. Springer et al. examined the effects of these whitewater contaminants on the strength properties of paper produced from secondary fiber.[6] Eight contaminant types were evaluated at three concentrations. The effects of the contaminants grouped as organic and inorganic were measured first, followed by separately determining the effect of individual contaminants. As a group, the organic contaminants had a greater effect in reducing sheet tensile index. For both groups, as concentration increased, sheet tensile index decreased. Taken individually, the organic contaminants – water-soluble wood extractives, NSSC lignosulfonate, kraft lignin, and defoamer – adversely affect sheet tensile index. Of the inorganic contaminants: sodium, calcium, iron and alum, only iron had a large adverse effect on sheet tensile index.

Two separate surveys were conducted to characterize manufacturing conditions and water use practices at zero-discharge paper board mills.[60] These surveys included a telephone survey of 16 US and Canadian mills conducted by BEAK consultants and mill visits by NCASI staff to nine zero-discharge non-deinking recovered fiber mills. A majority of the mills visited by NCASI were also included in the BEAK survey. Following conclusions were drawn: (1) Considerable whitewater surge capacity is needed for zero-discharge operation, especially where white-lined or colored grades are manufactured. (2) All but one of the mills visited by NCASI maintain zero discharge by providing either no external treatment or external treatment in the form of gravity clarification only. Only one of the mills used biological treatment. (3) Operational difficulties which may be encountered due to closure of the mill water cycle include (a) fogging in machine area, (b) wet-end chemistry changes, (c) reduced felt efficiency, (d) reduced vacuum pump efficiency, (e) development of slime holes or other imperfections in the sheet, and (f) odor, either in the sheet or emanating from external treatment process. Furthermore, accelerated corrosion rates associated with zero discharge dictate the need for stainless steal process equipment and piping.

Three major problems were encountered when Scott Paper and Crabtree municipality effluent treatment system were jointly conducted.[61] These problems were flotation on the primary sedimentation clarifier, foam on the areation basin and filamentous sludge bulking. The major source of Scott Paper mill effluent is from the deinking operations producing 310 t/day of recycled fiber. Ledgers, office waste and book stock are the main sources of waste paper.

The municipal effluent accounts for less their 5% of the BOD load to the effluent treatment system. The solutions to these problems were presented.[61] It was found that adequate sludge dewatering capacity is critical and must satisfy seasonal needs, that spill protection is essential and that treatment system cleanout strategies are required and also that deinking effluents need specific treatment plant configurations.

Conclusion

The optimum closure strategy for a given mill depends on the type of pulping process employed: unbleached kraft, bleached kraft, bleached sulfite, secondary fiber, mechanical, etc. In general, most unbleached mills probably will develop closure strategies based on minimization of water usage using conventional technology with treatment of effluent for recovery and reuse technology similar to that already in place at high-yield mills like Meadow Lake and Chetwynd. At present, evaporation is the only industrially proven technology. The strategy must (1) not have a negative effect on pulp quality, (2) not cause losses in pulp production, (3) have operating cost competitive to biological secondary effluent treatment system, and (4) minimize use of treatment chemicals to ensure purity of smelt and ash for potential recovery and to avoid effect on other unit's operations.

Extensive closure of a TCF bleach plant is very demanding and requires extremely good washing of the pulp between the stages. However, kraft mills utilizing TCF bleaching sequences will be able to close up their bleach plants by recycling the effluent directly to the chemical recovery loop. ECF-bleached kraft pulp mills will have much more difficulty achieving closure and, of course, effluent-free ECF mills will still carry the stigma of producing pulp containing traces of organic halides.

Most early closed-cycle implementations will be in existing mills in North America, Sweden, and Finland. Many mills are positioned to reduce bleach plant effluent to the level at which complete recycle is economically feasible and to consider recycling alkaline effluent to the recovery cycle. The closed-cycle pulp mill is a part of the ultimate strategy for a holistic approach to the environment. The driving forces are moving the industry in the long term to achieve environmental harmony and looking at the total impact on the environment. Coupled with sustainable, recyclable, biodegradable nature of the products, a closed cycle process where no process effluent are released into the receiving water is the long term solution.

Closed cycle bleached kraft mill appears to be technically feasible. The challenge is to optimize process choice to give good economics. The choice of closed-cycle should be based on market demands and sound economics and should not be imposed by regulations.

References

1. Rapson WH: The feasibility study of recovery of bleach plant effluent to eliminate water pollution by kraft mills. Pulp Pap. Mag. Can. 1967; 12: T635–T640.
2. Reeve DW, Rowlandson G, Kramer JD, Rapson HW: The closed-cycle bleached kraft pulp mill-1978. Tappi J 1979; 62(8): 51–54.
3. Gartz R et al.: The zero effluent mill. The 18th Intl. Mechanical Pulping Conference, Oslo 1993.
4. Nurmiaho-Lassila E-L, Salkinoja-Salonen MS, et al.: Electron microscopical analysis of biological slimes on paper and board machines. Inst. Phys. Conf. Ser. no. 98: Chap. 16
5. Welkener U, Hassler T, McDermott M: The effect of furnish components on depositability of pitch and stickies. Nord. Pulp Pap. Res J 1993; 8: 223–225, 232.
6. Springer AM, Dullforce JP, Wegner TH: The effects of closed white water system contaminants on strength properties of paper produced from secondary fiber. Tappi J 1985; 68(4): 78–82.
7. Young J: Meadow Lake enters third year of zero-effluent pulp production. Pulp Pap. 1994; March: 56–61.
8. Ricketts JD: Considerations for the closed-cycle mill. Tappi J 1994; 77(11): 43–49.
9. Young J: Chetwynd pioneers innovations in zero-effluent pulp production. Pulp Pap. 1994; March: 73–75.
10. Klinker RT: Successful implementation of a zero discharge program. Tappi J 1996; 79(1): 97–102.
11. Coghill RS, Dahl S, Thurley D: Wastewater management at Australian Newsprint Mills Albury: responding to community expectations. Appita 1995; 48(1): 25–29.
12. Lagace P, Stuart PR, Kubes GJ: Development of a physical-chemical zero discharge process for a TMP-newsprint mill. Pulp Pap. Can. 1996; 97(11): T414–T419.
13. Dorica J, Ramamurthy P, Elliott A, Crotogino R, Vrooman W, Chatterjee A, Wong P: Mill studies on closure of the effluent cycle in a newsprint mill. In: Proceedings of Tappi Minimum Effluent Mill Symposium, Tappi Press, Atlanta, USA 1996: 217–227.
14. Wilson RW, Purdue R: Water and effluent management at Bowater Mersey. Pulp Pap. Can. 1996; 97(8): T273–T275.
15. Jantunen E, Lindholm G, Lindroos C-M, Paavola A, Parkkonen U, Pusa R, Söderström M: The effluent-free newsprint mill. Paperi Ja Puu-Paper and Timber 1992; 74(1): 41–44.
16. Lindholm G, Jantunen E: Treatment and reuse of process water in pulp and paper industry, steps towards effluent free operation. In: Proceedings of Tappi Minimum Effluent Mill Symposium, Tappi Press, Atlanta, USA 1996: 195–208.
17. Elefsiniotis P, Hall ER, Johnson RM: Contaminant removal from recirculated white water by ultrafiltration and/or biological treatment. In: Proceedings of 1995 International Environmental Conference, Tappi Press, Atlanta, USA 1995: 861–867.
18. Schnell A, Dorica J, Ho C, Ashikawa M, Munnogh G, Hall ER: Anaerobic and aerobic pilot-scale effluent detoxification studies at an integrated newsprint mill. Pulp Pap. Can. 1990; 91(9): T347–T352.
19. Tardif O, Hall ER: Membrane biological reactor treatment of recirculated newsprint white-water. In: Proceedings of Tappi Minimum Effluent Mill Symposium, Tappi Press, Atlanta, USA 1996: 347–355.
20. Sundberg A, Ekman R, Holmbom B, Sundberg K, Thornton J: Interactions between dissolved and colloidal substances and a cationic fixing agent in mechanical pulp suspensions. Nord. Pulp Pap. Res. J. 1993; 8(1): 226–231.
21. Gaarder C, Englezos P: The use of clathrates for the concentration of mechanical pulp effluents. Nord. Pulp Pap. Res. J. 1995; 10(2): 110–114.
22. Sloan ED: Clathrate hydrates of natural gases. Marcel Dekker Inc., New York 1990.
23. Gerbasi B, Stuart PR, Arsenault F, Zaloum R: Techno-economic assessment of several closed cycle technology alternatives for an existing TMP-newsprint mill. Pulp Pap. Can. 1993; 94(12): 123–128.

24. Hinswey NW: EPA's proposed cluster rule: the end of end-of-pipe. Tappi J. 1994; 77(9): 65–74.
25. Anonymous: Mill closure has taken over the ECF vs. TCF debate. Pulp Pap. Can. 1996; 97(10): 11–15.
26. Johnson T, Gleadow P, Hastings C: Pulping technologies for the 21st century – a North American view. Part 4 Emerging technologies and mill closure. Appita 1996; 49(2): 76–82.
27. Albertsson V, Bergkvist S: Environmental protection techniques to be applied in a bleached kraft pulp mill in Sweden. Tappi J 1973; 56(12): 135–138.
28. Andersson KA: The non-polluting bleach plant. Tappi J 1997; 60(3): 95–96.
29. Lindberg S, Lund LB: The non-polluting bleach plant. Tappi J 1980; 63(3): 65–67.
30. Dorica J, Wong A, Garner BC: Complete effluent recycling in the bleach plant with ultrafiltration and reverse osmosis. Tappi J 1986; 69(5): 122–125.
31. Rapson HW, Reeve DW: The effluent free bleached kraft pulp mill: the present state of development. Tappi J 1973; 56(9): 112–115.
32. Reeve DW, Rowlandson G, Rapson WH: The effluent free bleached kraft mill Part VIII Bleach plant rennovation and design. Pulp Paper Can. 1997; 78(3): T27–33.
33. Reeve DW, Pryke DC, Lukes JA, Donavan DA, Valiquette G, Yemchuk EM: Chemical recovery in the closed cycle mill, Part I. Pulp Pap. Can. 1983; 84(1): T25–T29.
34. Reeve DW, Tran HN, Braham D: The effluent free bleached kraft mill. Part XI Morphology, chemical and thermal properties of recovery boiler superheater fire side deposits. Pulp Pap. Can. 1981; 82(9): T315–T320.
35. Reeve DW, Rapson WH: Recovery of sodium chloride from bleached kraft pulp mills. Pulp Pap. Mag. Can. 1970; 71(13): T274–280.
36. Gilbert AF, Rapson WH: Accumulation of potassium in a closed cycle mill. Pulp Pap. Can 1980; 81(2): T43–T47.
37. Reeve DW: The effluent-free bleached kraft mill Part XIII. The second fifteen years of development. Pulp Pap. Can. 1984; 85(2): T27–T30.
38. Maple GE, Ambady R, Caron JR, Stralton SC, Vega Canovas RE: BFR^{TM}-A new process toward bleach plant closure. Tappi J 1994; 77(11): 71–80.
39. Mannisto H, Mannisto E, Winter P: Technical and economic implications of converting bleached-kraft mills to low-effluent operation. Tappi J 1995; 78(1): 65–73.
40. Brooks TR, Edwards LL, Nepote JC, Caldwell MR: Bleach plant closeup and conversion to TCF: a case study using mill data and computer simulation. Tappi J 1994; 77(11): 83–92.
41. Prasad B, Kirkman A, Jameel H, Gratzl J, Magnotta V: Mill closure with high-kappa pulping and extended oxygen delignification. Tappi J 1996; 79(9): 144–152.
42. Shenassa R, Reeve DW, Dick PD, Costa ML: Chloride and potassium control in closed kraft mill liquor cycles. Pulp Pap. Can. 1996; 97(5): T173–179.
43. Covey GH, Nguyen KL: Chloride management in a closed ECF bleach plant. Appita 1996; 49(5): 332–336.
44. de Pinho MN, Geraldes V, Rosa MJ, Afonso D, Figueira H, Taborda F, Ganho R, Gaeta S, Creusen R, Hanemaaijer J, Amblard P, Gavach C, Guisard C: Water recovery from bleached pulp effluents. 1995 International Environmental Conference Proceedings, 883–892.
45. Kenny R, Yampolsky M, Goncharov A: An overview of a Russian zero discharge unbleached kraft pulp and paper mill-Selenga Pulp & Paper Company. Pulp Pap. Can. 1995; 96(5): T155–T157.
46. Saunamaki R: Treatability of wastewaters from totally chlorine-free bleaching Tappi J 1995; 78(8): 185–192.
47. Kenny R, Naumov A, Babinskiy GA, Yampolsky MD, Volsky O, Goncharov A: Experience with tertiary effluent treatment in three Russian mills. Tappi J 1995; 78(3): 191–196.
48. Barton DA, Buckley DB, Lee JW, Jett SW: Biotreatment of kraft mill condensates for reuse. In: Proceedings of Tappi Minimum Effluent Mill Symposium, Tappi Press, Atlanta, USA 1996: 277–289.
49. Nimmerfroh N, Suss HU, Bottcher HP, Luttgen W, Geisenheiner A: The German approach to the closed cycle sulphite mill development and implementation. Pulp Pap. Can. 1995; 96(12): T414–420.

50. Vendries E, Pfromm PH: Influence of closure on the white water dissolved solids and the physical properties of recycled linerboard. Tappi J. 1998; 81(9): 206–213.
51. Tyler MA, Stuart PR: Zero-discharge pulp and paper mills: the inevitable solution. Presented at the 80th CPPA Annual Meeting, Montreal, Canada, January, 1994.
52. Macleod M: Mill achieves maximum reuse of water with reverse osmosis. Pulp Pap 1974; 48(11): 62–66.
53. Young J: Green Bay packaging begins third year with closed water system. Pulp Pap 1994; 68(11): 105–111.
54. Evans J: Century-old board mill operates with zero effluent discharge. Pulp Pap 1981; 55(17): 98–103.
55. Anon: Development document for proposed effluent limitations, guidelines and standards for the pulp, paper and paperboard point source standard. US EPA-821-R-93-014, p610, October 1993.
56. Bowen W, Scogin R, Cobery J: Implementing municipal water reuse in a liner paper mill. Proceedings 1994 Int Environ. Conf. Portland OR April 1994; p.23–28.
57. Ramamurthy P, Dorica J, Rollin Y: Reduction in effluent discharge from recycled paperboard mills. In: Proc. Tappi Minimum Effluent mill Symp. Tappi Press, Atlanta, USA 1996; 335–340.
58. Rousseau S, Doiron B: Zero Process effluent discharge attained at St.-Laurent Paperboard Inc., Matane, Quebec. Pulp Pap. Can. 1996; 97(9): T302–T305.
59. Stuart P, Lagace P: Review of process modifications for implementation of zero discharge at recycled fiber board mills. In: Proc. Tappi Minimum Effluent mill Symp. Tappi Press, Atlanta, USA 1996; 327–333.
60. Barton DA, Stuart PR, Lagace P, Miner R: Experience with water system closure at recycled paperboard mills. Tappi J. 1995; 79(3): 191–197.
61. Jasmin G: Operating experience of the joint Scott Paper and Crabtee municipality effluent treatment system. Pulp Pap. Can. 1996; 97(2): T71–73.

10 Management of Wastewater Treatment Sludges

A variety of solid wastes are generated by pulp and paper mills that need final disposition. In addition to bark, wood waste, boiler ash, pulping residuals, and mill trash, it also generates wastewater treatment sludges like deinking sludge, primary clarifier sludge, and biological sludge. The quantity of sludges varies from site to site. On average, Canadian pulp and paper mills with activated sludge wastewater treatment system produce primary sludge of 31 kg(od)/ton of pulp while the secondary (biological) sludge generation is 16 kg/ton (Table 10.1).[1] A typical floatation deinking plant produces 80–150 kg of dry sludge/ton of recycled pulp.[2] Table 10.2 illustrates how the quantity of sludge generation varies with the type of pulping and paper making or both.[3] In the US, an NCASI study indicated that in 1988, 50 kg/ton of wastewater treatment sludge was generated averaged across the industry as a whole. Primary sludge accounted for 18%, secondary sludge 7%, and deinking sludge 2% of all the solid waste generation in US pulp and paper industry in 1986.[4] The US industry generated approximately 4.6 million tons of dry sludge solids in 1989.[4]

The amount of sludge generated by a recycled fiber mill is greatly dependent on the type of furnish being used and the end product being manufactured. For example, a recycled tissue mill and a recycled newsprint mill may use the same old newspaper (ONP) as furnish, but the higher brightness and lower dirt requirements of the tissue will result in a lower yield and higher sludge generation. Typical yields for various wastepaper recycling processes are summarized in Table 10.3.[5] The amount of sludge production by the use of these processes are given in Table 10.4.

Sludge composition is a function of the raw material, manufacturing process, chemicals used, final products, and the wastewater treatment technique. In the case of recycled papers, it also depends on the grade of paper fed to the recycling operation and the number and types of cleaning stages utilized. For example, sludge from mixed office wastepaper may contain high levels of clay and other types of fillers, printing inks, stickies from envelope adhesives, as well as fibers and paper fines. In fact, sludges from mixed office waste paper recycling operations may contain as much as 2% ash from fillers in the waste paper. Sludge solids produced by pulp and paper mills typically include a majority fraction of fiber. Depending on the mill ink, sand, rock, biological solids, clay/fillers, boiler ash, grits from recausticizing, etc. may make up the other fractions. Because of the constitutents that may exist, along with the water fraction, typical sludge analysis can vary widely. Therefore, it is

Table 10.1. Sludge generation for selected Canadian pulp and paper mills having activated sludge treatment facility

Mill type	Primary sludge (kg/ton)		Secondary sludge (kg/ton)	
	Mean	Range	Mean	Range
Bleached kraft	31	13–64	10	4–19
TMP[a]/Deinked	32	16–44	23	21–24
BCTMP[b]	35	21–67	24	12–44
All mills (average)	31	13–67	16	4–44

Based on data from Ref. 1.
[a] Thermomechanical pulp.
[b] Bleached chemi-thermomechanical pulp.

Table 10.2. Sludge generation rates by production category

Production category	Sludge (kg/ton)
Chemical pulping/low filler	9–36
Chemical pulping/high filler	23–68
Groundwood/newsprint and other	9–45
Semichemical/corrugating medium	9–27
Deinking/fine and tissue papers	36–136
Nonintegrated/fine papers	9–36
Recycled paper board	0–27

Based on data from Ref. 3.

Table 10.3. Waste paper recycling yields

Waste paper grade	Final product	Typical yield (%)
Old corrugated containers	Linerboard	87–92
Old newspapers	Deinked newsprint	80–85
Old newspapers	Deinked tissue	78–83
Mixed office wastepapers	Fine papers	65–73

Based on data from Ref. 5.

Table 10.4. Sludge generation from recycling operations

Waste paper grade	Final product	Yield (%)	Sludge generated (kg/ton)
Old corrugated containers	Linerboard	89	123
Old newspapers	Deinked newsprint	80	243
Mixed office waste paper	Fine paper	68	490

Based on data from Ref. 5.

critical to characterize the sludge carefully for determining the best method for sludge disposal.

The dumping of sludge in a company owned or municipal landfill is by far the most common method of disposal. However, alternative methods of disposal have been gaining acceptance as ways of reducing the volume of sludge that must be landfilled and tackling the environmental challenges posed by sludge disposal. Some of the alternative methods are given below:

1. ***Thermal Oxidation.*** This method of handling entails burning the organic material in the sludge and collecting the combustion ash for landfill disposal or for use as a feed stock for various products. To improve handling and storage characteristics, sludge can also be dewatered and pelletized for use as a fuel. It can be burnt in a refuse/bark boiler along with the hogged fuel in an existing stoker type grate fired boiler. In combination with coal or bark or 100% sludge can be utilized in a bubbling fluidized-bed boiler or circulating fluidized-bed boiler. The sludge can also be added to a cement kiln as a feed stock supplement, with ash becoming part of the cement composition.

2. ***Landfilling.*** This is the most common form of sludge disposal, but because of decreasing landfill space and new environmental regulations, use of this method will have to be reduced in the future.

3. ***Land Application.*** A number of sludge disposal methods are included in this category. All these methods involve processing of sludge either for agricultural use or for soil substitution. There are some risks involved with handling sludge in this manner due to the potential for buildup or runoff of potentially harmful substances. These methods are composting, land spreading (directly on farm or forest land where it adds to the organic matter in the sludge), and manufactured soil (sludge can be mixed with sand or poor soil and fertilizer to create a mixture for use in creating a top soil for use in land restoration).

4. ***Recycled Products.*** A number of innovative uses are being found for recycling sludge into other products and construction materials such as lightweight aggregates (pelletized sludge as a substitute for rock aggregate in concrete products) and cement additive (sludge is directly added to the concrete to develop certain desirable properties derived from its fibrous nature).

5. ***Reuse.*** Sludge can be cleaned and processed to recover good, long fibers for reuse in a mills furnish. The remaining sludge still requires disposal, but the volume can be reduced.

6. ***Miscellaneous.*** Sludge can also be utilized to produce alcohol, animal feed, or to recover clean energy by destructive distillation.

Dewatering of Sludge

Dewaterability of the sludge is very important because it determines the volume of waste that has to be handled. If a sludge is dewatered from 1 to 5% solids, four-fifths of the volume of material has been eliminated, 10% solids represents a nine-tenths volume reduction. The final waste volume is directly related to the cost of disposal. The nature of sludge differs greatly with the type of pulp and paper manufactured. Table 10.5 gives the ranking of various sludges on their resistance to dewatering.[6] The primary sludges that are high in fiber and low in ash are the easiest to dewater. The most difficult are the solids from the high-rate biological treatment systems. The primary sludge most difficult to dewater is that containing groundwood fines.

Those mills which are dewatering progressively more secondary sludges are faced with increasing challenges. Economic pressures are reducing the quantities of primary sludge (mainly fibers) discharged. Primary sludges are normally tertiary or quarternary pulp and paper mill rejects, but often consist of quality fibers having a high monetary value. As the percentage of secondary sludge increases, the dewatering characteristics deteriorate, resulting in decreased cake solid contents. A reduction in solids content adversely affects landfilling and incineration. Incineration of sludges with 28 to 50% solids content have been reported to impair boiler operation.[7] The dewatering properties of a 50:50 mixture are very poor. Some mills are approaching this 50:50 ratio and are therefore anticipating difficulties. Table 10.6 shows how the percentage of secondary sludge varies with production category. Tissue mills, NSSC plants, and recycle paper board plants have problems with dewaterability of combined sludges.

Sometimes, it may be desirable to dewater the primary sludge separately from the secondary sludge. One example is a situation in which the secondary sludge can be disposed of through land application. Blended sludges are not usually suitable for such disposal. Another example is a situation in which the

Table 10.5. Relative dewaterability of pulp and paper mill sludge

Sludge	Ranking of resistance[a] to dewatering
Primary sludge (>20% fiber or <30% ash)	1
Primary sludge (>20% fiber or >30% ash)	2
Hydrous primary sludge (groundwood, glassine, etc.)	3
Combined primary and secondary sludges	4
High-rate treatment system biological solids	5
Posttreatment alum or filter backwash solids	6

Based on data from Ref. 6.
[a] Relative resistance to dewatering increases with number.

primary sludge can be used to produce a by-product or can be reused within the production process, but the blended sludge cannot be used.

Sludges must be predewatered if the combined consistency is less than 4–5%. Predewatering assists the dewatering process by reducing solution volume while increasing solid content for further dewatering, absorbing fluctuations of inlet solids consistency while stabilizing the output consistency, increasing outlet solids content and solids capture efficiency, and reducing overall polymer consumption. Most prevalent predewatering technologies include rotary sludge thickeners (RSTs) and gravity thickeners. Other technologies in use include gravity table thickeners, dissolved air floatation (DAF) clarifiers, and belt presses. Although not common, centrifuges, V-presses, coil vacuum filters, and fabric vacuum filters are also used. The floatation thickener used on secondary sludge can achieve approximately 4% solids, giving it an advantage over the gravity thickener, which can achieve only about 2%.

Gravity thickener is a radial clarifier that is normally 11 to 12 m in diameter and 3 to 4 m in depth. The advantages of gravity thickeners include: simplicity, low operating costs, low operator attention, and a degree of sludge storage. Conditioning chemicals are not normally required and there is minimal power consumption. However, these advantages are often offset by potential septicity/odor, less dewatering capability (as compared to other technologies), and large space requirements. These disadvantages have limited the use of gravity thickeners in recent installations. An RST is a rotary screen where water is removed by gravity and tumbling action. RSTs have been installed in many mills as predewatering units before the screw presses. This type of predewatering device is capable of increasing consistency to between 4 to 10%, depending upon the proportion of secondary sludge and the percentage of solids from the secondary and primary clarifiers. In a gravity table filter, sludge is drained on a rotary wire. Drainage is assisted by moving paddles. The paddles are required to prevent wire pluggage. Gravity tables and RSTs produce sludges of similar consistency. Gravity tables are normally placed over screw presses to allow feeding by gravity. As with RSTs, polymers are applied before the table filter. With DAF clarifiers, secondary sludge is floated

Table 10.6. Combined sludge composition at pulp and paper mill treatment systems

Pulp/paper production	Secondary sludge (%)
Bleached kraft/paper, paper board	20–40
Groundwood/newsprint	25–35
Sulfite/tissue	50
NSSC/semichem. board	60
Nonintegrated fine paper	25
Recycle paper board	100

Based on data from Ref. 7.

with dissolved air, usually with the aid of some dawatering chemicals. Sludge is skimmed from the surface of the clarifier and the underflow retreated in the aeration pond or the primary clarifier. In the DAF process, solids can be increased to 3–6% for secondary sludges. The actual performance is frequently dependent upon the type of chemical applied and the dosage rate. DAF units also have the potential to eliminate odor problems. One American mill and one Russian mill rely solely on DAF units for sludge predewatering.[8] A few American activated sludge treatment plants use DAF units in combination with rotary sludge thickeners. One Canadian mill uses coil filters and V-presses to dewater primary sludge.[8] After predewatering of the primary sludge, the secondary clarifier sludge and predewatered primary sludge are mixed in a paddle mixer and then discharged for final dewatering on screw presses. No dewatering chemicals are required. Vacuum filter dewatering of biological sludges has been phased out of service in North America. Problems with poor capture rates, blinding, and landfilling difficulties have eliminated this option. The performance of different predewatering devices is summarized in Table 10.7.

Very often, conditioning of the sludges is required before dewatering. In general, the following chemicals are used for this, regardless of the type of equipment: lime, ferric chloride, and polymers. The three can be used separately or in combination. Ferric chloride has the disadvantage of being highly corrosive but is a very effective conditioning agent. The sludges that are difficult to dewater require high polymer addition rates. Wet air oxidation has also been used as a conditioning process to aid sludge dewatering and has been commercially applied in the paper industry,[9] where filler recovery was a side benefit. However, the brightness of the filler recovered was lower than that of the filler grade clay and the installation experienced considerable down time and high maintenance costs.

The commonly used dewatering devices used in the paper industry are rotary vacuum filters, centrifuges, V presses, twin-wire presses, and screw presses. In some situations, the primary dewatering device is followed by a press to further increase the solids content. The vacuum filter had been the most popular device. It operates on the same principle as a pulp decker. The key design parameter is the filter solids loading in kg/m^2/h, which is usually determined in bench-scale testing with a filter leaf test apparatus. The filter leaf test gives specific filtration resistance volumes. Solids capture in vacuum

Table 10.7. Performance of predewatering devices

Equipment	Expected solids (%)
Gravity thickeners	<3
RSTs	4–10
DAFs	3–6
Gravity tables	4–10

Based on data from Ref. 8.

filter is 90–95%, and the cake produced is about 20% solids. In order for the filter cake to discharge properly from the filter, 10–20% long fiber (>100 mesh) must be present in the sludge.[6] Vacuum filter cakes containing combined sludge solids can be further dewatered on V presses to approximately 35–40% consistency. A V press is just two disks providing a converging nip that applies pressure to the sludge to squeeze out the water. Vacuum filters can be equipped with either fabric media or steel coils. Fabric media are often used in situations when fiber content is low, the ash content is high, or the solids are otherwise difficult to dewater on a coil filter. The power costs for operating the large vacuum pump required by a vacuum filter are quite high. Vacuum filters are being replaced by belt presses, which seem to perform as well, if not better, at lower operating cost.

Disk centrifuges have found little application in the paper industry. They have been tried as thickening devices but experience has been unsatisfactory. Basket centrifuges have been used to a limited extent for sludges that are very difficult to dewater, but they operate in a batch mode rather than continuously. Usually, it is desirable to use the continuous decanter scroll centrifuge. Special scroll units have been developed for secondary sludge, and they are usually preferred over the basket centrifuge. Scroll centrifuges dewatering combined paper industry sludges generally produce cakes of 20–40% consistency at solids capture efficiencies of 85–98% from sludges conditioned with polymer. As the centrifuges operate on the basis of density difference separation, the sludges which are much denser than water, such as high-ash sludges, are the best for application of centrifuges. Specially designed scroll centrifuges can dewater secondary sludge from 2 to 11% solids with 99.9% capture efficiency.[10] However, 6–8 kg polymer per ton of sludge is required for conditionining. Centrifuges have a relatively low capital cost but can be expensive to operate due to the requirement of chemical conditioning agents, their high power requirements, and their maintenance costs. Dissatisfaction with centrifugation has been attributed to the following: (1) generation of poor-quality supernatant that could cause a buildup of fines in the treatment system, (2) susceptibility of centrifuges to plugging with pieces of bark, and (3) the severe screw conveyor abrasion experienced at many mills.

V presses have been applied successfully to the dewatering centrifuge and vacuum filter cakes containing as much as 30% biological solids. However, the combined sludges normally encountered require sufficient conditioning for vacuum filtration or centrifugation to render them amenable to V pressing.[6] V pressing can be performed to raise the solids content of the sludge high enough for incineration.[11] V presses generally produce primary sludge cake consistencies of 30–45%. Either a V press or a screw press would precede most bark boilers burning bark and sludge. The sludge would enter the press at 15–25% solids and be subjected to a pressure of 690 kPa to raise the solids to the 30–45% suitable for incineration.[12]

Pressure filters are the most powerful dewatering devices available. For combined sludge, cake of 30–35% consistency can be produced with solids capture efficiency of 95–100%.[6] However, it is necessary to precoat the filter cloth to facilitate cake discharge and minimize the frequency of media cleaning. Diatomaceous earth, fly ash, cement dust, etc. can be used for precoating. Media cleanliness has been indicated as a crucial parameter in determining the pressure filter cycle time. Pressure filtration also requires conditioning of the sludge before filtration. On pure secondary sludge, 35–40% cake solids can be achieved with a conditioning agent and a pressure of 200–250 psi. The main drawback of the pressure filter is that it is a batch operation and requires a lot of operator attention. Continuously operating automatic units have also been developed, but they are mechanically complex and therefore subject to many maintenance problems.

The moving-belt press (twin-wire press) has received intensive interest from industry in the past. Many paper mills have installed moving-belt presses. On primary or combined sludges, moving-belt presses have generated cakes of a consistency comparable to that of two-stage dewatering with V presses, with similar or somewhat higher conditioning costs and generally lower power consumption. Polymers are commonly used for the sludge conditioning, and some processes use dual-polymer systems. The cake solids are 20–50% for the primary sludge, and they are 10–20% solids for the secondary sludge. Capture efficiency is very high for belt presses, about 95–99% of the solids fed. Requirement of operator attention is low. These presses require power only to drive the belt; thus, they are energy-efficient. Another advantage is their ability to operate on secondary biological sludge. However, the major operating problem is belt life, which is only a few months. The usual cause of failure is puncture of the belt by incompressible objects in the sludge. The press is also subjected to corrosion due to hydrogen sulfide gas that is sometimes generated if there is any sulfur content in the sludge.

The latest development in sludge dewatering is a screw press of new design. These presses produce cake solids of 50–55% when operated as the only sludge dewatering device; solids capture ranges from 70–88% with no polymer addition on primary sludge.[13] Biological solids adversely affect solids recovery. Polymer can be used to improve efficiency but it has little or no effect on final sludge consistency; therefore is often not used on primary sludge. With secondary sludge, polymer is used. These presses appear to be energy-efficient. Screw presses are replacing twin-wire presses as the dewatering technology of choice for the pulp and paper industry.

Chemically precipitated tertiary sludges pose significant dewatering challenges. In a US mill, dewatering of activated sludge with tertiary chemical sludge substantially reduced the efficiency of belt presses. Belt press solids dropped from 25–30% to the 20% range when tertiary sludge was added. The bulk of tertiary sludge was shown to be lignin and not precipitated chemical (alum).[14]

It is estimated that up to 50% of the total cost of activated sludge treatment is due to sludge management.[15] For this reason, optimization of the sludge-dewatering process is very important. Implementation of instrumentation and improved sludge mixing are potential options to help increase the effectiveness of the sludge-dewatering process. Selection of predewatering and dewatering equipment can be based on capital costs, operating costs, and disposal method of the dewatered sludge. Some mills have installed unique systems for dewatering of sludge, some of which may be applicable to other operations also. The end disposal of sludge often dictates the type of sludge-dewatering equipment. Screw presses are mostly used for incineration of sludges.

Alternatives of Disposal

The pulp and paper industry disposes of its dewatered solids primarily by landfilling or by burning for power generation. Table 10.8 presents industrial experience as of 1989.[4] Landfills are still the dominant method of disposal but burning for energy is gaining.[3] Since landfill sites are increasingly more difficult to obtain, more mills are facing the need to incinerate. Land application of sludge as a soil amendment is also being practiced to a limited extent. One percent of total sludge produced is recycled. Sludge reuse is practiced by some waste paperboard mills that reuse the sludge as part of the furnish. Even some secondary biological sludge is reused in this way. Seven percent of the sludge produced is sent to the storage lagoons, where it is left to dry out slowly as the water drains away or evaporates. This practice requires a large land area and is gradually being phased out.[3] Other alternatives being investigated and put into practice for sludge disposal and utilization include: composting and synthetic soils, production of ethanol and animal feed, utilization as a hydraulic barrier layer in landfill cover systems and as fuel pellets, in building and ceramic materials, recovery of fibers and fillers, etc.[16]

1. Landfilling

Landfilling of sludge dominates the industry's sludge disposal practices.[17,18] Mills have typically favored landfilling whenever disposal sites are readily available and handling costs are low.[18] Landfilling has been the preferred

Table 10.8. Disposal of wastewater treatment sludges

Disposal alternative	1979 (%)	1988/89 (%)
Landfill or lagoon	86	70
Burning for energy	11	21
Land application	2	8
Recycle/reuse/byproducts	<1	1
Other	<1	<1

Reproduced with permission from Tappi Press, Ref. 3.

option because of the relatively low capital investment and the availability of mill-owned land. In recent years, however, regulatory agencies have recognized the potential for far-reaching adverse environmental effects from landfilling activities. This has resulted in the tightening of regulations and requirements for more monitoring, environmental impact assessments, closure plans, and public consideration. The average handling cost for land disposal in 1989 was \$12.8 per ton.[4] Normal sanitary landfill practices should be observed in constructing an industrial landfill. Some of the requirements that must be met are as follows:[12] (1) the disposal site should be at a minimum distance above groundwater, (2) all subsurface conduits – such as culverts, gas and water lines – should be removed, (3) the site should be above the flood plain and be protected from flooding, (4) the site should be at a minimum distance from a public well, highways, and watercourse, and (5) the nearest property line should be a certain distance away. After a site is chosen, according to the listed criteria, it should be used in accordance with good operating procedures for sanitary landfills.[19] Studies of the specific requirements for the design of papermill landfills are described by Wardwell,[20] Holt,[21] and Ledbitter[22] Modern landfill will require a liner design as shown in Fig 10.1. A leachate collection system is required plus FML liners and a clay liner. In daily use, intermediate cover is usually not required, but a final cover will be, and it must be impermeable, properly sloped, vented, and have the ability to support vegetation.

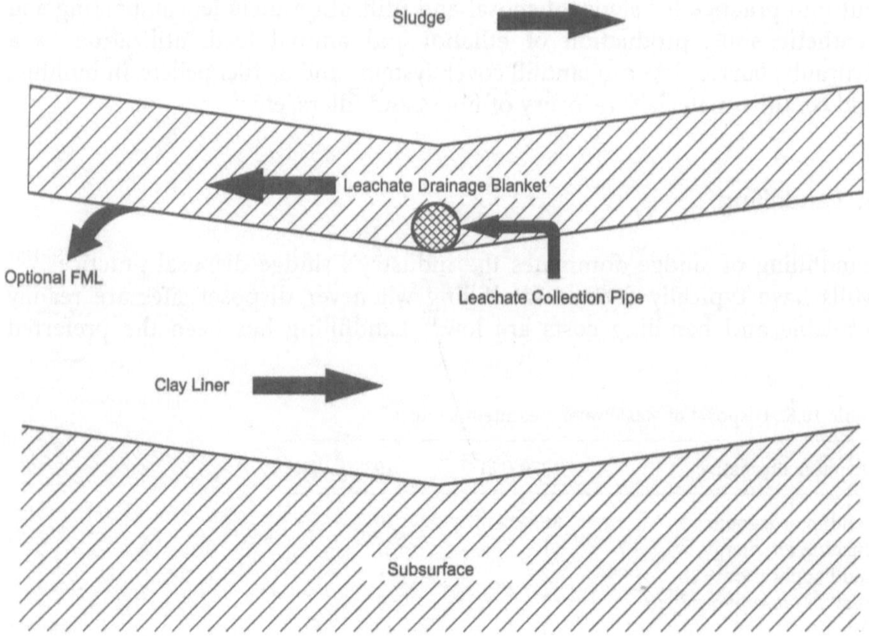

Fig. 10.1. Landfill liner design. Reproduced with permission from Tappi Press, Ref. 3

Most of the environmental effects from landfills arise from the runoff of liquid leached from the waste, that is, the leachate. Leachate is generated at solid waste landfills as a result of physical, chemical, and biological activity within the landfill. Leachate characteristics are effected by (1) precipitation, (2) runoff from and runon into the landfill, (3) groundwater flow into the landfill, (4) evapotranspiration, and (5) consolidation and water generated during the decomposition of the waste. These factors depend on local conditions such as climate, topography, soils, hydogeology, the type of cover on the filled sections, and the type of waste. Leachates from pulp and paper industry landfills are known to contain conventional pollutants as well as metals, volatile organic compounds, phenolic compounds, volatile fatty acids, and some base neutral compounds.[23] A 1992 NCASI study[23] found that metals were usually present at fairly low concentrations. Volatile organic compounds were detected, toluene being the most common with a median concentration of 35 µg/l, which is well below the Canadian Council of Resource and Environment Minister's goal of 300 µg/l for protection of aquatic life.[24] The only base/neutral compounds found more than once in detectable quantities were bis-(2 ethyl-hexyl)-pthalate and di-n-octyl pthalate. Pthalates are used in plasticizers, defoamers, and lubricating oils. Several kinds of phenolic compounds may be found in pulp mill landfill leachates including cresol isomers, phenols, and chlorinated phenols. Volatile fatty acids are produced from the decomposition of organic matter under anaerobic conditions and are common to leachates from many type of landfills. Acetic acid and propionic acid were found in the highest concentrations in pulp and paper mill landfills. A comparison of the average TOC and COD concentrations and the total VFA concentrations showed that VFAs contributed from 7 to 100% of the organic material in kraft mill landfill leachetes.[23] These leachates, if not properly collected and treated, may contaminate groundwater or surface water bodies. When landfills are on relatively permeable soils such as sand or gravel, leachate migration may cause contamination over areas many times larger than the area of the landfill. This can also occur over impermeable surfaces such as bed rock, where the leachate can flow quickly towards a receptor. Groundwater contamination is a concern if the groundwater is a drinking water source or if it flows to a surface water body. If groundwater contamination directly affects the drinking water supply, the liability implications for the landfill owner/operator may be enormous. In addition to impairment of drinking water quality, leachate contamination of ground or surface water may result in the impairment of biological communities, esthetics, and recreational uses. Recognition of these potential effects, together with public awareness of landfilling issues, dictates the necessity for a thorough EIA of new landfill sites.

In Canada, while the regulatory framework does not typically require an EIA for pulp and paper landfill proposals, many of the components of an EIA are fundamental to a successful permitting process. The key components include establishing a site development and approval plan, conducting effective public consultation throughout the process, and undertaking solid technical studies and impact assessment analysis in support of the project. The steps to achieving a successful environmental permitting process for a landfill development are illustrated in Fig 10.2.[18] The mill will need to decide on the specific scope of work based on the environmental conditions of the site, the community needs, and the input from local regulatory agencies. Regardless of scope or approach, the mill, as a proponent of a new landfill development, must recognize the long-term commitment associated with landfill effects and adopt a management approach which incorporates public involvement with solid technical design and assessment.

A cost-effective approach has been developed and recently applied to a landfill in Ontario.[18] Essentially, a control chart method is used where warning and control limits are established for selected leachate indicators. Leachate indicators are selected based on the ratios between background and leachate concentrations, with the highest ratios indicating the most appropriate indicators. The leachate indicators selected should also represent different chemical groups such as metals, nutrients, ions, and organic compounds. Before landfill operation, the selected leachate indicators (three to five chemical con-

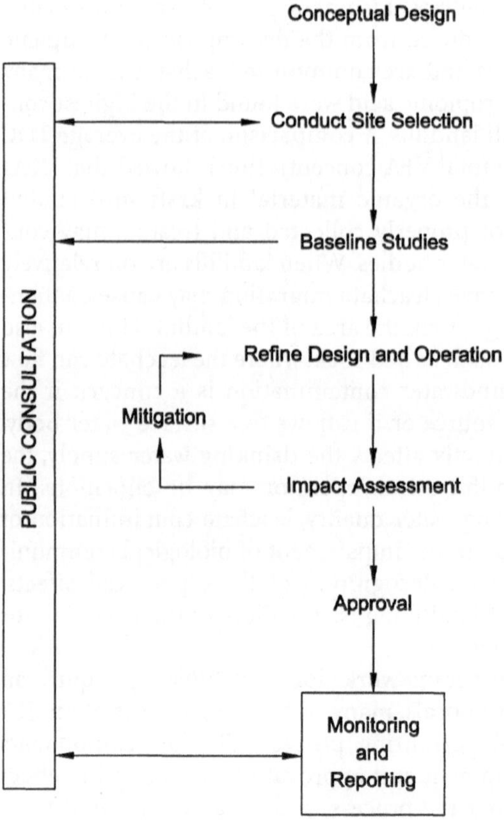

Fig. 10.2. Development of a landfill site from concept to closure-environmental considerations. Reproduced with permission from Pulp & Paper Canada, Ref. 18

stituents) are monitored monthly and the concentration differential is used to establish the warning and control limits. The landfill is monitored monthly during the operation and the concentration differential is plotted on a graph for each leachate indicator with the warning and control limits. If the value is within the warning limit, no action is required; however, if the value is above warning or control limits, an established response is implemented to determine the cause and, if necessary, initiate mitigative measures. The use of control charts for tracking water quality is beneficial as it is easily interpretable by the public and the mill's environmental managers.

2. Incineration

Disposal of sludge by combustion is becoming an alternative for many pulp and paper manufacturers due to constraints imposed on other disposal methods and the continuous pressure to reduce costs. Effective dewatering techniques have been and are being developed to produce a fuel with real potential. Traditional methods of firing these fuels have been nothing more than an incineration or disposal process with little energy gain for operation of the facility. The following three types of incineration are in practice in the industry: (1) burning in an incinerator specifically designed for the sludge, (2) burning in the bark boiler, and (3) burning in a power boiler that also burns fossil fuel. Burning the sludge in the bark boiler, which is a hogged fuel (combination fuel) boiler, seems to cause few problems except for reduced steam generation and reduced boiler efficiency.[25] Incineration in the bark boiler appears to be acceptable for sludge incineration if such a boiler is available on the mill site and if it can take the increased water load. Dewatering to higher levels, 45–50% solids, will make bark boiler incineration even more attractive and will minimize the effect on boiler operation. Table 10.9 summarizes some of the characteristics of burning sludge in combination fuel-fired boilers.

Combustion properties of a sludge are generally related to the amount of fiber present. Energy available is usually inversely related to the ash content. High ash values (up to 50% on dry basis) correlate with relatively low heating values. Sulfur values are important as related to emissions. Dewatering of the sludge stream will be required to increase solids up to some minimum level before combustion will be beneficial or even break even. Self-sustained combustion is available with some sludges generated depending on the moisture and organic levels. Cost and benefit evaluations can be made that will indicate the moisture level for optimum performance. Removal of additional water to increase solids above 50% requires a different method similar to paper passing from the press section to the dryer section on a paper machine.[26] Thermal drying with hot gases or air can be carried out in a conveyor drier, cascade system, or a stand-alone drying unit. Reduced water content obviously helps improve efficiency and also can improve long-term storage options through reduced microbial growth.

Table 10.9. Sludge burning practices in combination fuel-fired boilers

Characteristics	Value	
	Range	Median
Sludge consistency (% solids)	20–47	33.0
Heat content of sludge (BTU/lb volatile sludge solids)	6300–9500	7900
Sludge ash content (%)	2–40	8.5
Sludge solids feed rate (lb sludge solids/100 lb boiler fuel solids)	2.5–14	7.5
Sludge water feed rate (lb sludge water/100 lb boiler fuel water)	2–32	17.0
Sludge BTU feed rate (sludge BTU/100 lb BTU in boiler fuel)	1.5–11.5	5.5
Sludge ash feed rate (lb sludge ash/100 lb boiler fuel ash)	3–55	13.0

Based on data from Ref. 25.

The sludge product may be available in several forms, depending on the method of combustion and the boiler used. Dewatered sludge straight off a screw press will be lumpy and, after moving through several conveying operations, begin to break up into a fuel that is fine, uniform, and fibrous in nature. Sludge may also be processed further into briquettes or pellets[27–29] to improve handling, storage, or combustion characteristics. Blending dewatered sludge with other fuel (chip fines or saw dust) can help improve conveying characteristics.[30] Pelletizing has recently come to the forefront as a method to convert combustible solid waste into a usable fuel. Waste to energy via pellet fuels needs to be examined more closely and regarded more highly as a successful solution to landfill crisis. They are quickly becoming a very viable and profitable alternative.[31]

Various types of combustion methods are available which include:[32] travelling grate boilers, vibrating grate boilers, other hog fuel boilers,[33] bubbling bed combustors,[34] circulating fluidized boilers, stage combustors, rotary kilns, and pyrolysis/pulse combustors. The practicality of the above would be based on the sludge characteristics (contaminant contents, fuel size, volatility, ash characteristics, heat content, etc.) and to a great degree the volume to be fired.[26] Operating experiences with stoker firing of TMP clarifier sludge with wood waste[33] and combustion of the wastewater clarifier underflow solids in a hog fuel boiler with a new high energy air system[35] have also been reported. Combined-cycle fluidized-bed combustion of sludges and other pulp and paper mill wastes to useful energy has been suggested.[36] Pulp and paper companies can improve the cost of operation by using proven, readily available power plant and combustion equipment and systems to efficiently convert the energy available in mill wastes to useful thermal energy and electrical power. By using the combined-cycle concept, either as the combustion turbine combine cycle or the diesel combined cycle, the firing of wood waste and sludge provides net energy gain for the operation of facility rather than merely a means of disposal.

Other alternatives of recovering energy from the sludge have also been tried. A sludge gasification plant has been tested to generate the clean fuel.[37] Steam reforming as an alternative method for disposal of waste sludge has been suggested.[38] A novel method of thermal treatment of contaminated deinking sludge has been proposed which is based on the application of the low-high-low temperature (LHL) regions during the combustion.[39] The LHL approach allows for the simultaneous encapsulation of heavy metals within solid particles, removal of submicron particulate, and destruction of polycyclic aromatic hydrocarbons before they are emitted into the atmosphere. The encapsulation of the heavy metal layers surrounding the heavy metal-rich cores of the ash particles may prevent the metals from leaching under acidic conditions.

Sludge can be easy to burn with the right combustion technology. Knowing that the right technology is very fuel-specific and having the technology characterization customized for site-specific conditions is essential to make proper combustion technology choices. Incineration is not practical for high-ash sludges. Stringent air-pollution emissions requirements for combination boilers have diminished the amount of incineration practiced. One of the Finnish mills incinerates sludge if the solids content is over 32%, and landfills the sludge if it is less than 32%.[1] Operation of the boiler must also be considered when the sludge is not available as a fuel. Several points of consideration include the combustion temperature, fuel feed systems, and boiler rating. Older boilers burning sludge as an alternative fuel should be able to simply return to earlier operating states.

Some of the chlorinated organics not eliminated through process modifications could be trapped on the sludge from the external treatment process(es). The disposal of pulp and paper mill sludges, which may contain chlorinated organic compounds, represents an increasing problem. However, if those sludges could be dried to 90% dry content, in an energy-efficient manner, they could provide a flame temperature high enough upon combustion to destroy the organic chlorides entrapped in the sludges. In addition, this approach could improve a mill's fuel self-sufficiency.

3. Land Application

Much work has been done with land application of pulp and paper mill sludge in the last 10 years.[17] In 1991, 25 mills were routinely doing land application, 10 others had carried out a one-time application, while 30 more have conducted serious field trials in Canada.[3] QUNO Inc. Thorold, Ontario, Canada, has over 6 years' experience with land application of primary, secondary, and deinking sludges.[40] Primary and deinking sludges have been found to have similar characteristics – low nitrogen and high fiber content. Conversely, secondary sludges (biosolids) have relatively high nitrogen and phosphorus content and low fiber content, making them more suitable for land application. Tests at QUNO found that the heavy metal content of the combined paper mill sludge was equivalent to that of cattle manure, and about one-tenth that of municipal sludge. The sludge has been successfully used as a replacement for manure in agricultural applications, as well as for land reclamation projects of old sand pits, coal/clinker sites, and a former foundry site. Recently, work has been completed with Alberta pulp and paper mills in conjunction with the Alberta Research Council on land application.[41] Land-spreading trials have been completed on both agricultural and forest cut-block sites. Research is also being conducted by the Alberta Newsprint Corporation and Alberta Research Council to evaluate the environmental effect of land spreading conventional and deinking sludge.[17] Preliminary research indicated that the procedure should not present any problems in regard to soil quality or plant growth. Trials with land spreading around the mill site have been successfully completed by applying the sludge on top of a gravel base. Alberta Research Council has also completed research on ash and sludge land spreading in conjunction with the Slave Lake Pulp Corporation (SLPC).[17] Grass yield on the test plot site at SLPC indicated as much as five times the yield of control plots. SLPC has had favorable results with sludge application on the surrounding agricultural area. Previously, landfilled sludge had been reclaimed and distributed to the farming community and applied using manure spreaders.

There has been considerable interest in the use of paper mill biosolids and ink waste in agricultural land for many years.[40] Sludges function only as amendments and not as fertilizers because they do not contain the elemental analysis required of a fertilizer.[42] For a soil amendment, the carbon nitrogen ratio should be 20:1 to 30:1. An average composition of seven different paper mill combined sludges from ten different mill types was 26:1, so this criteria is being met. The calcium/magnesium ratio should be above 6:1; many combined sludges fail to meet this criteria but the addition of lime to the sludge fulfills it. Sludges are good soil amendments for sandy soils. Detailed analysis of the seven combined sludges did not indicate a heavy metal problem.[43] Trials have been conducted in which fly ash and either primary sludge or secondary sludge were applied to crop land.[44] The fly ash-sludge blends were as effective as commercial fertilizer. In these same trials, lime mud applied to agricultural land performed better than dolomite limestone used for the same purpose.[44]Australian Newsprint mills Ltd. (ANM) used small quantities of biosolids on vegetable and horticultural gardens with good results and no observed detrimental effects.[45] Several farmers have also used the material on small pasture areas and on orchards, but no objective evaluations have been carried out. Because of the high level of farmer interest, ANM carried out land-spreading trials on crops and pastures.[45] The biosolids were utilized on a farm land close to mill. For this, a desk study and a survey of local farmers was conducted. It was found that biosolids would be readily used by farmers, if it could be demonstrated that it was a viable fertilizer, that it was safe to apply to the environment, and the cost was competitive with existing practices. This study also confirmed that about 2000 ha/year of land would be required to dispose of the material. It identified the area of interest of land for economic disposal as areas of crops and pasture land within 20 km of the mill and luceme flats where disposal could take place in winter. In 1992, a field experimental program started with a large area experiment on oats at a location known as Waitara. Biosolids were found to be slow to release their nutrients and produced an effect similar to fertilizer without producing adverse environmental effects. Rates of 16 to 64 t/ha were required to substitute for normal rates of conventional fertilizer. The results of trials of using biosolids in oats production is shown in Table 10.10. ANM also conducted trials to spread the biosolids on forest land.[45] Trials started in the Carabost and Green Hills State Forests, near Tumbarumba. The results are not available so far. The major disadvantage with forest spreading over agricultural land

Table 10.10. Results of trials of using biosolids in oats production

	Dry matter yield (kg/ha)	Grain yield (t/ha)	Grain protein (%)
1992 trials			
Biosolids (4 t/ha)	2164	1.68	8.8
Biosolids (8 t/ha)	2387	1.89	8.8
Biosolids (16 t/ha)	2576	2.06	9.7
Biosolids (32 t/ha)	3051	2.86	10.0
Superphosphate (15 kg P/ha)	5212	2.37	8.5
Phosohorus and nitrogen (120 kg/ha)	11051	2.25	9.9
Biosolids (32 t/ha incorporated)	7127	3.55	-
Biosolids (32 t/ha direct drilled)	4887	3.85	-
Control	1875	1.47	8.8
1993 trials			
Biosolids (4 t/ha)	2830	1.44	8.0
Biosolids (8 t/ha)	2910	1.63	7.8
Biosolids (16 t/ha)	3080	1.97	8.9
Biosolids (32 t/ha)	3792	2.02	10.2
Biosolids (64 t/ha)	4680	3.08	11.8
Superphosphate (15 kg P/ha)	4310	1.74	8.3
Phosohorus and nitrogen (120 kg/ha)	7620	3.80	12.4
Control	2640	1.15	7.9

Reproduced with permission from Appita, Ref. 45.

spreading is the higher cost of transport to the disposal site, which would normally make forest spreading unattractive. However, if the solid could be backloaded on log trucks, then the economic disadvantage decreases. In Canada, Greater Vancouver Regional District (GVRD) and the University of British Columbia's Forest Sciences department embarked on a 3-year research program at UBC's Malcolm Knapp Research Forest in Maple Ridge to determine the environmental and silvicultural application of recycling pulp and paper sludge and treated sewage sludge as an organic forest fertilizer called Nutrifor.[17] The second phase of the program is currently underway to introduce Nutrifor as a viable fertilizer for forestry and other users.

Scott Paper Ltd. in New Westminster is currently participating in a full-scale land application project with GVRD in which paper mill sludge is combined with municipal sludge and then applied to a tree farm in the Fraser Valley.[17] In 1990, the GVRD, Western Forest Products Ltd. and the IBEC Aquaculture participated in a fertilization project in which various mixtures of pulp mill wastes, sewage sludge, and fish mort silage were applied to forest sites in Southern British Columbia near Port McNeil on Vancouver Island.[46] Initial results indicate a rapid response by young conifers to organic fertilization. In 1992, a project cosponsored by Nutrifor was completed at Malaspina College where 600 dry tons (2500 wet tons) of sludge were applied over an area of 26 ha in the Malaspina College Research Forest on Central Vancouver Island.[47] Full-scale projections were made using data obtained from the trials to determine cost per ton of sludge for each of three application methods.[47] The lowest cost method of spreading the sludge was found to be dry application. Projected cost could be reduced to $56/wet ton to apply approximately 36 000 wet tons onto 400 ha.

Seattle, Washington, has a sludge management plan which calls for the development of a number of alternative methods.[40] Since halting ocean disposal in 1972, the system has made compost, undertaken strip mine reclamation, and is said to have been one of the first to use biosolids in forestry. An innovative application is the growing of hops for the beer industry. Seattle is making use of about 101 000 dry tons/year at 20% moisture. The effects of lands spreading wastewater sludges from pulp and paper mills were investigated by examining (1) the fate of chlorinated organic materials in landspread sludge and (2) the impact of sludge on plant growth and wild life.[48] The results indicated that high molecular weight chlorolignins were rapidly absorbed by soil or humic matter and organic chlorine was slowly released as inorganic chloride. There was no detectable release of new monomeric chlorolignin-related chlorocompounds. Even under severe extraction conditions, the extractability of low molecular weight chloroaromatic compounds decreased rapidly (half lives of 6–70 days), apparently the result of biodegradation and biologically mediated chemical binding into the soil humic structure. No persistent biotransformation products were detected. Sludge applications produced an increase in plant growth (grass, hay, corn, trees). Studies of wildlife on sludge-amended soils did not detect any adverse effects on the health of individuals or on reproductive parameters. Criteria have also been proposed for the land spreading of solid waste.[3] Briefly, the proposed criteria are: (1) the soil sludge mixture must not have a high content of heavy metal that can be taken up by growing plants, (2) the soil-waste pH should be 6.5 or higher, (3) excess nitrogen should not be applied beyond that normally taken up by the crop in one season, (4) the sludge applied should be free of living pathogenic organisms, and (5) solids must be applied in such a manner that they are not available for direct ingestion by domestic animals or humans. Land application is not a trouble-free technology, however.[3] The most commonly noted problems are odors, groundwater contamination, heavy metals, and specific organic toxics. Other problems are noise, surface water contamination, pathogens, and excessive nitrogen application. The process of applying sludge is dirty and noisy, so if there are houses in the vicinity, potential difficulties will arise. Actually, public and user acceptance has been very good because sludge is applied mostly to rural areas close to the mill and in some cases on mill-owned land.

4. Composting and Synthetic Soils

Research into the feasibility and potential benefits of composting pulp and paper mill sludges was most prevalent between about 1975 and 1985. Pulp and paper mill sludges are usually amenable to well-controlled composting techniques. Markets for compost include land application for agriculture, horticulture, land reclamation, landscaping and individual consumer use. One mill has had considerable success with marketing its composted sludge.[50] This mill presently composts about 50% of its sludge. The mill sells the compost to a limited number of distributors who market the material in an area within a 250-mile radius from the mill.[51] Initiation of new composting operations within the industry has slowed considerably since the mid-1980s. Lack of sufficiently large, locally available markets for compost, and regulatory concerns about the possible presence of chlorinated dioxins and furans in industry sludges are two common reasons for the limited utilization of this management alternative. Recent industry initiatives to reduce the presence of dioxin in sludges are likely to relieve some regulatory concerns about land application of sludges.

Recently, a mill in the northeastern United States began working with a third-party company to produce synthetic topsoil using sludge.[16] The process involves the homogenization of sludge with varying proportions of sand, gravel, and fertilizer to produce a synthetic soil. More than a dozen landfills have used the soil as part of the final cover. It also has use in other applications requiring vegetative cover. The pulp fiber content of the synthetic soil probably allows for an increased resistance to erosion before the establishment of vegetative cover.

5. Recovery of Raw Materials

Paper industry sludges usually contain significant percentages of both cellulose fiber and paper-making fillers such as clay and titanium dioxide. Attempts have been made to reduce sludge volume by reclaiming the fiber or filler or both for reuse.[16] Although reclamation of usable materials from sludge does occur, the industry more commonly uses in-mill loss control measures as the primary means of recovering raw materials. There are several methods to recover raw materials from sludge. One of the most common is to recycle primary sludge back into the mill's fiber-processing system. Recycled paperboard mills commonly use this technique. Some manufacturers of unbleached and bleached pulp and paper have also practiced recycling primary sludge back to the mill, with limited success.[52] Segregated effluents from paper machines, bleach plants, and various cleaning and screening operations can be good targets for fiber reclamation because they usually lack contaminants such as bark or causticizing waste solids. Using some fractionation scheme for the sludge may also provide recovery of fiber alone. The complexity of fiber-recovery systems varies widely and depends on the nature of the constituents in the sludge. Mills producing bleached pulp sometimes add recovered fiber to the unbleached pulp entering the bleach plant. This strategy allows for both the reclamation of unbleached fiber and the brightening of previously bleached fiber which may have dirtied by exposure to contaminants in the wastewater. Some mills have associated the reuse of fiber recovered from sludge with increased deposits of pitch on equipment. Use of a fractionation system helps to recover filler. Most systems for which pilot- or full-scale data are available have employed a thermal oxidation technique for destroying the organic fraction of the sludge to yield filler in the form of ash.[16] Experiments with calcination systems have revealed that controlling the kiln temperature at 816 and 843 °C helps to avoid formation of fused agglomerates which can cause the recovered filler to be excessively abrasive. Wet-air oxidation can be also used to recover filler materials from sludge. One US mill is presently practicing this process an a full scale.[16] Wet-air oxidation is an oxidation reaction carried out in a liquid environment under high temperature and pressure. This process is capable of reducing sludge volume through oxidation of the organic fraction to yield an ash composed of inert materials, e.g., filler clay, titanium dioxide, and calcium carbonate for reuse in the paper-making process. Initial experience with the operation of WAO unit for filler recovery revealed problems with Ca sulfate and Ca oxalate scale deposition. Both pilot- and full-scale systems have demonstrated some problems with low brightness of the recovered filler. In Turkey, primary sludge has been successfully used in the manufacture of hardboard.[53] Full-scale studies using sludge at a 1:4 ratio indicate that the use of 28 bdt/d of waste primary sludge mill save $455 000/year on wood costs and $130 000/year on electricity costs.

6. Production of Ethanol and Animal Feed

Ethanol is a common additive in automobile gasoline. Traditionally, it is produced by fermentation of starches and syrups. Recently, interest has been shown in producing ethanol from agricultural waste, municipal solid waste, and pulp and paper mill sludge in order to reduce production cost and to make ethanol more widely available. Laboratory and pilot-scale studies to produce ethanol from wood-based feedstocks have used acid and enzymatic hydrolysis followed by fermentation of the resulting sugars into ethanol.[54–56] Primary sludges can be used as feedstock for ethanol production because they are widely available in sufficient quantity and have little economic value. In the University of Florida, Dr. Ingram's group has conducted research on the conversion of cellulose and hemicellulose fractions of wood-based feedstocks into hexose and pentose sugars followed by fermentation to ethanol using a genetically engineered strain of *Escherichia coli*.[57] The advantage of this process is that it can ferment both the pentose and hexose sugars into ethanol, thereby increasing the overall yield. It is expected that this technology will be commercialized soon.[58]

Sludge has also been used for production of animal feed, for which there are two methods. One involves production of single-cell protein. Cell protein is present in secondary sludge and derives from the fermentation of fibrous sludge. It is possible to dry these proteins and incorporate them into feed mixtures. In the USA, one mill used a process to convert secondary sludge into saleable protein product for use in animal feed.[59] Mechanically dewatering secondary sludge to 12% solids with further dewatering by feeding a mixture of sludge and oil to specially designed multiple-effect falling film evaporators produced a 45% protein material. Centrifugation of the evaporator discharge gave 83% dry solids, 1% water, and 16% oil. Targeted markets for the finished product included feed for cattle and poultry and use in agricultural composting.[59] However, acceptance of this product on the market was not sufficient to support continued production.

The second method incorporates sludge directly into animal feed mixtures.[16] This method exploits the presence of carbohydrates, which are primarily in the form of cellulose and other nutrients present in primary or combined sludges. Research in the early 1970s included experiments on the palatibility and digestibility of sludge-augmented feed mixture on goats, sheep, and cattle. It was found that the digestibility of sludge relates directly to the carbohydrate content and inversely to the ash and lignin content. Hardwood pulp residues were found to be more digestible than softwood residues.[60]

7. Pelletization of Sludge

The reasons for producing sludge pellets are: (1) volume reduction, (2) odor control, (3) recovery of fuel value, and (4) by-product applications. The most common reason for production of pellets is for use as an alternative fuel. One mill transports dewatered sludge to an off-site pellet mill for drying and formation into pellets. The mill purchases the finished pellets as a fuel supplement. The finished pellets contain 15–20% moisture and 10% ash. They have a heating value of 14.7×10^6 Joules/kg.[16]

Two companies are now manufacturing pellets by using mixtures of sludge and nonrecyclable paper.[61] These pellets are being marketed as an alternative fuel compatible for use in most stoker and some pulverized coal boilers. The amount of sludge in these pellets can range between 10 and 66%. It is possible to control the fuel value of the pellets by manipulating both the sludge content and the grade of nonrecyclable paper used. The fuel values of the finished pellets are in the range of $14–23 \times 10^6$ J/kg. The regulatory agencies require evaluation of alternative fuels for by-products of combustion before widespread use of the fuel. Companies involved in both production and use of sludge and NRP fuel pellets have indicated that regulatory reaction to trial run data has generally been positive. NCASI has developed a proprietary process to convert combined sludge from a recovered paper deinking mill into a granular product. The product has been used as a carrier material for agricultural as well as home and garden pesticides and can compete with other common pesticide-carrier materials composed of clay, vermiculite, diatomaceous, and cob products. Claims for the product indicate that it is superior to some of these conventional carriers because it is dust-free and attrition-resistant.[16] The company's production facility has a capacity of 180 tons/day of the granular product.

Kitty litter, poultry litter, and large animal bedding have all used pelletized sludge. One US mill processes all of its primary sludge into several varieties of animal litter sold to a distributor for marketing. The litter production process is proprietary. It involves sanitizing and deodorizing primary sludge followed by drying and pelletization. Kitty litter is the primary product manufactured, but other products include large animal bedding, pet bedding, and bedding for laboratory animals. Grocery stores market kitty litter and feed stores market bedding products. Bedding sells in 25- and 50-lb bags and 1000-lb totebins.[16] Several other companies have studied the feasibility of using sludge to produce kitty or poultry litter. In these cases, they have usually demonstrated production of a quality litter product from primary sludge. Initial capital costs, distribution, and marketing issues and incompatibility with company business strategies have inhibited some companies from pursuing this by-product alternative.

8. Manufacture of Building and Ceramic Materials and Light-weight Aggregate

Sludge use in building products has followed three general techniques. One method is the use of sludge as a feedstock to a cement kiln. Raw materials used to produce cement can include calcium carbonate, clay, silica, and smaller amounts of aluminum and iron. Some sludges contain significant quantities of these materials. Two companies have extensively investigated this alternative and one mill currently practices this full scale.[62] The mill sends all its primary sludge and all its coal boiler ash to the cement manufacturer. This is a combined total of approximately 100 tons/day. For the kiln involved, this amount of material represents only about 2% of the total feed stock.

Another alternative is the use of sludge in cementitious products. A lot of work has been done on the use of organic fibers, including wood pulp, in cementitious composites. The advantages include increased durability and pumpability as well as reduced shrinkage-related cracking.[63] Two studies undertaken to assess the performance characteristics of composites, which included paper industry sludge, concluded that a composite material potentially useful in building blocks, wallboards, panels, shingles, fire retardants, and filler materials for fireproof doors could result from combining Portland cement with sludge from a deinking mill.[63] It was found that mixtures including Portland cement, ash, sand, and sludge yielded a compressive strength comparable to conventional concrete and superior flexural strength.[63]

Sludge has also been used in the production of lightweight aggregates (LWA).[16] Aggregate is a term describing a collection of materials used as a filler in construction materials. Aggregates find use in cementitious products such as concrete, masonary, building blocks, and asphalt. Sand and gravel, or both, are typical aggregate materials mixed with cement to produce concrete. Lightweight aggregate refers to a select group of materials which allow for reductions in final density while maintaining acceptable strength properties. Products which sometimes incorporate LWA include concrete block, architectural panels, and decorative stone.

9. Landfill Cover Barrier

Paper industry sludges have been found to show low hydraulic conductivity (permeability). This finding has led to research by many groups on the potential utilization of sludge as hydraulic barrier layer in landfill cover systems. In 1987, NCASI completed construction of four pilot-scale field test cells designed to allow investigation and comparison of the performance of hydraulic barrier layers made from sludge and made from clay.[16] Results obtained from these cells during the first 5 years of operation indicate that the sludge barriers perform as well or better than the clay barriers. Experience with the use of paper industry sludge as daily, interim, and final cover for paper industry and municipal landfills is available. Worthy of special mention is the experience of one organization. To demonstrate the utility of paper mill sludge as landfill-capping material, this recovered-fiber processing mill constructed six test cells to compare the performance of primary sludge, combined sludge, and clay as hydraulic barriers.[16] Data from these test cells sufficiently supported a petition to the Massachusetts Department of Environmental Protection for a full-scale demonstration project. The project involved capping a 2-ha municipal landfill with combined mill sludge. To date, monitoring of cap performance indicates that the demonstration has been successful. The company has received a "beneficial use determination" which allows use of the sludge in other landfill cover systems in Massachusetts.

10. Other Uses

Pyrolysis, gasification, and supercritical water oxidation have been studied as a way of reducing sludge volume. During pyrolysis, oil-like liquids and gases are formed which have fuel value. Study has been conducted on pyrolysis of cellulose-based waste materials but there is not much published information on experience with pyrolysis of pulp and paper industry wastes.[16] Pilot studies have been conducted on the application of this technology to wood chips, recycle mill sludge, and bleached kraft mill sludge, There is no report on a full-scale experience with the pyrolysis of sludge. Supercritical water oxidation has undergone research as a waste management technology for approximately 10 years. The process involves the decomposition of organic and some inorganic material in the aqueous phase above the critical point of water (374°C and a pressure of 22×10^3 kPa). In this state, organic materials become much more soluble in water and oxidize readily. The principle of supercritical water oxidation is similar to that of wet-air oxidation except that wet-air oxidation maintains subcritical conditions. No full-scale supercritical water oxidation units are currently in operation. Laboratory-scale research has been conducted on supercritical water oxidation of pulp and paper mill sludge. This work used an 80 cm^3/min benchtop system. Operating limits for the reactor were 600°C and 25.5×10^8 kPa. Residence time in the reactor ranged from 10 s to 10 min. In the experiments, a 99% reduction of total organic carbon was possible.[16] The problems anticipated with large-scale and or full-scale systems involve (1) corrosion of equipment particularly for low pH and high chloride concentration wastes and (2) deposition of salts or pyrolytic chars which could lead to plugging or increased cleaning requirements.

In Canada, Ensyn Technologies has developed a Rapid Thermal Processing (RTP) reactor which heats biomass to an extremely high temperature (400 to 900°C) for 0.5 s at atmospheric pressure with no oxygen.[64] RTP is also called fast cracking, and is similar to the catalytic cracking process used by the oil industry. The rapid heating of the biomass cracks the chemical bonds and produces a liquid bio-oil. Rapid cooling prevents the completion of chemical reactions. The feed stock can vary: pulp sludge, wood waste, rice husks, agricultural residue. The bio-oil created from the process has been used as a fuel-oil substitute. Destructive distillation as a resource recovery process for solid waste was evaluated during 1982–1984 at Marcel Paper Mills, Elmwood Park, New Jersey (USA).[65] The results indicate that the process is environmentally friendly and has the abilility to provide substantial energy savings utilizing organic solid waste as its sole source of fuel. The technology is able to fractionate the biomass content of municipal and industrial wastewater sludge to a combustible gas and inert char in an environmentally safe manner, as shown in Fig. 10.3. Full-scale operation of the process was carried out on sewage and deinked paper mill sludge at installations in California and New Jersey.

The expense of solids disposal could be eliminated by destroying the microorganisms in the excess secondary sludge and recycling the material through the treatment process. Springer et al. used a simple mechanical device – a kady mill to break down the microorganisms in the excess sludge, allowing all the material to be recycled to the treatment process.[66] The kady mill combines the effects of high shear and temperatures, both of which are required for efficient cell distruction. Based on 60 days of operating data, it was found feasible to operate an activated sludge plant in extended aeration mode by recycling sludge that has been lysed in a kady mill. This process could be an alternative wastewater-treatment system for use in the pulp and paper industry. The system would be most suitable for use in mills operating well within EPA permit discharge limits for BOD. This system operated with an average COD-removal efficiency of 80%, compared with 87% removal for the conventional system. Both systems operated with an influent COD of 260 mg/l. The sludge-lysis-and-recycle process operated free of bulking problems. This process appears to be an economically attractive alternative to conventional treatment if higher BOD values can be accommodated.

The biosolids generated by the activated sludge process can also be anaerobically digested to reduce its volatile solids and generate energy in the form of methane gas.[67]

Conclusion

Environmental impact will be the major factor to be addressed in planning for the stabilization and ultimate disposal of paper mill sludge. In the near future, we can expect even more stringent environmental regulations to control air and solids emissions coming from ultimate sludge disposal processes. The methods that produce unacceptable air emissions or produce solid residues containing toxic metals or chemicals, especially dioxins and furans, will be facing costly retrofit in order to comply. However, paper mill sludges have a net environmental advantage over sewage sludges in that they are nearly pathogen-free; handling and use pose lower health risks.

The challenge to find efficient methods for firing sludge still exists today and is becoming increasingly important as pulp and paper mills strive to be competitive. So far, incineration has been the primary alternative to landfill. However, incineration is associated with environmental pollution problems. The emission of gaseous NO_x and SO_2 are the major precursors of acid rain. The residue ash contains various toxic metals which need to be landfilled and

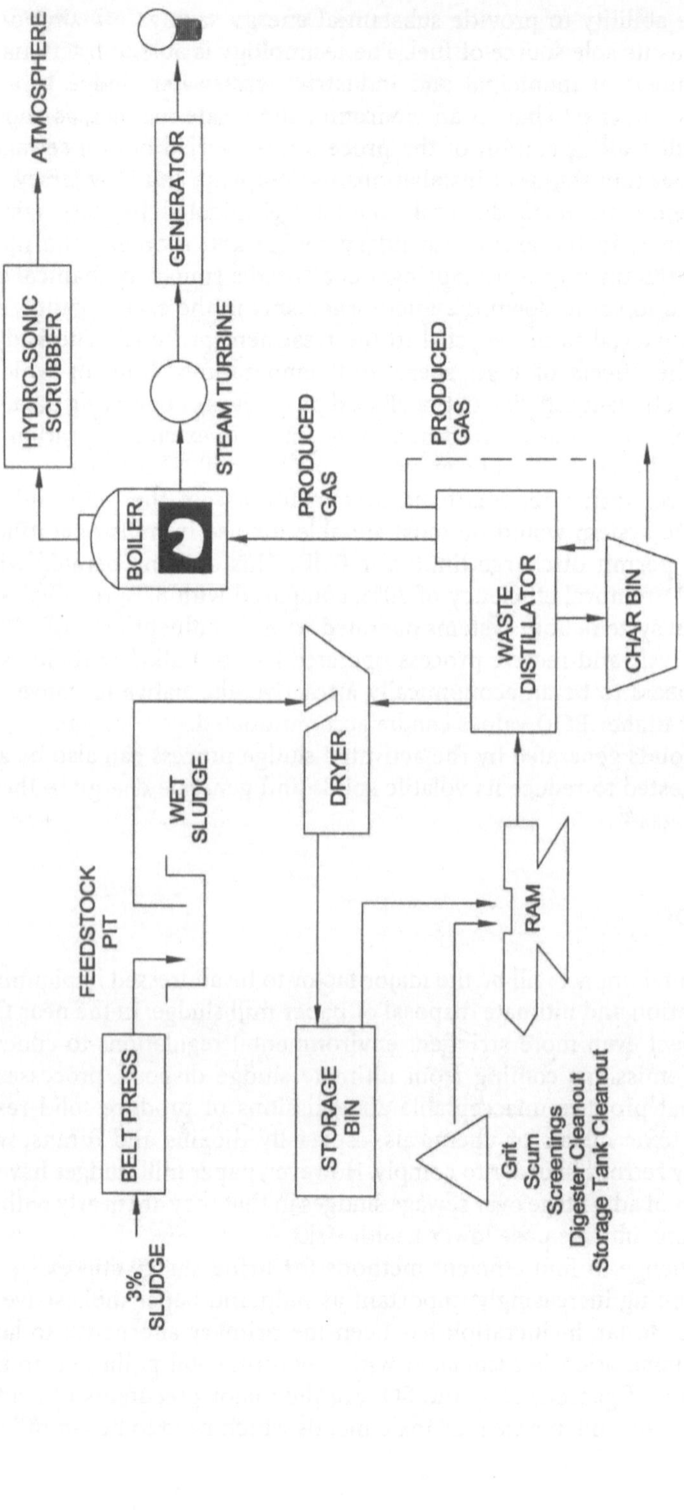
3% SLUDGE
BELT PRESS
FEEDSTOCK PIT
WET SLUDGE
STORAGE BIN
Grit
Scum
Screenings
Digester Cleanout
Storage Tank Cleanout
RAM
DRYER
WASTE DISTILLATOR
CHAR BIN
PRODUCED GAS
PRODUCED GAS
BOILER
STEAM TURBINE
GENERATOR
HYDRO-SONIC SCRUBBER
ATMOSPHERE

hence result in groundwater contamination. The plastics and glue found in the sludge are the sources of chlorinated compounds such as HCl, dioxins, and furans, which are a major threat to the environment.

Landfilling is becoming less of a viable option as environmental problems and restrictive legislation are making landfills a buried liability. Also, landfill operating cost has increased. Sites and permits for new landfills have become more difficult to obtain in many countries. Incineration and the production of steam and power with sludge will continue to be an option for the foreseeable future. Improvements in sludge-drying techniques and boiler configuration are making sludge more of an asset than a liability for heat production. Incineration in beehive burners will soon be eliminated, as these burners are being phased out. Land application is gaining momentum in the pulp and paper industry as well as with municipal waste treatment systems. Trials are quickly developing into permanent land-spreading programs for forestry application, land reclamation, and agricultural application, and research is demonstrating that the hurdles of environmental legislation and disposal costs could be overcome. Recently, there has been interest in the use of sludge for production of light-weight aggregate and granules to carry agricultural chemicals, in pelletization of sludge for use as a fuel, production of ethanol and single-cell protein, use in cement kiln feedstock, and as hydraulic barrier material in landfill capping systems. The interest in these particular waste management opportunities probably relates mostly to their potential for using significant amounts of sludge. With the exception of ethanol and light-weight aggregate production from sludge, full-scale operations have successfully demonstrated each of these alternatives.

References

1. Kenny R, Coghill R, Almost S, Easton C: CPPA international sludge dewatering survey. Proc. 1995 Tappi Environmental Conference, Atlanta, GA, April 1995.
2. Latva-Somppi J, Tran HM, Barham D: Characterization of deinking sludge and its ashed residue. Pulp Pap. Can. 1994; 95(10): 31–35.
3. Springer AM: Solid waste management and disposal. In: Industrial environmental control-pulp and paper industry (Springer AM, ed). Tappi Press, Atlanta 1993; Second Edition: 458–493.
4. Miner RA, Unwin J: Solid waste management and disposal practices in US paper industry. Technical bulletin no. 641, National Council of the Paper Industry for Air and Stream Improvement, New York, Sept. 1992.
5. Yale JR: Paper recycling impact on effluent sludge. In: Environmental issues and technology in the pulp and paper industry – a Tappi Press anthology of published papers 1991–1994 (Joyce TW, ed). Tappi Press, Atlanta 1995; 325–333.
6. Miner RA, Marshall DW: Sludge dewatering practice in the pulp and paper industry. Technical bulletin no. 286, National Council of the Paper Industry for Air and Stream Improvement, New York, 1976.
7. Pickell J, Wunderlich R: Sludge disposal-current practices and future options. CPPA Pacific & Western Branches Conference, Jasper, Alberta, Canada, May 1994.
8. Kenny R, Almost S, Coghill R, Easton C, Osterberg F: CPPA/International review of pulp and paper activated sludge dewatering practices. Pulp Pap. Can. 1997; 98(8): T277–T281.

9. Mertz HA, Jayne TG: Start up and operating experience with Zimpro high pressure wet oxidation system for sludge treatment and clay reclamation. Proc. of Tappi Environmental Conference, Savannah, GA, April 9–11, 1984.
10. Reilly MT, Krepps WE: A case study – trials with a mobile unit demonstrate centrifugation of secondary sludge. Tappi J. 1982; 65(3): 83–85.
11. Stovall JH, Berry DA: Pressing and incineration of kraft mill primary clarifier sludge. Tappi J. 1969; 52(11): 2093–2097.
12. McKeown JJ: Sludge dewatering and disposal. A review of practices in the U.S. paper industry. Tappi J. 1979; 62(8): 97–100.
13. Toole NK, Kirkland JH: Pilot studies of screw presses for dewatering primary sludges. Proc. Tappi Environmental Conference, Savannah, Georgia, April 9–11, 1984.
14. Acker S, Malloy W: Color removal practices at a bleached kraft mill. Proc. Tappi Environmental Conference, Seattle, WA, 1990; 92.
15. Saunamaki R: Sludge handling and disposal at Finnish activated sludge plants. Water Sci. Technol. 1988; 20(1): 171–180.
16. Weigand PS, Unwin JP: Alternative management of pulp and paper industry solid wastes. Tappi J. 1994; 77(4): 91–97.
17. Pickell J, Wunderlich R: Sludge disposal: Current practices and future options. Pulp Pap. Can. 1995; 96(9): T300–306.
18. Russel C, Odendahl S: Environmental considerations for landfill development in the pulp and paper industry. Pulp Pap. Can. 1996; 97(1): T17-T22.
19. Tchobanoglous R. Solid waste management, McGraw-Hill, New York, 1975.
20. Wardwell RE, Cooper SR, Charlie WA: Disposal of paper mill sludge in landfills. Tappi J. 1978; 61(12): 72–76.
21. Holt WH: Solid waste landfills at paper mills. Tappi J. 1983; 66(9): 51–54.
22. Ledbetter RH: Design considerations for pulp and paper mill sludge landfills. U.S. Environmental Protection Agency EPA-600/3-76-11, December 1976.
23. National Council of the Paper Industry for Air and Stream Improvement (NCASI), New York. Chemical composition of pulp and paper industry landfill leachates. Technical Bulletin No. 643, Sept 1992.
24. Canadian Council of Resource and Environment Ministers: The Canadian Water Quality Guidelines prepared by the task force on water quality guidelines, March 1987.
25. Miner RA: A review of sludge burning practices in combination fuel-fired boilers. National Council of the Paper Industry for Air and Stream Improvement, New York, November 1981; Technical Bulletin No. 360.
26. Busbin SJ: Fuel specifications – sludge. Environmental issues and technology in the pulp and paper industry – a Tappi Press anthology of published papers 1991–1994 (Joyce TW, ed). Tappi Press, Atlanta 1995; 349–355.
27. David PK: Converting paper, paper mill sludge and other industrial wastes into pellet fuel. Environmental issues and technology in the pulp and paper industry – a Tappi Press anthology of published papers 1991–1994 (Joyce TW, ed). Tappi Press 1995; 365–367.
28. Nichols WE, Flanders LN: An evaluation of pelletizing technology. Environmental issues and technology in the pulp and paper industry – a Tappi Press anthology of published papers 1991–1994 (Joyce TW, ed). Tappi Press, Atlanta 1995; 357–363.
29. Sell NJ, McIntosh TH: Technical and economic feasibility of briquetting mill sludge for boiler fuel. Tappi J. 1988; 71(3): 135–139.
30. James BA, Kane PW: Sludge dewatering and incineration at Westvaco, North Charlston, SC. Tappi J. 1991; 74(5): 131–137.
31. Bezigian T: Alternative solutions to landfilling paper mill packaging waste. Environmental issues and technology in the pulp and paper industry – a Tappi Press anthology of published papers 1991–1994 (Joyce TW, ed). Tappi Press, Atlanta 1995; 459–473.
32. Kraft DL, Orender HC: Considerations for using sludge as a fuel. Tappi J. 1993; 76(3): 175–183.
33. King B, McBurney B, Barnes TW, Cantrell M: Operating experience with stoker firing TMP clarifier sludge with wood waste. Environmental issues and technology in the pulp and paper

industry – a Tappi Press anthology of published papers 1991–1994 (Joyce TW, ed). Tappi Press, Atlanta 1995; 393–403.

34. Fitzpatrick J, Seiler GS: Fluid bed incineration of paper mill sludge. Environmental issues and technology in the pulp and paper industry – a Tappi Press anthology of published papers 1991–1994 (Joyce TW, ed). Tappi Press, Atlanta 1995; 369–378.
35. La Fond JF, Lantz D, Ritter LG: Combustion of clarifier underflow solids in a hog fuel boiler with a new high energy air system. Environmental issues and technology in the pulp and paper industry – a Tappi Press anthology of published papers 1991–1994 (Joyce TW, ed). Tappi Press, Atlanta 1995; 385–392.
36. Davis DA, Gounder PK, Shelor FM: Combined cycle-fluidized bed combustion of sludges and other pulp and paper mill wastes to useful energy. Environmental issues and technology in the pulp and paper industry – a Tappi Press anthology of published papers 1991–1994 (Joyce TW, ed). Tappi Press, Atlanta 1995; 379–384.
37. AghaMohammadi B, Shekarchi S, Durai-Swamy K, Steedman W, Dauber R: Testing of a sludge gasification plant at Inland Containers Ontario (California) Mill. Environmental issues and technology in the pulp and paper industry – a Tappi Press anthology of published papers 1991–1994 (Joyce TW, ed). Tappi Press, Altanta 1995; 431–443.
38. AghaMohammadi B, Durai-Swamy K: A disposal alternative for sludge waste from recycled paper and cardboard. Environmental issues and technology in the pulp and paper industry – a Tappi Press anthology of published papers 1991–1994 (Joyce TW, ed). Tappi Press, Altanta 1995; 445–458.
39. Kozinski JA, Zheng G, Saade R, DiLalla S: On the clean and efficient thermal treatment of deinking solid residues. Can J. Chem. Eng. 1997; 75(1): 113–120.
40. Pridham NF, Cline RA: Paper mill sludge disposal: completing the ecological cycle. Pulp Pap. Can. 1988; 89(2): T73-T75.
41. Macyk T: Research relative to land application of BCTMP mill waste in Alberta. Preprints 1993 Pacific Paper Expo 1993; 91–95.
42. Atwell JS: Disposal of boiler ash. Tappi J. 1981; 64(8): 67–70.
43. McGovern JN, Berbee JG, Bockheim TG, Baker AJ: Characteristics of combined effluent treatment sludges from several types of pulp and paper mills. Tappi J. 1983; 66(3): 115–118.
44. Simpson GG, King LD, Corlile BL, Blickensderfer PS: Paper mill sludges, coal flyash, and surplus lime mud as soil amendments in crop production. Tappi J. 1983; 66(7): 71–74.
45. Hoffman, R, Coghill R, Sykes J: Solid waste management at ANM, Albury – from waste problems to resource opportunity. Appita 1995; 48(1): 12–14.
46. Taylor BR, Mcdonald MA, Kimmins JP, Hawkins BJ: Combining pulp mill sludges with municipal sewage to produce slow-release forest fertilizers. Pac. Pap. Expo 1992; 63–65.
47. Braman JR: Forest fertilization with sludge in Malaspina College research forest. Operations report on Malaspina project 1992, February 1993; 1–46.
48. Garvey D, Guarino C, Davis R: Sludge disposal trends around the globe. Water/Eng. Manage. 17–20 December 1993.
49. Sherman WR: A review of the Maine 'Appendix A' sludge research program. Tappi J. 1995; 78(6): 135–150.
50. Fitzpatrick GE: Biocycle 1989; 30(9): 62.
51. Smyser S: Biocycle 1982; 23(3): 25.
52. Rosenqvist GV: The use of primary waste water treatment sludge in the manufacture of printing paper at Kymi Kymmene. Paperi Ja Puu 1978; 60(4a): 205–217.
53. Ozturk I, Eroglu V, Basturk A: Sludge utilization and reduction experiences in the pulp and paper industry. Water Sci. Technol. 1992; 26(9–11): 2105–2108.
54. Goldstein IS, Easter JM: An improved process for converting cellulose to ethanol. Tappi J. 1992; 75(8): 135–140.
55. Alterthum F, Ingram LO: Ethanol production from glucose, lactose, and xylose by recombinant *E. coli.* Appl. Environ. Microbiol. 1989; 55(8): 1943–1948.
56. Lee YY, McCaskey TA: Hemicellulose hydrolysis and fermentation of resulting pentoses to ethanol. Tappi J. 1983; 66(5): 102–107.

57. Ingram LO, Conway T: Expression of different levels of ethanologenic enzymes from *Zymomonas mobilis* in recombinant strains of *E. coli*. Appl. Environ. Microbiol. 1988; 54(2): 397–404.
58. Brannigan M: Wall Street J. Nov. 3, 1992.
59. Evans JC: Pulp Pap. 1983; 57(3): 124.
60. Millet MA, Baker AJ, Satter LD, McGovern JN, Dinius DA: Pulp and papermaking residues as feedstuffs for ruminants. J. Anim. Sci. 1973; 37(2): 599–607.
61. Nichols W: Solid waste options Am. Papermaker 1992; 55(10): 41.
62. Huston B, Hardesty KL, Beer HE: Pollut. Prevent. Rev. 1992; 2(4): 453.
63. Thomas CO, Thomas RC, Hover KC: Wastepaper fibers in cementitious composites. J. Environ. Eng. 1987; 113(1): 16–31.
64. Rodden G: The new alchemy: turning waste into oils and chemicals. Can. Chem. News 1993; 45(8): 45–48.
65. FioRito WA: Destructive distillation – paper mill sludge management alternative. Environmental issues and technology in the pulp and paper industry – a Tappi Press anthology of published papers 1991–1994 (Joyce TW, ed). Tappi Press, Altanta 1995; 425–429.
66. Springer AM, Dietrich-Velazquez, Higby CM, Digiacomo D: Feasibility study of sludge lysis and recycle in the activated-sludge process. Tappi J. 1996; 79(5): 162–170.
67. Krogmann U, Boyles LS, Martel CJ, McComas KA: Biosolids and sludge management. Water Environ. Res. 1997; 69(4): 534–550.

11 Biofiltration of Exhaust Gases

People living in or near a kraft pulp mill often complain of the odor nuisance associated with the mill's operations. These complaints are directly related to the production of odorous compounds during the cooking of wood chips with white liquor and subsequent points of gaseous release to the atmosphere. Even when pure sodium hydroxide is used to treat wood and straw, odors are produced. The cause of these odors is to be found in the residual sulfur-containing protoplasm which reacts with the alkali to form mercaptans and organic sulfides during the digestion phase. It was found that the mercaptans are formed by the saponification of lignin methoxyl groups by sulfide ions.

Emission from Kraft Pulping

The evil-smelling gases emitted from the kraft process include H_2S, methyl mercaptan (CH_3SH) and various organic sulfides such as dimethyl sulfide (CH_3-S-CH_3), and dimethyl disulfide (CH_3-S-S-CH_3), collectively referred to as total reduced sulfur (TRS). They are formed during kraft pulping by reaction of sulfides with methoxy groups of lignin via nucleophilic substitution reactions. The major sources of TRS emissions include digester blow and relief gases, multiple-effect evaporator vent and condensates, the recovery furnace with direct-contact evaporators, smelt dissolving tank and slacker vents, brown-stock washers and seal tank vents, and the lime kiln exit vents. Typical characteristics of the gaseous emissions are given in Table 11.1.[1,2] It is apparent that the source of largest volume of potential emissions is the recovery furnace, followed closely by the digester blow gases and the washer hood vents. However, the most concentrated emissions come from the digester blow and relief gases. Overall, the three most important sources of odor production are black liquor combustion, weak black liquor concentration, and the digestion process.

About 0.1–0.4 kg of TRS is emitted per ton of pulp at 5 ppm in the recovery boiler flue gases. The total TRS production has decreased from 3 kg/ton pulp in 1960 to less than 0.5 kg/ton in the early 1990s. The principal difficulty with TRS emission is their nauseous odor, which is detected by the human nose at very low concentrations. Table 11.2 lists the odor threshold (odor detectable by 50% of the subjects) concentrations of the principal TRS compounds emitted by kraft mills which are only few parts per billion by volume.[3] At low concen-

Table 11.1. Typical off-gas characteristics of kraft pulp mill

Emission source	Off-gas flow rate (m^3/ton pulp)	Concentration (ppm by volume)			
		H_2S	CH_3SH	CH_3SCH_3	CH_3SSCH_3
Digester batch:					
Blow gases	3–6 000	0–1000	0–10 000	100–45 000	10–10 000
Relief gases	0.3–100	0–2000	10–5 000	100–60 000	100–60 000
Digester, continuous	0.6–6	10–300	500–10 000	1500–7 500	500–3 000
Washer hood vent	1500–6 000	0–5	0–5	0–15	0–3
Washer seal tank	300–1 000	0–2	10–50	10–700	1–150
Evaporator hotwell	0.3–12	600–9000	300–3 000	500–5 000	500–6 000
BLO tower exhaust	500–1 500	0–10	0–25	10–500	2–95
Recovery furnace	6000–12 000	(after direct-contact evaporator)			
		0–1500	0–200	0–100	2–95
Smelt-dissolving tank	500–1 000	0–75	0–2	0–4	0–3
Lime kiln exhaust	1000–1 600	0–250	0–100	0–50	0–20
Lime slacker vent	12–30	0–20	0–1	0–1	0–1

Based on data from Refs. 1 and 2.

Table 11.2. Odor threshold concentration of TRS pollutants

Reduced sulfur compound	Odor threshold concentration (ppb)
Hydrogen sulfide (H_2S)	8–20
Methyl mercaptan (CH_3-SH)	2.4
Dimethyl sulfide (CH_3-S-CH_3)	1.2
Dimethyl disulfide (CH_3-S-S-CH_3)	15.5

Based on data from Ref. 3.

trations, TRS is more of a nuisance than a serious health hazard. Thus, odor control is one of the main air-pollution problems in a kraft mill.

Oxides of both sulfur and nitrogen are also emitted in varying quantities from a few points in the kraft system. The dominant source of SO_2 emission is the recovery furnace, due to the presence of sulfur in the spent liquor used as a fuel. SO_3 is sometimes emitted when fuel oil is used as an auxiliary fuel. The lime kiln and smelt-dissolving tank also emit some SO_2. The emission of nitrogen oxides is more general because nitric oxide is formed whenever oxygen and nitrogen, which are both present in air, are exposed to high temperatures. A small part of the nitric oxide formed may further oxidize to nitrogen dioxide. These two compounds, nitric oxide and nitrogen dioxide, are termed total oxide of nitrogen. Under normal operating conditions, the temperature in the recovery furnace is not high enough to form large quantities of oxides of nitrogen (NO_x). The main source of NO_x emissions is the lime kiln. Table 11.3 summarizes SO_x and NO_x emission rates from various kraft mill sources.[1,4] Large

Table 11.3. Typical emissions of So_x and NO_x from kraft pulp mill combustion sources

Emission source	Concentration (ppm by volume)			Emission rate (kg/ton[a])		
	SO_2	SO_3	NO_x (as NO_2)	SO_2	SO_3	NO_x (as NO_2)
Recovery furnace						
No auxiliary fuel	0–1200	0–100	10–70	0–40	0–4	0.7–5
Auxiliary-fuel added	0–1500	0–150	50–400	0–50	0–6	1.2–10
Lime kiln exhaust	0–200		100–260	0–1.4		10–25
Smelt-dissolving tank	0–100	–	–	0–0.2		
Power boiler	–	–	161–232	–	–	5–10[b]

Based on data from Refs. 1 and 4.
[a] Kg/ton of air dried pulp.
[b] Kg/ton of oil.

variations in the emission rates are due to the variations in operating conditions at different mills. Large amounts of NO_x are produced if the flame temperature is above 1300 °C and oxygen concentration greater than 2%. Modern recovery boilers should have SO_x emissions below 100 ppm when properly operated. Sulfur emissions from power boilers are controlled by using fuels of low sulfur content.

Another type of odorous emission of nonsulfur compounds is produced by the hydrocarbons associated with the extractive components of wood, such as terpenes and fatty and resin acids, as well as those from materials used in processing and converting operations, such as defoamers, pitch control agents, bleach plant chemicals, etc. These hydrocarbon emissions are small compared to TRS emissions, but they may be odorous, or act as liquid aerosol carriers contaminated with TRS, or undergo photochemical reactions.

Some organic compounds that are not odorous are also emitted from the kraft process, These compounds include some terpenes, hydrocarbons alcohols, and other miscellaneous materials released from the wood upon pulping. The significance and emission rates of these compounds are not well characterized.

Emissions from Neutral Sulfite Semichemical (NSSC) Pulping

In general, the emissions from NSSC are much less than those from the kraft process. Both methyl mercaptan and dimethyl sulfide are absent from the gaseous emissions because no Na_2S is present in the pulping liquor. A very low amount of reduced sulfur is emitted.[5] The sulfur emissions from the Na_2CO_3 (sulfur-free) process have been traced to sulfur in the fuel oil and process water streams used. The emissions of SO_2 and NO_x are similar to those of a kraft mill.

Emissions from Sulfite Pulping

The sulfite process mainly operates with acidic SO_2 solutions and, as a consequence, SO_2 is the principal emission. Organic reduced sulfur (RS) compounds are not produced if proper conditions are maintained in the process.[1] Because the odor threshold is about 1000 times higher for SO_2 than for RS compounds, sulfite mills generally do not experience the odor problem of a kraft mill. The method of attack on lignin by sulfite liquor is quite different from that by kraft liquor. The sulfite process involves sulfonation, acid hydrolysis, and acid condensation reactions.[6] Volatile compounds such as methyl mercaptan and dimethyl sulfide are not produced in sulfite pulping.

Typical emissions in the sulfite process are SO_2 with special oxides of nitrogen problems arising in the ammonium-base process, SO_2 is also emitted during sulfite liquor preparation and recovery. Very little SO_2 emission occurs with continuous digesters. However, batch digesters have the potential for releasing large quantities of SO_2, depending on how the digester is emptied. Digester and blow-pit emissions in the sulfite process vary, depending on the type of system in operation. These areas have the potential for being a major source of SO_2 emission. Pulp washers and multiple-effect evaporators also emit SO_2.

Treatment Techniques

Air pollution is pervasive in its effects. It affects human health, plants, animals, building materials, and finally the physical properties of the atmosphere itself. Even small emissions of odorous compounds may be the cause of nuisance or, at worst, may directly endanger health.

A number of methods are available to eliminate odorous components from gaseous emissions. The important ones are:[7]

Gas Phase Methods. One of the gas phase methods is the masking of odorous components, which in fact is no real elimination process but consists of masking the undesired odor by a component pleasing to smell. The undesired component is not eliminated in this way, which is, of course, the main disadvantage of this technique and makes it, therefore, unsuited to the purification of waste gases. Another gas phase method is making waste gases odorless by oxidation with ozone, which was applied for a few years but discontinued due to harmful effects on the bronchial tubes and the cost of the process.

Liquid Phase Method. The waste gases are contacted with a liquid phase in which the pollutants are absorbed. The loaded liquid effluent needs an aftertreatment to eliminate the absorbed components. This offers the possibility of recovering valuable absorbed components. Mere absorption processes are generally applicable only in the case of high solubility of the compounds to be eliminated from the waste gas. Biological regeneration processes have also been introduced. Under aerobic conditions many odorous compounds, of organic as well as of inorganic nature, are oxidized by suitable microorganisms. The microbial population can be either dispersed in the liquid phase or immobilized on a carrier material.

Solid Phase Methods. The waste gas is contacted with a solid phase. The molecules of the adsorbate condense at the surface of the adsorbent, where they are bound by physical adsorption or chemisorption and eliminated from the waste gas. Active carbon is frequently used as an appropriate adsorbent mainly for traces of odorous components. A disadvantage of this procedure is the saturation of the adsorbent after some time of operation, which necessitates regeneration of the active carbon. Biological filtration methods have also been introduced into this field.

Combustion. Organic compounds can be burned to carbon dioxide and water at sufficiently high temperature. However, energy costs are considerable due to the generally low concentration levels of the organic components present in waste gases, which necessitates combustion in an external flame. The combustion temperature can be considerably decreased when an appropriate catalyst is used. This reduces fuel consumption. However, this method can only be used with well-defined waste gases because of poisoning of the catalyst by certain compounds.

Biological Methods. A broad spectrum of compounds can usually be converted with chemical method, but energy consumption is relatively high. Consumption of chemicals (oxidants) is another disadvantage of these methods. Physical methods, such as absorption in liquids and adsorption on solids, have the disadvantage that the pollutants are not converted, and that the sorbents have to be regenerated. In this way, a new polluted mass flow is often created, although these methods sometimes offer the possibility of recovering valuable components. On the other hand, the biological methods generally have the specific advantage that the pollutants are not concentrated in another phase, but are converted to harmless or much less harmful oxidation products (e.g., CO_2, H_2O, etc.). These processes do not generally give rise to new environmental problems, or if they do, these problems are minimal. An exhaust air problem should preferably not become a solid waste or wastewater problem. Another advantage of biological treatment is the possibility of carrying out the process at normal temperature and pressure. Moreover, the process is reliable and relatively cheap, while the process equipment is simple and generally easy to operate.

The elimination of volatile compounds present in waste gases by microbial activity is due to the fact that these compounds can serve as an energy source and/or a carbon source for microbial metabolism. Hence, a broad range of compounds of organic as well as of inorganic origin can be eliminated by microbiological processes.

As microorganisms need a relatively high water activity, these reactions generally take place in the aqueous phase and, as a consequence, the compounds to be degraded as well as the oxygen required for their oxidation first have to be transferred from the gas phase to the liquid phase. Therefore, mass transfer processes play an important role in this methodology. The microbial population can either be freely dispersed in the water phase or is immobilized on a packing or carrier material. The first-mentioned operation is carried out in bioscrubbers, the second in trickling filters and biofilters. Bioscrubbers and tricking filters are more energy-intensive than biofilters, as water circulation in these two systems requires relatively much more energy than gas transport through a biofilter. Also, the reliability of operation of bioscrubbers is relatively low due to possible washing away of active microorganisms, while the presence of a large amount of packing material with a buffering capacity diminishes the sensitivity of biofilters to different kinds of fluctuations. Therefore, biofiltration technology is receiving significant attention.[8]

Biofilters

Biofilters are reactors in which a humid polluted air stream or contaminated exhaust gas is passed through a porous packed bed on which pollutant-degrading microbial cultures are naturally immobilized. Biofilters excel in two main domains: in the removal of odoriferous compounds[9] and in the elimination of volatile organic chemicals,[7,9–11] primarily solvents, from air. Under optimum conditions, the pollutants are fully biodegraded without the formation of aqueous effluents. As gases pass through a biofilter, odorous compounds are removed by processes thought to include sorption (absorption/adsorption) and biooxidation.[12] The odorous gases adsorb onto the surface of the biofilter medium and/or are absorbed into the moisture film on the biofilter particles. Given a sufficient rate of biological activity in the filter, the sorbed compounds are then oxidized (degraded) by microorganisms. End products from the complete biooxidation of the air contaminants are CO_2, water, mineral salts, and microbial biomass. The elimination of a gaseous pollutant in a biofilter is the result of a complex combination of different physicochemical and biological phenomena.

Biofilters are commonly constructed in a vessel packed with loose beds of solid material, soil, or compressed cakes with microbes attached to their surface. Waste gases are passed through these units via induced or forced draft. Biofilters are capable of handling rapid air flow rates and volatile organic carbon (VOC) concentrations in excess of 1000 ppm. These units are gaining importance in bioremediation also and are timely in that they are a cost-effective means by which to deal with the more stringent regulations on VOC emission levels.

There are essentially two types of biofilters. The first and simplest is the soil filter. Contaminated air from a small waste stream or other treatment process is passed through a soil-compost type design, so-called open system, which is schematically illustrated in Fig. 11.1.[7] Sometimes, nutrients are preblended into the compost pile to provide conditions for microbial growth and biodegradation of the waste by indigenous microorganisms. Being usually installed in the open air and partly underground, these systems are exposed to many weather conditions: rain, frost, temperature fluctuations, etc. These filters are usually overdesigned; they require a very large area. To increase the reliability of these filters, a number of other types (closed type) of systems have been developed which house the treatment beds or disks of different packing materials/media. In the treatment bed, the waste air stream and the filter are humidified as the waste is passed through one, two, or more beds. In this approach, a series of humidified disks or beds are placed inside a reactor shell (Fig. 11.2).[13] These layered disks contain packing material/media, nutrients, microbial cultures, and/or compost material. The waste air stream organics undergo biodegradation as they pass through the system. Any collected water condensate from the process is returned to the humidification system for reuse. Biofilters have reportedly been built to handle up to 3000 m^3/min of air flow using filters up to 6500 metre in wetted area.[14] The filters can be customized with specific carriers, nutrient blends, or microbial cultures. Some biofilters can endure up to 5 years before replacement is necessary.[15] Spent filters can be utilized as fertilizer since they present no hazard.

Microorganisms in Biofilters

For the purification of waste air containing a few known chemicals, the degrading microflora may be restricted to a few species of microbes. In these cases, it has become common practice to inoculate the filter bed using pure cultures of microorganisms known to actively degrade the pollutants. Waste air streams polluted with numerous chemicals require many various microbes possessing different capabilities of catabolism. Compost material generally has such versatile microflora. In addition, activated sludge originating from biological wastewater treatment plants may be advantageously used as an inoculum.

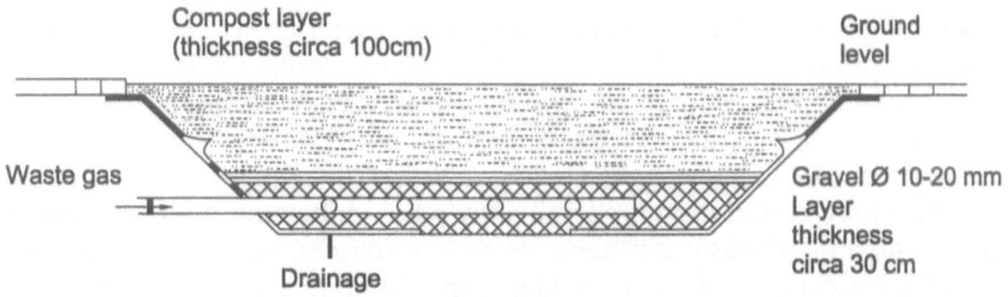

Fig. 11.1. Cross-section of a conventional open compost filter. Reproduced with permission from VCH, Germany, Ref. 7

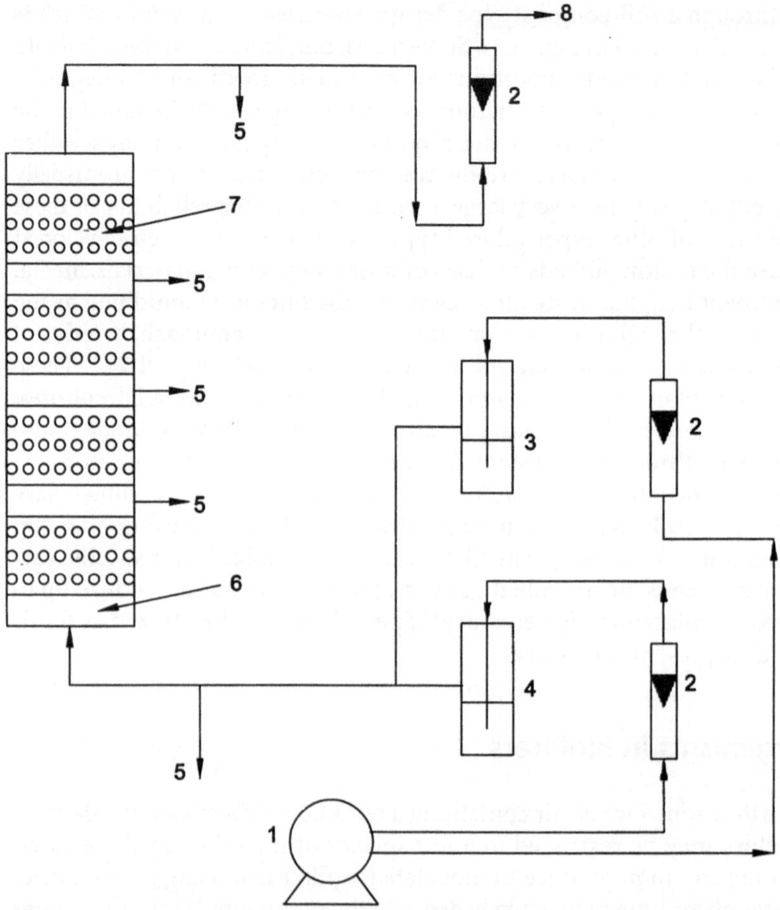

Fig. 11.2. Scheme of the experimental fixed-bed biofilter unit: *1* air pump; *2* air flow meters; *3* water; *4* methanol; *5* sampling ports; *6* polyurethane segments used as bottom support for the packing, as spacers around on column sampling ports, and at the top of the biofilter; *7* peat/perlite packing material in four segments of the column; *8* treated air (exhaust). Reproduced with permission from John Wiley & Sons Inc., Ref. 13

Several groups of microorganisms, including bacteria (*Actinomyces globisporus, Micrococcus albus, Micromonospora vulgaris, Proteus vulgaris, Bacillus cereus, Streptomyces* sp., *Thiobacillus, thioparus, Pseudomonas* sp., *Hyphomicrobium* sp., etc.), actinomycetes and fungi (*Penicillium, Cephalosporium, Mucor, Circinella, Cephlotecium, Ovularia, Stemphilium* sp., etc.)[8] are utilized in biofiltration. Biofiltration predominantly relies on heterotrophic organisms that use organic off-gas constituents as energy sources. As a result, introduction of these compounds into the filter material upon startup will generally shift the distribution of existing microbial populations towards strains that metabolize the largest pollutants. Acclimation takes around 10 days.

Table 11.4. Microbial cultures used for degradation of pollutant

Culture	Pollutant(s)	Reference
Soil (indigenous microbes)	Sewage odors – H_2S	16, 18, 19
Compost (indigenous microbes)	Various VOCs	17, 19, 20
Aerobically digested sludge of night soil	Sulfur compounds (H_2S, DMS, methanethiol)	9, 21
Peat (indigenous microbes)	H_2S	22
Sludge from sewage treatment	H_2S, C_2H_5SH, $(C_2H_5)_2NH$, C_4H_9CHO	23
Thiobacillus sp. strain MS_1	Methyl sulfides (DMS, DMDS)	24
Thiobacillus thioparus TK-m	H_2S, DMS, DMDS, Methanethiol	25, 26
Thiobacillus thioparus strain E_6	DMDS	27, 28
Hyphomicrobium sp. *strain S*	DMS, DMSO	29
Hyphomicrobium sp. *strain* EG	Methylated sulfur compounds	30, 31
Pseudomonas fluorescens (DSM50 090)	Methanol, isopropanol, butanol, etc.	32
Bacterial consortium (8 members) consisting of *Pseudomonas, Methylomonas, Aeromonas, Achromobacter, Flavobacterium, Alcaligenes*	Methanol	13

Different types of representative microorganisms/cultures used by various investigators have been given in Table 11.4. Soil and compost contain a large variety of indigenous microorganisms which degrade the odorous compounds in air. The common soil bacteria, *Bacillus cereus* var. *mycoides*, and strains of *Streptomyces* are most frequently identified in the soil samples. Autotrophic bacteria like *Thiobacillus* are also present in the soils which grow on thiosulfate medium; but the counts of heterotrophic bacteria are much higher. They have been demonstrated to reduce the sewage odors, especially by eliminating the hydrogen sulfide present in the waste air stream.[16–18] Following bacteria and microfungi: *Actinomyces globisporus, Penicillium* sp, *Cephalosporium* sp, *Mucor* sp., *Micromonospora albus, micrococcus albus, Ovularia* sp., etc. are the most frequently occurring microorganisms in the compost cultures. Compost has been a common choice of microbial source in biofiltration.[11,17,19,20] In addition to the source of microorganisms, the soil and compost provide both a physical support for the microorganisms and also water-holding capacity and some amount of minor and trace nutrients. Aerobically digested sludge of night soil has also been used as a source of microbial cultures in the biofilters for the removal of H_2S, DMS and methanethiol.[9,21] The digested sludge of night soil is supposed to contain several types of microorganisms, useful in biooxidation of air pollutants. The indigenous microorganisms in the peat have been tried for biooxidation of H_2S in a biofilter.[22] In a few cases, sludge from sewage treatment works is used as the source of microorganisms.[23] Classical microbiological techniques have revealed the presence of mixed populations of bacteria, yeast, fungi, and higher organisms in the biofilters.

Bacterial species of *Thiobacillus* and *Hyphomicrobium* degrade many sulfur compounds like H_2S, methyl sulfide, DMS, DMDS, DMSO, methanethiol, etc.[24–31] For methanol bio-oxidation, *Pseudomonas fluorescens*[32] and a bacterial consortium[13] consisting of *Methylomonas, Aeromonas, Achromobacter, Flavobacterium, Alcaligenes*, and *Pseudomonas* have been used.

Packing Materials for Biofilters

The traditional bed materials in a biofilter have been soil,[16,18,19] compost,[7,11,17,18,23] peat moss,[9,13,22] bark,[24,33,34] or other material that contains a large variety of indigenous microorganisms, but compost is a common choice. In addition to providing a physical support for the microorganisms, these materials also provide water-holding capacity and some amount of minor and trace nutrients. Most operating systems in Europe use compost. It is cheap, with a large existing microbial population capable of degrading various pollutants. Buffer capacity is adequate and inorganic nutrients are available. However, compost is prone to aging phenomena, including compaction and lumping of material, that decrease the effectiveness of the biofilter. Every 2 or 3 years, part of the compost must be replaced to preserve the structure of the filter, When the pressure drop becomes too high, the whole filter bed has to be changed. Sometimes, the bed material is amended with bulking agents such as wood chips,[34,35] saw dust,[34] bark,[24,34] sand,[43] bagasse,[35] etc. to improve air flow, or with other additives such as limestone[7,10,36,37] for pH control in systems removing sulfur-based odors. Peat has the advantages over soil or compost of maximum permeable value of the moisture content and a lower pressure drop due to its fibrous structure.[9,22] Peat has been reported to possess a unique combination of chemical and physical properties, such as adsorbency, which could be employed in environmental protection applications.[38]

Aging phenomena in the natural packing material can also be caused by local loss of moisture due to a low relative humidity of the air and/or the occurrence of temperature gradients in the filter bed. The occurrence of temperature gradients is inherent in microbial activity, which may differ locally. Where this activity is high, the temperature will be slightly higher than elsewhere as a result of the liberated energy of oxidation. As a result of these temperature gradients, gradients in the maximum water vapor pressure will also develop. Thus, the biologically active zones have a tendency to dry up, while less active zones take up this excess water by condensation. Usually, this develops into shrink cracks in the carrier material in which drying occurs and often anaerobic zones in the wet region. Such aging phenomena are mostly irreversible. This means that a shrink crack once developed will not disappear on its own. Application of narrow sieve fraction of the compost, use of bark compost, and mixture of bark compost with inert materials to increase the degradation capacity and to decrease the pressure drop of the filter bed have been suggested.[7] Additional porous materials with high internal porosity and hydrophilic properties are advantageous since such materials may also function as a buffer for excess moisture in the carrier material. Use of activated carbon has been suggested for this.[36] The combination of activated carbon and the natural packing materials may be very favorable when a biofilter is subjected to varying loads, for instance, with varying influent gases or for discontinuous processing. This enables a greater fraction of the VOCs entering the bed during peak pollution load to be captured. This captured material subsequently desorbs from the activated carbon when loads are low. Much of it is degraded by the microbes present. This permits, in principle, a considerable reduction of the filter volume required for the plants that discharge discontinuously. Granulated activated carbon has also been tried as a sole packing material in a biofilter for treatment of ethanol vapors.[39] An activated carbon fabric, developed by Otsuka Ind. Co. Ltd. (Japan) as a heat-resistant material, has been studied as a carrier for microorganisms in biofilters for deodorization of gas containing methyl mercaptan and DMDS.[21]

In addition, several other materials like a textile carrier (technical fabrics),[40] structured ceramic media,[41] peat and perlite,[13] sintered diatomaceous earth,[13] and a commercial packing material Bioton (compost plus polystyrene spheres and lime stone)[37] have been reported to be used in biofilters. Rock, sand, ceramic saddles, charcoal, kaolin particles, bentonite, anthracite, etc. are also used in different combinations with other materials in the biofilter as packing beds. A representative list of packing material is given in Table 11.5.

The role of the packing is to support the biofilm. In addition, it serves as a reservoir for water, pollutants, and nutrients by adsorption on its matrix and absorption in porewater.[42] The packing material generally comprises a fair

Table 11.5. Packing materials for biofilter

Material	Reference
Soil	16, 18–20
Compost	7, 11, 17, 18, 23
Peat	9, 13, 22
Bark	24, 33, 34
Activated carbon	36, 39
Activated carbon fabric	21
Textile	40
Structured ceramic media	41
Peat + perlite	13
Polyurethane foam	13
Sintered diatomaceous earth	13
Granulated activated carbon + sintered diatomaceous earth	39
Bioton	37

amount of organic residues that may be utilized as nutrients by the process culture. It is known that use of high surface area per unit volume structured media results in higher contaminant treatment rates per unit biofilter volume. Peat/compost biofilters exhibit lower removal efficiencies at high (more than 100 ppm) inlet contaminant concentrations and require control of media moisture content. Use of structured ceramic media allows effective control of biomass buildup by continuous removal of biomass from the biofilter media, and the biomass removal rate depends on nutrient flow rates.[40] The choice of the packing material is determined by the structure, the void fraction, the area per unit volume, the specific flow resistance, the ability of the filter material to hold water, and its working life.

Development of Biofiltration Technology

The principle of filtering air by contacting it with a microbial population has been known for quite some time. Where and when this principle was first used on a technical scale is difficult to say. It is known that around 1953, a soil system was employed for the treatment of odorous exit sewer gases in Long Beach, California (USA); in Europe, the first attempts, also with soil filters, were made in Geneva-Villete for the treatment of exhaust air from composting works.[7] Since then, the biological purification of waste gases has made considerable progress.

From the soil filtration experiments with H_2S-containing waste gases, it was concluded in 1966 that the odor reduction is effected by microorganisms present in the soil.[16] For the removal of H_2S, *Thiobacillus* and a number of heterotrophic bacteria are responsible. It was also concluded by these investigators that environmental conditions in the soil bed should be optimal for growth of these bacteria (e.g., temperature, moisture content, pH, etc.). From their experimental results, a maximum H_2S elimination rate of 110–120 mg/m^3/h could be estimated.

Biofilter systems for the elimination of volatile organics have been explored both on the experimental and mathematical modeling levels primarily in The Netherlands by the pioneering contributions of Ottengraf and his associates.[7,23,43,44] Research on and commercial use of biofilters has been less extensive in the United States, although the need for cost-effective air emission technology is clearly acute.[11] Bohn,[19] Pomeroy,[20] and Bohn and Bohn[18] described the use of soil and compost beds for biological treatment of malodorous emissions, but land area requirements and lack of process control restricted the industrial use of these systems.

Furusawa et al. used a packed bed of fibrous peat for the removal of hydrogen sulfide from air.[22] H_2S was almost completely removed irrespective of its inlet concentration when the loading was less than 0.44 g sulfur/day/kg dry peat. The removal rate of hydrogen sulfide by the acclimatized (in 15 days) peat was fairly constant under a constant inlet concentration but the reaction rate constant was proportional to the influent concentration of H_2S. In another study, the elimination of H_2S from odorous air using a wood bark filter to improve the low permeability of soil beds has been reported.[33] Lee and Shoda reported the biological deodorization of methyl mercaptan using an activated carbon fabric as a carrier of microorganisms for the biofilters.[21] The activated carbon fabric seeded with digested night soil was found to be the best packing material amongst the five materials evaluated. The critical load of methyl mercaptan, in which the gas can be completely removed, was determined as 0.48 g S/kg activated carbon fabric/day. About 80% of methyl marcaptan removed in the biofilter was converted into the sulfate ion. Effluent gas concentrations of methyl mercaptan and dimethyl disulfide were not detected below 50 ppm inlet concentration at a space velocity of 50/h. Fibrous materials which are flexible, light, and less microbially degradable may become significant as carriers of microorganisms.

The kinetics of removal of three kinds of odorous sulfur compounds – H_2S, methanethiol (MT), and dimethyl sulfide (DMS) – in acclimatized peat (seeded with aerobically digested sludge of night soil) were compared by Hirai et al. by supplying single or mixed odorous gases.[9] H_2S and MT were found to be degraded on peat irrespective of the acclimatizing gas, and their maximum removal rates were unaffected by the presence of DMS, whereas DMS was degraded only in DMS acclimatized peat. It has been reported that peat has the advantages over soil or compost of broadness of maximum permeability of the moisture content and a lower pressure drop due to its fibrous structure. The same laboratory reported earlier on the characteristics of peat as a packing material in a deodorization device with the following results: zero-order kinetics in complete H_2S removal by peat biofilters,[22] characteristics of isolated H_2S-oxidizing bacteria inhabiting a peat biofilter,[45] and biological removal of organosulfur compounds by peat biofilters.[46] Gradual increase of load was better for obtaining a high removal rate than high load at the start of the experiment. Acclimation periods for H_2S, MT, and DMS were 19, 17, and 24 days, respectively. During this period, the pH of the peat gradually decreased due to accumulation of sulfate ions.

The maximum removal rate of H_2S in its acclimatized peat was 1 order larger than those in MT- and DMS-acclimatized peat. The removability of DMS was affected by the mixed gases. Although the removal of DMS decreased when present with MT, the existence of H_2S will weaken the effect of MT on DMS removal to a certain extent. Thus, it would be better to maintain the SV (space velocity) value lower in order to guarantee DMS removal.[9] At a high SV, two stage columns in series are recommended. In the first column, most of the H_2S and MT can be removed, while the second column will be exclusively for DMS removal. This method is also appropriate for the maintenance of the operation, including the washing of accumulated ions and the exchange of packing material.

Shareefdeen et al. used an eight-membered bacterial consortium, obtained from methanol-exposed soil, and a peat-perlite column for the biofiltration of methanol vapors.[13] The biofilter was found to be effective in removing methanol at rates up to 112.8 g/h/m^3 packing. They also derived a mathematical model and validated it. Both experimental data and model predictions suggested that the methanol biofiltration process was limited by oxygen diffusion and methanol degradation kinetics. Bench-scale experiments and a numerical model were used by Hodge and Devinny to test the effectiveness of biofiltration in treating air contaminated with ethanol vapors.[39] Out of the three different packing materials used, viz., GAC, compost, and a mixture of compost and diatomaceous earth, the GAC supported the highest elimination rates, ranging from 53 to 219 g/m^2/h for a range of loading rates. Partitioning coefficients for the contaminant on the biofilter packing material had a strong effect on the efficiency of the biofilters. Several studies on the removal of volatile solvents like ketone mixtures,[10,32,36] toluene,[36,47,48] and ethyl acetate[36,48] by biofiltration have also been reported.

The performance of biofiltration to remove odors (about 40 compounds) from animal rendering plant's gaseous emissions was investigated by Luo and Oostron using pilot-scale biofilters containing different media (sand, sawdust, bark, bark/soil mixture).[34] Biofilter odor removal efficiencies of 75–99% were obtained at various air loading rates (0.074–0.057 m^3/m^3 medium/min) and medium moisture contents. Bio-Reaction Industries Inc., Tualatin, OR, USA, has reported developing a modular vapor phase biofilter which is capable of treating extremely high concentrations of VOC in low air volumes.[49] These systems are more suitable for point source industrial process air streams, storage tanks, and other vent emissions.

Biofiltration of NO_x is reported to be enhanced by the addition of an exogenous carbon and energy source.[50] pH control is found to be an important operating parameter due to the acidic nature of the gas. Addition of calcite to the biofilter bed provided an effective internal buffer and the optimum temperature was found to be 50–60 °C. The biofilter using activated carbon or anthracite as packing material was reported to be the most acceptable process for the removal of malodorous compounds containing nitrogen or sulfur,[51] because it produced no oxidized organics noticed with ozonation, and it had an equally high removal efficiency of both sulfur- and nitrogen-containing odorous compounds.

Recently, biofiltration has been successfully applied to remove α-pinene, a very hydrophobic VOC discharged in pulp and paper and wood products emissions, from a contaminated air stream.[52] Two identical bench-scale biofilters were utilized for more than 4 months of experiment. The biofilter medium consisted of a mixture of wood chips and spent mushroom compost that was amended with higher perlite for the first filter and with granulated activated carbon (GAC) for the second biofilter, The experiment was conducted at loading rates between 5 and 40 g α-pinene/m^3 bed medium/h. Under steady-state operating conditions, both biofilters, amended with perlite and GAC, performed similarly and provided removal rates of up to 30–35 g α-pinene/m^3 bed medium/h with gas retention times as low as 30 s. The adsorption characteristics of GAC were significant only during the startup period, where the GAC biofilter had a significantly better performance than perlite biofilter. When the biofilters were subjected to a sudden increase in the loading rate, the performance of the biofilters decreased significantly. The reacclimation period, however, was not long and biofilters reached more than 99% removal within less than 48 h of the spike load.

Studies on the transient behavior of a laboratory-scale compost-based biofilters have been reported recently.[53] This included startup, carbon balances, and interactions between pollutants in the aerobic biodegradation of VOC mixtures from effluent air streams. The study of transient behavior offers a genuine basis for the development of a conceptual explanation of the complex phenomena that occur in biofilters during pollutant elimination, thereby providing an opportunity for further progress in establishing fundamental understanding of such reactors.[54–56] During long-term operation of a biofilter, the mandatory absence of net cell growth forces the cells into maintenance metabolism, which is of relatively low rate compared to substrate consumption during the active growth of the acclimation phase. Postacclimation nutrient addition increases activity primarily by allowing a return to the high substrate consumption rate of active growth, and only secondarily helps raise bed activity because of the ultimately higher amount of biomass in the bed.[57] The biomass content of a biofilter during the acclimation phase can be estimated using two approximate methods. The first follows the cumulative amount of substrate converted and uses the yield of cells from substrate during active growth to estimate the total biomass created. The second method follows a rate constant for conversion of substrate in the bed. This number is proportional to the amount of biomass as long as the conditions in the bed (e.g., temperature, pH, substrate concentration) are relatively constant.[57]

Generally, the empirical knowledge dictates the design and scaleup of biofiltration plants, even though substantial performance improvement could be expected from a more comprehensive knowledge of the individual processes involved in pollutant elimination. For improved design and performance, an appropriate model for the whole process is required. Deshusses et al. have developed a novel diffusion reaction model for the determination of both the steady-state and transient-state behavior of biofilters for waste air treatment,[56] and experimentally evaluated/verified the same.[37] Although this model deals with the aerobic biodegradation of methyl ethyl ketone (MEK) and methyl isobutyl ketone (MIBK) vapors from air, similar mathematical treatment can be given to other biofilters degrading H_2S, organosulfur compounds, and other volatile organics. Most of the mathematical models have been developed mainly to correlate a particular set of experimental data, to explain the influence of selected parameters on the efficiency of the process, and sometimes to seek a better fundamental understanding of the phenomena occurring in a biofilter.[13,39,56] More promising quantitative structure activity relationships for biofiltration have been presented by Choi et al.[58] However, none of the above has resulted in either a universal theory of biofiltration or some sort of a predictive model, allowing a priori estimation of detailed pollutant elimination in biofilters to be achieved. Recently, a model allowing the maximum performance and the range of optimum removal for VOCs in waste air biofilters to be described in a very broad manner has been presented.[59] This model used only those parameters which do not require experimental determination, so that it might prove useful for design purpose. Although very good agreement between experimental data and model computations was achieved, the predictability of the model was not fully verified, nor were model parametric sensitivities performed.

Present Status

Biofiltration is now a well-established air pollution control (APC) technology in several European countries, most notably The Netherlands and Germany. In Europe, biofiltration has been used successfully to control odor, and both organic and inorganic air pollutants that are toxic to humans, as well as volatile organic compounds (VOCs) from a variety of industrial and public sector sources.[7,8] In The Netherlands and Germany, biofiltration has developed since the early 1960s into a widely used APC technology which is now considered the best available control technology in a variety of VOC and odor-control applications. More than 500 biofilters are known to be active in Germany and The Netherlands. Successful biofilter applications in Europe include the following: chemical manufacture, chemical storage, adhesive production, coating operations, iron foundries, waste oil recycling, flavors and fragrances, tobacco processing, industrial waste treatment plants, composting facilities, other food-processing industries, oil mills, beer yeast drying, etc.,[8] with odor control efficiency of 91–99% and organic removal efficiency of 71–95%. Compounds that are typically well degraded include alcohols, ethers, aldehydes, ketones, amines, sulfides, and inorganic compounds like ammonia and hydrogen sulfide. Higher chlorinated organics show a relatively lower ratio of biodegradation.

More than 40% of New Zealand animal rendering plants now use biofilters, which are usually effective.[34] Commercial use of biofilters has been less extensive in the United States, although the need for cost-effective air emission technology is clearly acute.[13] In the recent past, biofiltration technology has started picking up in the US also. Few examples are given below.

Rivana Water and Sewer Authority operates a 6-MGD sewage pumping station at Charlottesville, Virginia. A gas flow of 80 m^3/min is exhausted which contains 1–4 ppm H_2S and other odorous compounds. Based on the extremely favorable cost analysis and performance results of other installations, a full-scale biofilter system has been installed.[60] ABTco Inc. operates a hardboard manufacturing facility in a residential area of Alpena, Michigan. A full-scale biofilter on a press and cooler exhaust stream has been operating since 1 May 1996, for odor control[61] as required by the Michigan Department of Environmental Quality. Wastewater treatment process and distribution system odors are reported to be due to reduced sulfur (RS) compounds like H_2S, methyl mercaptan, dimethyl sulfide, and dimethyl disulfide.[62] In a single biofiltration system, removal of H_2S (99%), methyl mercaptan (90%), dimethyl sulfide (50%), and no removal of dimethyl disulfide have been obtained. These results indicate that H_2S can inhibit complete removal of RS compounds; sequential biofiltration may be an effective design for complete odor control when mixed RS compounds are involved. Application of two full-scale biofiltration systems is providing odor control at two wet-well locations.[62] The first application is a wet well where H_2S concentration is more than 400 ppm, the second is a wet well where H_2S concentration, though predominating, is less than 100 ppm and methyl mercaptan and dimethyl sulfide are also present in measurable quantities.

Although very little information is available in the literature about the application of biofilters in the pulp and paper industry for odor removal, substantial information is available for the removal of various compounds similar to those generated in the pulp and paper industry. This information could be very useful in installing biofiltration systems in pulp and paper mills.

Parameters Affecting the Performance of a Biofilter

In addition to the microbial culture and packing materials, several other parameters are also important which affect the performance of a biofilter. In order to avoid deposits in the filter layer, dusts and aerosols are to be removed to a great extent from the waste gas by means of appropriate separators. Before it enters the filter, the waste gas should be humidified to saturation. The raw gas is humidified in a spray humidifier or by adding steam to it. The dust separation and humidification can be combined in wet scrubbers in which scrubbing is done by water. Sometimes, the biofilters can be poisoned by the presence of off-gas constituents that are toxic to the microorganisms. Elimination of these substances or changing the vent system can make the off-gas suitable for biofiltration. High particulate loads in the raw gas can adversely affect the operation of a filter in different ways. Clogging of the air distribution system and the filter material itself by grease and resin can also occur. The deposition of dust in the humidifier will generate sludge and can result in improper humidification. In such cases, the installation of a particulate filter is required.[11]

Pollutant concentration and pollutant loading rates affect the performance. For example, in cases of H_2S removal by a compost biofilter, the efficiency does not change as long as the H_2S loading rate is less than the maximum acceptable value for the compost. A concentration of H_2S as high as 4000–4500 ppm can be treated with an efficiency of 99%, but if the concentration increases drastically, say more than 100 000 ppm, then fresh air can be introduced to reduce the H_2S concentration and increase the oxygen concentration.[63] The maximum elimination capacity is a function of the biofilter material and the operating conditions. The pollutant loading should be applied accordingly.

To ensure the maximum pollutant elimination capacity of the biofilter system, the gas should stay on the bed for a sufficient time, e.g., 30–40 s for H_2S elimination in a compost bed.[63] There is no significant increase in the efficiency if the time is greater than 25 s, but when it is decreased to 10 s, for example, the efficiency decreases by about 80%. The reduction of H_2S removal efficiency at shorter residence time is not necessarily due to the insufficient reaction time between the H_2S molecule and the biomass, but may be due to the slow step involved in the overall process. This slow step comprises H_2S diffusion from the gas phase into the liquid phase where the microorganisms exist.[63]

The moisture content and pH of the packing bed are other important parameters. For compost, the moisture level should be held between 40 and 60%. If the moisture content is reduced below 30%, the H_2S removal efficiency decreases proportionately. Proper moistening equipment like sprinklers should be installed and operated in such a way that moisture content stays in the prescribed limits. Since the dominant active species present in this biofilter are primarily acidophiles which prefer an optimum pH value near 3, maximum H_2S removal occurs at a compost pH of 3.2. Sulfur-oxidizing bacteria can live in environments having a wide range of pH (1–8). At a pH below 3, the efficiency decreases drastically. At the higher pH range, chemical reaction between H_2S and the compost material or reaction products can significantly enhance its removal, in addition to biological oxidation.[63]

For high pollutant removal efficiency, the temperature of the filter bed should be in the optimum range. The optimum range is 35–50 °C for H_2S removal. The efficiency drops rapidly with decreases in temperature. For example, if the temperature reduces to, e.g., 7 °C, the H_2S removal efficiency decreases by about 80%. The decrease in H_2S removal at the higher temperature is less significant than that at lower temperature.

The removal of H_2S at higher temperatures is probably due to increased chemical oxidation reactions in addition to biological oxidation. Normally, the temperature of biofilter is 10–15 °C higher than the ambient temperature. This is due to the biological respiration of the microbes and the exothermic reactions in the filter. Thus, the biofilter can function properly even if the ambient temperature is low.

Since sulfate is the final product of the biofiltration involving sulfur compounds, it may accumulate in the filter bed if not removed. Accumulation of sulfate can easily reach a level that can significantly reduce the biological function of the biofilter. Therefore, sulfate should be washed off periodically before it reaches the toxic level. A sulfate content of 25 mg/g is a critical level for the microbial environment.

The pressure drop increases approximately linearly with packing height. It increases in significantly larger increments with packing height for smaller particles than for larger ones. It also depends on the water content of the packing. If the water content is increased, the coagulation of small viscous particles is enhanced and the pressure drop increases sharply.

However, the rapid buildup of pressure can be suddenly released by channeling, i.e., a breakdown of filter bulk with much less resistance caused by a separation of packing materials. This situation is undesirable because it allows pollutants to exit the system without treatment. To prevent the high back pressure buildup, the surface load of up to 300 m^3 off gases/h/m^2 of filter should be maintained for proper functioning of the compost filter. Mineralization and compaction of the compost packing during extended operation may eventually increase the bed pressure drop. Practically, the bed needs to be repacked or the compost replaced when the overall pressure drop is greater than 25 kPa.[63]

Advantages, Limitations, and Future Prospects

Since biofilters compete with incineration and carbon adsorption in many situations, they are attractive in terms of not having to deal with landfilling costs or regeneration headaches. This has already been recognized in Europe, and some biofilter technology has found its way to the US.[14] Also, the thought of not simply transferring contaminants from one medium to another is particularly appealing. The biofilter creates a truly destructive process.

The use of microbial filter techniques in the treatment of air effluents containing organic pollutants can offer a number of advantages. They are inexpensive, efficient at ambient temperature, self-generating, maintenance-free with low running cost, have a long life, are kind to the environment, and oxidize most common volatile organic compounds to carbon dioxide and water, producing virtually no byproducts. The microbial flora survive for a fairly long period during which the filter bed is not loaded (periods of a fortnight are easily spanned with hardly any loss of microbial activity). This is important in view of the dynamic behavior of the filter bed at discontinuous operation, and means a very short starting time after longer periods of not operating the filter bed.[23] Moreover, the presence of a large amount of packing material with a buffering capacity diminishes the sensitivity of biofilters to different kinds of fluctuations.

Although such methods have long been known to be cost-effective, they have not found general acceptance in practice, even when the exhaust gas components to be removed are biodegradable. Long adaptation periods of the biomass (in particular with large exhaust gas flow discontinuities) or low space velocities, i.e., low specific purification capacities, are the reasons often cited. Bed compaction problems, specially with soil and compost biofilters, have also been noticed. This results in high pressure drop across the filter. However, the help of granulated activated carbon and other synthetic packing materials, individually or in combination with soil-peat-compost materials, has solved these problems to a great extent.

While biooxidizing H_2S and organic sulfur compounds in a filter, accumulation of sulfate can easily reach a level that can significantly reduce the biological activity of the biofilter. Therefore, sulfate should be periodically washed off before it reaches the toxic level. The removal of DMS decreases considerably if methanethiol (MT) is also present in the exhaust gas.[9] However, the existence of H_2S weakens the effect of MT on DMS removal rate to a certain extent. In this case, it would be desirable to maintain a low space velocity to ensure DMS removal. At high space velocity, two-stage columns in series are recommended, so that, in the first column, most of the H_2S and MT can be removed, while the second column will be exclusively for DMS removal. This method may also be appropriate for the maintenance of the operation, including washing accumulated ions and replacement of packing material. Multistage operation of biofilters may also be necessary when the waste gases contain components which require different conditions for their microbial degradation. In this way, optimal growth conditions for the different microbial population can be provided in separate stages. Also, more stages may be necessary when the waste gases include one component in a concentration so high that the capacity of one stage is inadequate for a sufficient degradation.

Depending on the nature of the organic compounds present in the waste, the filter sometimes needs inoculation with appropriate microorganisms to start biological activity.

In recent years, there has been significant maturation of biological waste air treatment research. This has resulted in a large number of studies concerning the performance and operation of the biofilters. Biofilter technology has a high potential for exhaust gas cleanup, but as with many biological processes, the design requirements have not been fully appreciated. Interestingly, the fundamental processes involved during the elimination of a pollutant in a gas phase bioreactor are still very poorly understood.

Biofilter technology was utilized in the field well before there was a basic understanding of its fundamental principles. This has resulted in several cases of unsuccessful or suboptimum operation of large-scale bioreactors. Today, with recent advances in the understanding of the fundamental principles underlying biofiltration, promise exists for better reactor design with optimal operating conditions. However, a number of fundamental questions remain unanswered or require further clarification, e.g., the quantification of biomass turnover, biodegradation kinetic relationships, and factors influencing these relationships, ecology of biofilter microflora, and the determination of the availability and cycles of pollutant, oxygen, and essential nutrients. The above factors have been found to significantly influence the performance and long-term stability of biofilters, and thus require further investigation in quantitative terms. The expanding use of modern tools of biotechnology should be able to make it easier. The largest problem to overcome will be the translation of recent and future basic advances into real process improvements for biofiltration technology to mature from the mysterious black box reactor to a well-engineered process based on solid science rather than on trial and error.

Biofiltration technology for purification of exhaust gases from the pulp and paper industry has a great potential. Very little information directly related to the industry is available, although reasonably good information can be found on the biofiltration of organic compounds similar to those found in the exhaust gases of the pulp and paper industry. More studies are needed to obtain a better understanding of the heat transfer, mass transfer, and reaction processes occurring within the biofilter beds. Comprehensive long-term studies of full-scale biofilter systems would also be valuable in improving our understanding of biofilters used to remove VOCs from off-gases generated in the paper industry. Extended studies of the transient behavior of biofilters are also needed to provide the basic empirical knowledge necessary for plant design, scaleup, and performance evaluation under real conditions.

References

1. Environmental pollution control pulp and paper industry, Part 1, Air, U.S. EPA Technology Transfer Series, EPA-625/7-76-001, October 1976.
2. Andersson B, Lovblod R, Grennbelt P: Diffuse emissions of odorous sulfur compounds from kraft pulp mills, 1 VLB145, Swedish Water and Air Pollution Research Laboratory, Gotenborg, 1973.
3. Springer AM, Courtney FE: Air pollution: a problem without boundaries. In: Industrial environmental control pulp and paper industry, 2nd Edition (Springer AM, ed) Tappi Press, Atlanta, Georgia 1993; 525–533.
4. Someshwar AV: Impact of burning oil as auxiliary fuel in kraft recovery furnaces upon SO_2 emissions. NCASI Technical Bulletin No.578, December 1989.
5. Dallons V: Multimedia assessment of pollution potentials of non-sulfur chemical pulping technology. EPA-600/2-79-026, January 1979.
6. Rydholm SA: Pulping process. John Wiley & Sons. Inc., New York, 1965., 452.
7. Ottengraf SPP: Exhaust gas purification. In: Biotechnology Vol. 8 (Microbial degradations), Schonborn W – vol. ed., Rehm H-J, Reed G – series eds., VCH, Weinheim, Germany 1986; chap. 12: 425–452.
8. Singhal V, Singla R, Walia AS, Jain SC: Biofiltration – an innovative air pollution control technology for H_2S emissions. Chem. Eng. World 1996; 31(9): 117–124.
9. Hirai M, Ohtake M, Soda M: Removal of kinetics of hydrogen sulfide, methanethiol, and dimethyl sulfide by peat biofilters. J. Ferment. Bioeng. 1990; 70: 334–339.
10. Deshusses MA, Hammer G: The removal of volatile ketone mixtures from air in biofilters. Bioprocess Eng. 1993; 9: 141–146.
11. Leson G, Wikener AM: Biofiltration: an innovative air pollution control technology for VOC emissions. J. Air Manage. Assoc. 1991; 41: 1045–1054.
12. Williams TQ, Miller FC: Odor control using biofilters. BioCycle 1992; 33(10): 72–77.
13. Shareefdeen Z, Baltzis BC, Oh Y-S, Bartha R: Biofiltration of methanol vapor. Biotechnol. Bioeng. 1993; 41: 512–524.
14. Anon: Air pollution control may be reduced with biotechnology. RMT Network, Madison, WI 1991; 6(1): 5–8.
15. Holusha J: Using bacteria to control pollution. The New York Times 1991; March 13: C6.
16. Carlson DA, Leiser CP: Soil beds for the control of sewage odors. J. Water Pollut. Control Fed. 1966; 38: 829–840.
17. van Lith C, Leson G, Michelson R: Evaluating design options for biofilters. J. Air Waste Manage. Assoc. 1997; 47: 37–48.
18. Bohn H, Bohn R: Soil beds weed out air pollutants. Chem. Eng. 1988; 95(4): 73–76.

19. Bohn H: Soil and compost filters of malodorant gases. J. Air Pollut. Control Assoc. 1975; 25: 953–955.
20. Pomeroy D: Biological treatment of odorous air. J. Water Pollut. Control Fed. 1982; 54: 1541–1545.
21. Lee S-K, Shoda M: Biological deodorization using activated carbon fabric as a carrier of microorganisms. J. Ferment. Bioeng. 1989; 68(6): 437–442.
22. Furusawa N, Togashi I, Hirai M, Shoda M, Kubota H: Removal of hydrogen sulfide by a biofilter with fibrous peat. J. Ferment. Technol. 1984; 62(6): 589–594.
23. Ottengraph SPP, Van Denoever AHC: Kinetics of organic compound removal from waste gases with a biological filter. Biotechnol. Bioeng. 1983; 25: 3089–3102.
24. Sivela S, Sundman V: Demonstration of *Thiobacillus* type bacteria which utilize methyl sulfides. Arch. Microbiol. 1975; 103: 303–304.
25. Kanagawa T, Kelly DP: Breakdown of dimethyl sulfide by mixed cultures and by *Thiobacillus thioparus* FEMS Microbiol. Lett. 1986; 34: 13–19.
26. Kanagawa T, Mikami E: Removal of methanethiol, dimethyl sulfide, dimethyl disulfide, and hydrogen sulfide from contaminated air by *Thiobacillus thioparus* TK-m. Appl. Environ. Microbiol. 1989; 55(3): 555–558.
27. Smith NA, Kelly DP: Isolation and physiological characterization of autotrophic sulfur bacteria oxidizing dimethyl disulfide as sole source of energy. J. Gen. Microbiol. 1988; 134: 1407–1417.
28. Smith NA, Kelly DP: Mechanism of oxidation of dimethyl disulfide by *Thiobacillus thioparus* strain E6. J. Gen. Microbiol. 1988; 134: 3031–3039.
29. DeBont JAM, vanDijken JP, Harder W: Dimethyl sulfoxide and dimethyl sulfide as a carbon, sulfur and energy source for growth of *Hyphomicrobium* S. J. Gen. Microbiol. 1981; 127: 315–323.
30. Suylen GM, Stefess GC, Kuenen JG: Chemolithotropic potential of a *Hyphomicrobium* species capable of growth on methylated sulfur compounds. Arch. Microbiol. 1986; 146: 192–198.
31. Suylen GM, Large PJ, vanDijken JP, Kuenen JG: Methyl mercaptan oxidase, a key enzyme in the metabolism of methylated sulfur compounds by *Hyphomicrobium* EG. J. Gen. Microbiol. 1987; 133: 2989–2997.
32. Kirchner K, Hauk G, Rehm HJ: Exhaust gas purification using immobilized monocultures (biocatalyst). Appl. Microbiol. Biotechnol. 1987; 26: 579–587.
33. Van Langenhove H, Wuyts E, Schamp N: Elimination of hydrogen sulfide from odorous air by a wood bark biofilter. Water Res. 1986; 20: 1471–1476.
34. Luo J, van Oostrom A: Biofilters for controlling animal rendering odor – a pilot-scale study. Pure Appl. Chem. 1997; 69(11): 2403–2410.
35. Chou M-S, Chen W-H: Screening of biofiltering material for VOC treatment. J. Air Waste Manage. Assoc. 1997; 47(6): 674–681.
36. Campbell HJ, Connor MA: Practical experience with an industrial biofilter treating solvent vapor loads of varying magnitude and composition. Pure Appl. Chem. 1997; 69(11): 2411–2424.
37. Deshusses MA, Hammer G, Dunn IJ: Behavior of biofilters for waste air treatment. 2. Experimental evaluation of dynamic model. Environ. Sci. Technol. 1995; 29: 1059–1068.
38. Martin AM: Peat as an agent in biological degradation: peat biofilters. In: Biological degradation of waste (Martin AM, ed). Elsevier Applied Sciences, New York 1992; 341–362.
39. Hodge DS, Devinny JS: Biofilter treatment of ethanol vapors. Environ. Prog. 1994; 13(3): 167–173.
40. Eisenring R: Technical fabrics as novel carrier materials for biofilters and biological trickling-bed reactors. WLB, Wasser, Luft Boden 1997; 41(9): 57–61.
41. Govind R, Bishop DF: Overview of air biofiltration – basic technology, economics and integration with other control technologies for effective treatment of air toxics. Emerging Solutions VOC Air Toxics Control, Proc. Spec. Conf. Pittsburgh, Pa 1996; 324–350.
42. Deshusses MA: Biological waste air treatment in biofilters. Current Opin. Biotechnol. 1997; 8(3): 335–339.

43. Ottengraf SPP, Meesters JPP, van den Oever, AHC, Rozema HR: Biological elimination of volatile xenobiotic compounds in biofilters. Bioprocess Eng. 1986; 1: 61–69.
44. Ottengraf SPP, Konings SPP: Emissions of microorganisms from biofilters. Bioprocess Eng. 1991; 7: 89–96.
45. Wada A, Shoda M, Kubota H, Kobayashi T, Katayama FY, Kuraishi H: Characteristics of H_2S oxidizing bacteria inhabiting a peat biofilter. J. Ferment. Technol. 1986; 64: 161–167.
46. Hirai M, Terasawa M, Inamura I, Fujie K, Shoda M, Kubota H: Biological removal of organosulfur compounds using peat biofilter. J. Odor Res. Eng. 1988; 19: 305–312.
47. Bibeau L, Kiared K, Leroux A, Brzezinski Viel G, Heitz M: Biological purification of exhaust air containing toluene vapor in a filter-bed reactor. Can. J. Chem. Eng. 1997; 75: 921–929.
48. Deshusses M, Johnson CT, Hohenstein GA, Leson G: Treating high loads of ethyl acetate and toluene in a biofilter. Air & Waste Management Association 90th Annual Meeting & Exhibition, June 8–13, 1997, Toronto, Canada pp.13.
49. Stewart WC, Thom RC: High VOC loading in biofilters industrial applications. Emerging Solutions VOC Air Toxics Control, Proc. Spec. Conf. Pittsburgh; Pa 1997; 38–65.
50. Apel WA, Barnes JM, Barrett KB: Biofiltration of nitrogen oxides from fuel combustion gas streams. Proc. Annu. Meet-Air Waste Manage. Assoc. 1995; 88th (vol.4A): 95-TP9C.04.
51. Hwang Y, Matsuo T, Hanaki K, Suzuki N: Identification and quantification of sulfur and nitrogen containing odorous compounds in wastewater. Water Res. 1995; 29(2): 711–718.
52. Mohseni M, Grant AD: Biofiltration of α-pinene and its application to the treatment of pulp and paper air emissions. Tappi Environ. Conf. Exhib. 1997; 2: 587–592.
53. Deshusses MA: Transient behavior of biofilters: start-up, carbon balance, and interactions between pollutants. J. Environ. Eng. 1997; 123: 563–568.
54. Shareefdeen Z, Baltzis BC: Biofiltration of toluene vapor under steady-state and transient conditions – theory and experimental results. Chem. Eng. Sci. 1994; 49: 4347–4360.
55. Tang HM, Hwang SJ, Hwang SC: Dynamics of toluene degradation in biofilters. Haz. Waste and Haz. Mat. 1995; 12(3): 207–219.
56. Deshusses MA, Hamer G, Dunn IJ: Behavior of biofilters for waste air biotreatment. I: Dynamic model development. Environ. Sci. Technol. 1995; 29(4): 1048–1058.
57. Cherry RS, Thompson DN: Shift from growth to nutrient – limited maintenance kinetics during biofilter acclimation. Biotechnol. Bioeng. 1997; 56(3): 330–339.
58. Choi DS, Webster TS, Chankg AN, Devinny JS: Quantitative structure – activity relationships for biofiltration of volatile organic compounds. In: Proc. 1996 Conf. on Biofiltration (Reynolds Jr. FE, ed.), The Reynolds Group, Tustin, CA USA 1996; 231–238.
59. Johnson CT, Deshusses MA: Quantitative structure – activity relationships for VOC biodegradation in biofilters. In: Proc. Fourth In-Situ and On-Site Bioremediation Symp. New Orleans, Battelle Press, Columbus, Ohio April 28 – May 1, 1997; Vol. 5: 175–180.
60. Williams TO, Boyette RA: Biofiltration for odor control at a sewage interceptor pumping station in Charlottesville, Virginia. Proc. – WEFTEC'96 Annu. Conf. Expo., 69th, Water Environment Federation, Alexandria, Va. 1996; 6: 445–449.
61. Allen PJ, Van Til TS: Operating Experience with a full scale biofilter at a hardwood mill. AIChE Symp. Ser. 1997; 315: 124–129.
62. Singleton B, Milligan D, Blanchard J: An effective sequential biofiltration design applied to control odors caused by reduced sulfur compounds at a wet well. Proc. – WEFTEC'96 Annu. Conf. Expo., 69th, Water Environment Federation, Alexandria, Va. 1996; 6: 405–411.
63. Yang Y, Allen ER: Biofiltration control of hydrogen sulfide design and operational parameters. J. Air Waste Manage. Assoc. 1994; 44: 863–868.

Subject Index